PETRI NETS IN FLEXIBLE AND AGILE AUTOMATION

PETRI NETS IN FLEXIBLE AND AGILE AUTOMATION

edited by

MengChu Zhou

New Jersey Institute of Technology

KLUWER ACADEMIC PUBLISHERS
Boston / Dordrecht / London

Distributors for North America:
Kluwer Academic Publishers
101 Philip Drive
Assinippi Park
Norwell, Massachusetts 02061 USA

Distributors for all other countries:
Kluwer Academic Publishers Group
Distribution Centre
Post Office Box 322
3300 AH Dordrecht, THE NETHERLANDS

Library of Congress Cataloging-in-Publication Data

A C.I.P. Catalogue record for this book is available
from the Library of Congress.

Printed on acid-free paper.

Printed in the United States of America

Contents

Contributors

Numbers in parentheses indicate the pages on which the authors' contribution begin.

Reggie J. Caudill (265), Department of Mechanical Engineering, Center for Manufacturing Systems, New Jersey Institute of Technology, Newark, NJ 07102-1982, USA.
Electronic-mail address: *caudill@admin.njit.edu*

Haoxun Chen (239), Systems Engineering Institute, Xian Jiaotong University, Xian 710049, P. R. of China.

Frank DiCesare (149), Department of Electrical, Computer and Systems Engineering, Rensselaer Polytechnic Institute, Troy, NY 12180-3590, USA.
Electronic-mail address: *dicesare@ecse.rpi.edu*

Luca Ferrarini (71), Dipatimento di Elettronica e Informazione, Politecnico di Milano, Piazza L. da Vinci, 32, 20133 Milano - ITALY.
Electronic-mail address: *ferrarin@elet.polimi.it*

Baosheng Hu (239), Systems Engineering Institute, Xian Jiaotong University, Xian 710049, P. R. of China.

Mohsen Jafari (337), Department of Industrial Engineering Rutgers University, P. O. Box 909, Piscataway, NJ 08855, USA.
Electronic-mail address: *jafari@princess.rutgers.edu*

Shengbing Jiang (207), Department of Automatic Control, Nanjing University of Science and Technology, Nanjing 210014, P. R. of China.

Doo Yong Lee (149), Department of Precision Engineering and Mechatronics, School of Mechanical Engineering, Korea Advanced Institute of Science and Technology, 373-1 Kusong-dong Yusong-gu, Taejon, 305-701, Korea.

Shifang Li (189), Takamori Laboratory, Department of Computer and System Engineering, Kobe University, Rokkodai 1-1, Nada, Kobe 657, Japan.
Electronic-mail address: *lee@in.kobe-u.ac.jp*

Ioannis Minis (109), Institute for Systems Research, University of Maryland, College Park, MD 20742, USA.
Electronic-mail address: *minis@src.umd.edu*
Electronic-mail address: *murata@sdl.hitachi.jp*

Tomohiro Murata (43), Systems Department Laboratory, Hitachi, Ltd. 1099 Ohzenji, Asao-ku, Kawasaki 215, Japan.

Kazuo Nose (93), Electronics Research Laboratory, Kobe Steel, Ltd., 5-5, 1-chome, Takatsukadai, Nishi-ku, Kobe 651-22, Japan.

Jean-Marie Proth (109), INRIA-Lorraine Technopole Metz 2000, 4 rue Marconi, F-57070 Metz,France.
Electronic-mail address: *proth@ilm.loria.fr*

Abd-El-Kader Sahraoui (305), Laboratoire d'Architecture des Systemes du C.N.R.S, 7, Avenue Colonel Roche 31077 Toulouse, France.
Electronic-mail address: *kader@laas.fr*

Takashi Sato (93), Industry Machinery Group, Kobe Steel, Ltd. 3-1, 2-chome, Shinhama, Arai-cho, takasago 676, Japan.
Electronic-mail address: *GFD03414@niftyserve.or.jp*

Satoshi Tadokoro (189), Takamori Laboratory, Department of Computer and System Engineering, Kobe University, Rokkodai 1-1, Nada, Kobe 657, Japan.

Toshi Takamori (189), Takamori Laboratory, Department of Computer and System Engineering, Kobe University, Rokkodai 1-1, Nada, Kobe 657, Japan.

Kurapati Venkatesh (265), Department of Mechanical Engineering, Center for Manufacturing Systems, New Jersey Institute of Technology, Newark, NJ 07102-1982, USA.
Electronic-mail address: *vxk8876@hertz.njit.edu*

Jiacun Wang (207), Department of Automatic Control, Nanjing University of Science and Technology, Nanjing 210014, P. R. of China.

Keyi Xing (239), Systems Engineering Institute, Department of Applied Mathematics, Xidian University, and Xian Jiaotong University, Xian 710049, P. R. of China.

MengChu Zhou (1, 265), Laboratory for Discrete Event Systems, Department of Electrical and Computer Engineering, Center for Manufacturing Systems, New Jersey Institute of Technology, Newark, NJ 07102-1982, USA.
Electronic-mail address: *zhou@njit.edu*

Richard Zurawski (1), Laboratory for Robotics and Intelligent Systems, Department of Electrical and Computer Engineering, Swiburne University of Technology, P. O. Box 218, Melbourne, Vic. 3122, Australia.
Electronic-mail address: *rzz@stan.xx.swin.oz.au*

Preface

Companies around the world are competing for tomorrow's opportunities. One such opportunity is the development and implementation of flexible and agile manufacturing systems to maintain and improve their competitive edge in the world market. Both researchers and engineers are constantly seeking advanced and unifying methodologies for modeling, design, analysis, performance evaluation, scheduling, and control of these systems whose complexity is kept increasing rapidly. One methodology resulting from this effort is based on Petri nets and related graphical and mathematical tools. This book is dedicated to the applications of this methodology to industrial automation of the growing importance.

A series of volumes published by Springer-Verlag has been dedicated to advances in the applications and theory of Petri nets. This series, however, is oriented towards readers in the computer science community as its contents are often theoretical with few industrial applications. This makes it very difficult for new researchers and practicing engineers to understand the potential applications for Petri nets due to mathematical complexity and the various notation and interpretations presented by different authors. Consequently, a primary motivation of this volume is to present a unifying approach to the potential applications of Petri nets in flexible and agile automation.

Not only does this volume unify the notation and terminology of Petri net methodology, but it presents many practical applications in addition to theoretical development. It presents the latest research results and industrial applications of Petri nets, particularly in discrete-event control design, scheduling, performance evaluation, and deadlock avoidance. It also presents comparative studies between Petri nets and other approaches. The book complements several other Petri net-related titles in manufacturing automation.

The research interest in Petri nets is expanding from Europe to the U.S., Asia, and elsewhere in both academic and industrial communities. For example, Grafcet, a Petri net-like standard for sequence control specification, was developed in France in 1977 and has become an international standard called Sequential Function Charts since 1990. Evidence of the rapid development and progress in using Petri nets as a graphical and mathematical tool for automation research is the growing number of papers in journals and conference proceedings. For example, over the past several years, there were a growing number of technical sessions on Petri nets in manufacturing automation in IEEE International Conferences on Systems, Man and Cybernetics, IEEE International Conferences on Robotics and Automation, IEEE Workshops on Emerging Technologies and Factory Automation, and International Conferences on CAD/CAM, Robotics and Factories of the Future. Many papers have appeared in IEEE Transactions and other manufacturing research journals. A single volume involving the most important advances in the use of Petri nets in the area of automation will help researchers and industrial engineers better disseminate the new ideas and research results and forge new areas of collaboration between academia and industry.

The current studies in Petri net applications to industrial automation have focused on two critical objectives for modern manufacturing systems: flexibility and agility. To compete in today's world market, manufacturing companies are required to design and manufacture products in a flexible and agile manufacturing system. "Flexible" signifies the manufacturing system's ability to adjust to customers' preferences and "agile" describes the system's speed in reconfiguring itself to meet changing demands. Meanwhile, flexible machine and robotics technology, information technology, and computer integration have been introduced into factories in an increasing speed. Applications of these advanced technologies provide tremendous challenges to engineers, analysts, and managers whose responsibility ranges from system design and analysis, planning, scheduling and dispatching to coordination, synchronization and operation of the production facilities. Design and implementation of flexible and agile manufacturing systems require advanced methodologies which are powerful, flexible, mathematically sound, and easy-to use.

The last two-decade research and development of the theory and graphical tools based on Petri nets has yielded one of the most powerful methodologies and tools. This volume has compiled some important results on their applications to the modeling, specification, discrete event control, planning and scheduling, performance analysis, supervisory control of flexible and agile manufacturing systems. It also presents comparative studies between Petri nets and other approaches, and several industrial application examples. This volume shall allow readers to evaluate the benefits and applicability of state-of-the-art Petri net methods and CAD tools from the best-known institutions, laboratories and companies for a particular operating environment and apply the tools to a problem of interest. This volume is not only a necessary reference book for researchers and Petri net experts but also directly applicable for engineers, analysts, and managers responsible for the design, implementation, and efficient operation of the next-generation manufacturing systems which are flexible and agile, and preferably computer-integrated.

Organization

Chapter 1 "Introduction to Petri Nets in Flexible and Agile Automation" written by Zhou and Zurawski introduces Petri nets as a basic methodology and tool for flexible and agile automation. Their application scope ranges from specification, modeling, performance analysis to simulation, planning, scheduling, real-time control, and implementation of various industrial systems. Among these systems are production lines, robotic assembly systems, material handling systems, flexible manufacturing systems (FMS), agile manufacturing and test facility, intelligent machines, computer networks, and traffic control systems. Both mathematical descriptions and intuitive ideas behind them are given. Many simple and illustrative examples are provided for a reader to obtain the basic ideas about Petri nets. This chapter presents a Petri net modeling and synthesis method and two examples including a production line and an AT&T's Printed Circuit Board assembly station. The Petri net properties and their relevance to manufacturing systems are presented. Analysis methods including reachability tree, invariant analysis, reduction, and simulation are briefly discussed and simple analysis examples are given. This chapter discusses two basic timed Petri nets: deterministic and stochastic. The basic solution methods are given and illustrated through a production line example. This chapter concludes with the discussion of industrial system-driven development of extended Petri net theory and applications and the remaining research issues to achieve a wide scope of industrial applications. This chapter serves as an introductory one which could be particularly helpful for new researchers and practicing engineers.

Chapter 2 "Application of Petri Nets to Sequence Control Programming" by Murata introduces one of the first successful applications of Petri nets for their sequence controller development in Hitachi, Ltd. The resultant Hitachi's Station Controller is the first commercial proof on how Petri nets can be used to control discrete manufacturing and assembly system while reducing system development time drastically. This chapter discusses hierarchical control structure of flexible manufacturing systems; several models for sequence control including state-transition diagrams, statecharts, rule-based models, Petri nets, Grafcet, and predicate/event nets; and C-nets for sequence control programming. Examples are given to illustrate the usefulness of C-nets. The differences between C-net and Grafcet are highlighted. Implementation and performance evaluation of C-net based sequence controllers are presented for several industrial systems. Among them are automatic assembly sequence control, material flow control and tracking, and information process control systems.

Chapter 3 "Computer Aided Design of Logic Controllers with Petri Nets" by Ferrarini concentrates on the CAD tool development for logic controller design. Computer-aided control system design plays a significant role in engineering design of numerous continuous and/or discrete-time control systems. A challenge in discrete-event systems design is to design CAD environment and tools to achieve systematic development of discrete-event or logic controllers for advanced industrial systems. This chapter accepts this challenge and presents important contributions along the direction of CAD design of discrete event systems. It proposes and discusses a generic standard control unit which includes an analog-to-binary converter, input combinatorial net, sequential part, and output combinatorial net. Strongly-connected state machines as a special class of Petri nets are used to model elementary control tasks. They are then connected to express an overall system controller. The connection rules are defined and discussed. Both bottom-up and top-down modular methods can be effectively used for complex system design in the developed CAD environment. The chapter explores the issues related to real-time communication in PLC and microprocessor-based implementation. A super-elevated conveyor is used to demonstrate the application of the proposed design methodology and CAD tool. The chapter presents an in-depth treatment of the development of CAD tools for discrete event systems and lays a foundation for the future industrial applications.

Chapter 4 "Automatic Generation of Sequence Control Programs via Petri Nets and Logic Tables for Industrial Applications" by Sato and Nose combines Petri nets and logic tables to generate sequence control programs automatically. It describes the application to a press machine of two thousand I/Os in Kobe Steel, Ltd. A novel method is presented, which uses Petri nets to represent sequences of motions combined with the on/off readings of sensors and logic tables to represent the on/off switches and actuators. The rules are then derived to transform Petri nets and logic tables into ladder logic programs for Programmable Logic Controllers (PLCs). The program generator is implemented as a software package in MS-DOS-based personal computers and PLCs. The package consists of a sequence flow and table editor, modules for conversion, linkage, transfer, and document-making, and motion simulator. The chapter presents its successful application for large-scale industrial systems. Compared with the conventional program development method, the use of the new method and developed tool allows to reduce the programming time by 75% - a very significant saving. From this industrial example, one might be convinced that traditional approaches become ineffective to cope with an industrial discrete-event system with thousands of I/Os. It shows another success story of using modern tools such as Petri nets and related system engineering approaches to shorten the system development cycle and thus to increase the system's agility.

Rapid determination of a feasible, incrementally better, near-optimal, and finally optimal schedule for manufacturing operations is necessary and has aroused numerous interests in both theory and practice. Can one derive such a schedule more efficiently and effectively in the framework of Petri nets? The following three chapters are devoted to answering this question.

Chapter 5 "Planning and Scheduling Based on Petri Nets" by Proth and Minis discusses planning and scheduling problems and solution techniques for both cyclic and non-cyclic production systems using Petri nets and algorithmic methods. For cyclic systems, marked graphs as a special class of Petri nets are used as basis to model the physical and decision-making systems. The proposed scheduling algorithm is able to generate a near-optimal schedule which maximizes productivity while minimizing work-in-process in deterministic cases. It should be noted that for large-size scheduling problems, there exists no efficient algorithm which guarantees to generate an optimum solution. Non-cyclic production systems arise to satisfy the short-term (e.g., one day) planning requirement. Petri net models are constructed and several new concepts are formulated to facilitate the discussion of their scheduling problem. A scheduling algorithm is proposed to minimize the sum of inventory and backlogging cost for non-cyclic production systems. It can generate a near-optimal schedule. In addition, a modular approach is proposed to develop Petri net models to facilitate the formulation of and solution to management problems in manufacturing systems. This chapter presents extensive research results toward practical solutions to general job-shop scheduling problems.

Chapter 6 "Petri Net-Based Heuristic Scheduling for Flexible Manufacturing" by Lee and DiCesare investigates job-scheduling problems for flexible manufacturing systems by using Petri net models and heuristic search approaches. First a brief discussion on scheduling methods in the literature is given. The use of Petri nets for scheduling problems is motivated by their ability to concisely capture the dynamics of flexible manufacturing with a coherent formulation, and their graphical feature which allows one to better understand and formulate the problems in a consistent way. Then a method is introduced to construct a Petri net model which incorporates routing flexibility, shared resources, process times, and lot sizes. Next, this chapter proposes a scheduling algorithm which is based on the heuristic search to find the optimal or near-optimal solution in the reachability graph of the model. Different heuristic functions used in the algorithm are discussed. Four FMS examples are used to illustrate the algorithm and the computational results are reported. Finally, this chapter compares the results between the proposed method and three heuristic rule-based dispatching methods, i.e., shortest queue length, shortest process time, and shortest waiting time methods. The Petri net-based heuristic method outperforms all these rule-based methods by more than 10% for the FMS example from the literature.

Chapter 7 "Scheduling and Rescheduling of AGVs for Flexible and Agile Manufacturing" by Li, Takamori and Tadokoro focuses on the scheduling and rescheduling of a typical flexible and agile manufacturing system. In their system, parts are transported by a number of Automated Guided Vehicles (AGVs) and processed by the machines. Engineers have to schedule AGVs to transport parts cooperatively to satisfy a machining schedule which is free from jamming and collision, and generate quickly a new schedule in case of AGV troubles. Two deterministic timed Petri net models are constructed for machining and transporting events, respectively, resulting in a hierarchical net model for the entire system. The chapter defines machining and transporting priority functions in order to resolve the conflicts on line. They are determined to achieve job balancing, the least idle time, etc. A simulation study is performed for the system consisting of three machines, four AGVs, and three input and output buffers to produce four job types in a required quantity. The simulation shows that excellent scheduling and rescheduling results are obtained with the proposed priority functions.

Most scheduling problems discussed above involve deterministic timed operations. When an industrial system is better represented as a stochastic model, stochastic timed Petri nets can be well utilized to evaluate system performance. In these systems, the duration of an event/operation is described by a random time variable. Chapter 8 "Stochastic Petri Net Models of Communication and Flexible Systems" by Wang and Jiang gives a comprehensive presentation on this important application area of Petri nets. The use of this class of stochastic models proves cost-effective to evaluate the performance of a system under varying architectures and operational policies including routing and scheduling at the design stage, and helps determine optimal one(s) among several design alternatives. They can be used to investigate the impact of the operational parameters on the system performance without changing drastically themselves in structure while this may be the case for a state-machine approach. When each transition has exponentially distributed time delay in which only one parameter, i.e., the average firing rate, is needed to characterize the random variable, one obtains stochastic Petri net (SPN). Allowing in the above SPN immediate transitions, i.e., ones with no time delay, results in generalized SPN. Allowing arbitrarily distributed time delays for certain transitions leads to extended SPN. Permitting deterministic time delays for certain transitions in the SPN yields Deterministic SPN. Arbitrary SPN allows any distribution and cannot be analytically solved for general cases. This chapter presents all these models and solution methods based on semi-Markov chain or moment generating function approaches. They are illustrated through a number of application examples. Among them are a semi-duplex communication protocol, local area network based on carrier sense multiple access with collision detection (CSMA/CD) protocol, flexible manufacturing system, robotic assembly system, and microprocessor-based system.

Deadlock is regarded as one of most critical issues which have to be resolved before a full implementation and operation of an automated manufacturing system. Due to the high capital investment, many pieces of FMS equipment may need to be shared by several job types. Resource-sharing generates the possibility for a system to be deadlocked. How to avoid the deadlock while maximizing the resource utilization is a challenging problem. Chapter 9 "Deadlock Avoidance Policy for Flexible Manufacturing Systems" by Xing, Hu and Chen presents their solution to this deadlock avoidance problem by applying the feedback control principle to the Petri net models of a class of manufacturing systems. The deadlock structure concept in Petri nets is proposed; and necessary and sufficient conditions are proved to avoid deadlock for the systems under consideration. The Petri net realization of their proposed deadlock avoidance policy is also presented. This chapter applies their research results successfully to a flexible manufacturing system. It consists of four workstations, four input/output buffers, and a transportation system and deals with two types of products. Their proposed policy is significantly better than that in the literature.

Chapters 2 and 4 demonstrated in practice that using Petri nets and related CAD tools could greatly reduce the system development time compared with conventional methods such as ladder logic diagrams. Is there any fundamental reason for this? Chapter 10 "Discrete-Event Control Design for Manufacturing Systems via Ladder Logic Diagrams and Petri Nets: A Comparative Study" by Venkatesh, Zhou and Caudill explores this very basic reason from the viewpoint of design complexity when ladder logic diagrams and Petri nets are used to design discrete-event controllers for automated systems. This chapter extends Petri nets to real-time Petri nets to design discrete-event control of manufacturing systems. It then identifies the comparison criteria, i.e., graphical complexity and response time. The chapter focuses on graphical complexity and particularly on design changes with respect to specification changes for both ladder logic diagrams and Petri nets. A case study on a typical industrial system, i.e., an electro-pneumatic system, is performed. It is shown that Petri net designs have less complexity, better adaptability, and greater maintainability than ladder logic diagrams do for the considered system. The chapter also presents an estimation method to compute the graphical complexity of designs with a ladder logic diagram and Petri net prior to their actual designs. This allows an engineer to select between them before the design and implementation. The chapter concludes with the importance of performing benchmark studies for the existing approaches, e.g., ladder logic diagrams, sequential function charts, Petri nets, structured text, and high-level programming languages. Thus engineers can justify the use of modern methodologies for their factory automation projects.

Most engineering systems particularly automated manufacturing systems can be viewed as reactive systems which have to maintain interaction with their environment. It is a well-known fact that the errors made in the requirement specification stage can heavily increase system development cost if they cannot be revealed in the early design stage. It is noted that often these errors are not found until the later stages. Chapter 11 "From State Transition Models to DFD Extended Methods for Specifying Reactive Systems" by Sahraoui investigates this important topic and presents a variety of requirements specification tools and their related features. It starts with the requirements of reactive systems and current approaches including state-based, algebraic and functional, temporal logic methods to specify reactive systems. The chapter then discusses in detail the existing state transition models, e.g., Petri nets, Grafcet, and statecharts. These methods are proved insufficient for complex systems such as computer-integrated manufacturing systems. Next, the chapter explores the requirements specification methods used in software engineering. The methods based on structured analysis, later extended and called real-time structured analysis, are the focus of the discussion. The basic models in these methods are data flow diagrams which are an industry standard. An FMS case study is performed, in which statecharts are used to specify the synchronous model and Petri nets for the asynchronous model. The chapter demonstrates how to combine real-time structure analysis and Petri nets to model and specify effectively a reactive system. It concludes with a discussion about the limitation of the proposed method and further research issues in the area of reactive system requirements specification.

The final chapter "Supervisory Control Specification and Synthesis" by Jafari discusses the supervisory control and specification for discrete event dynamic systems. It starts with the discussion on the supervisory control concept from several perspectives. Then the chapter focuses on the formal language or finite automaton-based supervisory control. This discrete-event control theory is relatively mature. However, more successful industrial-scale applications remain to be developed. The chapter discusses the fundamentals of this control theory through several examples. Parallel with the finite automaton-based supervisory control, Petri net-based supervisory control theory is not fully established although Petri nets have gained a few discrete-event control applications. This chapter provides engineers with essential knowledge for potential industrial applications of formal language and related methodologies which take system requirements specifications as their input and produce supervisory controllers as their output.

Acknowledgments

Many people have helped me in completing this book. My relationship with Kluwer Academic started with Ken Tennity on my co-authored book *Petri Net Synthesis for Discrete-Event Control of Manufacturing Systems* published in 1993, and with Carl Harris brought this book to its completion. They helped me obtain comments on the proposal from anonymous referees to whom I owe a great deal.

I wish to thank all the contributing authors. Without them the book would have been impossible. Their hard work kept the writing of the book on schedule. I would like to thank some colleagues of mine in Department of Electrical and Computer Engineering, and the Center for Manufacturing Systems at New Jersey Institute of Technology (NJIT), who have given me many helps and encouragement, especially Profs. Nirwan Ansari, Reggie J. Caudill, Bernard Friedland, Andrew Meyer, Anthony D. Robbi, Sol Rosenstark, and Kenneth Sohn. I would also like to acknowledge the assistance provided by Kurapati Venkatesh, Henry H. Xiong, and Yingshi Fu. Finally and most importantly, I sincerely appreciate the support of my wife, parents, and mother-in-law who have taken care of my lovely eighteen-month old son, Albert, since his birth.

The preparation of this book volume has been financially supported by the New Jersey State Commission on Science and Technology through the Center for Manufacturing Systems at NJIT and partially by the National Science Foundation under Grant No. DMI-9410386.

MengChu Zhou
New Jersey Institute of Technology
Newark, New Jersey
Zhou@njit.edu

1

Introduction to Petri Nets in Flexible and Agile Automation

MengChu Zhou
Discrete Event Systems Laboratory
Department of Electrical and Computer Engineering
New Jersey Institute of Technology
Newark, NJ 07102-1982, USA
zhou@njit.edu

Richard Zurawski
Laboratory for Robotics & Intelligent Systems
Department of Electrical and Computer Engineering
Swiburne University of Technology
P. O. Box 218, Melbourne, Vic. 3122, Australia
rzz@stan.xx.swin.oz.au

Abstract

As customers' demands change rapidly, more and more manufacturers recognize the importance of flexible and agile production systems. To achieve high flexibility and agility, computer and related information technologies have to be introduced into their daily practice. The design and implementation of such manufacturing systems requires new methodologies and computer-aided design tools. Petri nets and the related graphical tools represent a promising solution technique for many challenging problems arising from the endeavor to design flexible and agile automated systems. This chapter introduces fundamentals of Petri nets and their related applications in industrial automation. It serves as a tutorial for researchers and engineers who are newly interested in these technologies to familiarize themselves with the basic terminologies, concepts, and approaches of Petri nets.

1. Introduction

Mass production arose from the need to supply a large volume of uniform products to satisfy the market at the end of last century. The early production and assembly systems were rigid and expensive, thus impossible to change to handle any variations in product types. Development of electrical devices such as motors and electrical switches led to better control of machines and resulted in machines with certain flexibility. Computer technologies allowed researchers and engineers to develop computer numerically controlled machines and programmable logic controllers to automate manufacturing operations. Recent development in information technology including computer hardware/software and networking techniques makes it feasible to achieve the purposes of rapid product prototyping, concurrent engineering, flexible and agile manufacturing, and computer-integrated manufacturing.

Flexibility signifies a manufacturing system's ability to adjust to customers' preferences and agility means the system's speed in reconfiguring itself to meet changing demands. Both together make it possible for manufacturers to respond instantly to the market. To achieve high flexibility and agility, a company has to introduce information technology into all of their departments. Thus it can operate a factory in a market-oriented way with high emphasis on flowline production and Just-In-Time manufacturing [40, 49]. The central problems associated with design of flexible and agile automated manufacturing systems are related to process planning, scheduling, real-time control, and monitoring. At the initial design stage, modeling, requirements specification, and performance estimates are very important. It is well known that Computer Integrated Manufacturing (CIM) has not achieved the manufacturing potential it had promised more than a decade ago. One major reason for this is that CIM control software is very hard made reusable, flexible and agile. Along with the time, the requirements and specifications are kept changing. This makes the development of a manufacturing system's control software very difficult. Meanwhile, it also motivates development of new methods and tools which are powerful, mathematically sound, and easy-to-use.

Petri nets originated from the computer science studies promise as a way to address many difficult problems associated with the design of manufacturing systems as well as their related control and information systems. This chapter tries to lay a foundation of Petri nets for the following chapters which discuss their related graphical techniques to challenge the traditional approaches to designing manufacturing systems. Thus the main objective of this chapter is to introduce the concepts, terminologies, and basic methods of Petri nets to the researchers and practitioners, both from academia and industry, who are actively involved in the work in the areas of modeling, analysis, and design of manufacturing systems, as well as those who may potentially become involved in these areas in the future. The presentation focuses on the ordinary Petri nets, although other types of Petri nets are also introduced in the context of application driven developments.

This chapter is organized as follows. Section 2 describes the application scopes of Petri nets as a tool and methodology in the area of flexible and agile automation. The important contributions made by Petri net researchers and practitioners in this area are outlined. This section motivates the next section on the detailed mathematical and graphical description of Petri nets. Following the basic Petri net definitions, Section 4 presents several simple modeling examples and emphasizes the importance of the Petri net modeling or synthesis in the development of Petri net applications. Then Petri net properties are defined and related to their modeled systems' behavior in Section 5. Their analysis methods are briefly discussed in Section 6. Section 7 discusses deterministic and stochastic timed Petri nets. Examples are given to illustrate the basic solution techniques. Section 8 concludes the chapter by discussing briefly other extended Petri net models for convenience or other purposes in modeling, simulation, control, and performance analysis of manufacturing systems.

2. Petri Nets as a Tool and Methodology in Automation

The growth in the complexity of modern manufacturing systems and information systems creates numerous problems for their developers. In the planning stage, one is confronted with increased capabilities of these systems due to the unique combination of hardware and software, which operate under a large number of constraints arising from the limited system resources. In view of the capital intensive and complex nature of modern industrial systems, the design and operation of these systems require modeling and analysis in order to select the optimal design alternative, and operational policy. It is well known that flaws in the modeling process can substantially contribute to the development time and cost. The operational efficiency may be affected as well. Therefore special attention should be paid to the correctness of the models that are used at all planning levels [73].

Petri nets, as a graphical and mathematical tool, provide a uniform environment for modeling, formal analysis, and design of discrete event systems, particularly flexible and agile manufacturing systems. One of the major advantages of using Petri net models is that the same model is used for the analysis of behavioral properties and performance evaluation, as well as for systematic construction of discrete-event simulators and controllers. Petri nets were named after Carl A. Petri who created in 1962 a net-like mathematical tool for the study of communication with automata [48, 51]. Their further development was facilitated by the fact that Petri nets can be used to model properties such as process synchronization, asynchronous events, sequential operations, concurrent operations, and conflicts or resource sharing. These properties characterize discrete-event systems whose examples include industrial automated systems, communication systems, and computer-based systems. This, and other factors - discussed in this chapter, make Petri nets a promising tool and technology for the application to flexible and agile automation.

Petri nets as a graphical tool provide a powerful communication medium between the user, typically requirements engineer, and the customer [73]. Complex requirements specifications, instead of using ambiguous textual descriptions or mathematical notation difficult to understand by the customer, can be represented graphically using Petri nets. This combined with the existence of computer tools [12, 14, 22, 50, 53, 55, 69] allowing for interactive graphical simulation of Petri nets, puts in hands of the development engineers a powerful tool assisting in the development process of complex engineering systems [73].

As a mathematical tool, a Petri net model can be described by a set of linear algebraic equations, or other mathematical models reflecting the behavior of the system [43-44, 48, 51, 73]. This opens a possibility for the formal analysis of the model. This allows one to perform a formal check of the properties related to the behavior of the underlying system, e.g., precedence relations amongst events, concurrent operations, appropriate synchronization, freedom from deadlock, repetitive activities, and mutual exclusion of shared resources, to mention some. The simulation-based model validation can only produce a limited set of states of the modeled system, and thus can only show presence (but not absence) of errors in the model, and its underlying requirements specification. The ability of Petri nets to verify formally the model is especially important for the real-time safety-critical automated systems.

Petri nets have been used extensively to model and analyze manufacturing systems. In this area, Petri nets were used to represent simple production lines with buffers, machine shops, automotive production systems, flexible manufacturing systems, automated assembly lines, intelligent machines and robots, resource-sharing systems, and recently just-in-time and kanban manufacturing systems. Some of the most recent developments involving modeling and qualitative analysis were reported in [1-2, 10, 18-20, 23-24, 38, 58-59, 63, 70-72]. The deadlock avoidance was studied in [4, 31, 57, 65].

The application of Petri nets to modeling sequence controllers is another success story. Programmable Logic Controllers are commonly used for the sequence control in automated systems. They are designed using ladder logic diagrams, which are known to be very difficult to debug and modify. Petri net-based sequence controllers, on the other hand, are easy to design, implement, and maintain. In the early 80's, Hitachi Ltd developed a Petri net-based sequence controller which was successfully used in real applications to control parts assembly system, and automatic warehouse load/unload system [45]. The use of Petri nets, it was claimed, substantially reduced the system development time compared with the traditional approach. Numerous approaches to the synthesis and implementation of Petri net-based sequence controllers as well as reusable control software have been reported in the past few years [7, 16-19, 23-24, 30-32, 36, 46, 56, 66].

Petri nets, as a mathematical tool, allow for the performance evaluation of the modeled systems. Both deterministic and stochastic performance measures can be evaluated by using a broad class of Petri net models incorporating in their definitions deterministic and/or probabilistic time functions. Petri nets allowing deterministic times can be used to derive production cycle time and identify bottleneck workstations in a production system [26, 28-29, 42, 47, 67]. The use of Petri net models which incorporate time functions having probabilistic distributions allows one to obtain production rates or throughput, average delays, and critical resource utilization and reliability measures for the modeled systems. The stochastic nets with certain constraints can be solved by mapping onto and solving the underlying (semi)-Markov processes [12, 25, 41] while for general cases they can be analyzed via Petri net-driven discrete-event simulation [21, 53, 55, 69]. In recent years, this class of Petri net models has been extensively used to model and study analytically performance of production systems, involving simple production lines, job shops, robotic assembly cells, flexible manufacturing systems [8, 18, 64]. When a state explosion problem arises, or the underlying stochastic models are not amenable for tractable mathematical analysis, simulation may be conducted for the analysis of both qualitative and quantitative properties. The discrete event simulation can be driven from the model, sometimes using complex algorithmic strategies representing real-time scheduling and control policies for production systems [69].

Petri nets with time extensions, combined with heuristic search techniques, were used to model and study the planning and scheduling problems for manufacturing systems [9, 11, 35, 54, 62]. The robotic assembly and trajectory planning using Petri nets was presented in [39]. Petri nets with time extensions were also used to model and analyze dynamics of continuous chemical processes [28]. The continuous time and discrete-event process control was modeled and analyzed in [30, 61].

Another area of application was communication networks [25, 41, 60]. Work was conducted on Fiber Optics Local Area Networks such as Expressnet, Fastnet, D-Net, U-Net, Token Ring and Fiber Distributed Data Interface (FDDI) network [6, 13, 37]. Fieldbuses, such as FIP and ISA-SP50, have attracted much attention in the last two years [21]. This is not surprising, since they are very important networks for factory automation.

A recent overviews of industrial applications of Petri nets can be seen in [73]. Other tutorial and survey papers can also be seen in [19, 52, 69, 73]. The monographs and textbooks related to manufacturing systems design using Petri nets include [17-18, 51, 68].

3. Description of Petri Nets

A Petri net may be identified as a particular kind of bipartite directed graphs populated by three types of objects. These objects are places, transitions, and directed arcs connecting places to transitions and transitions to

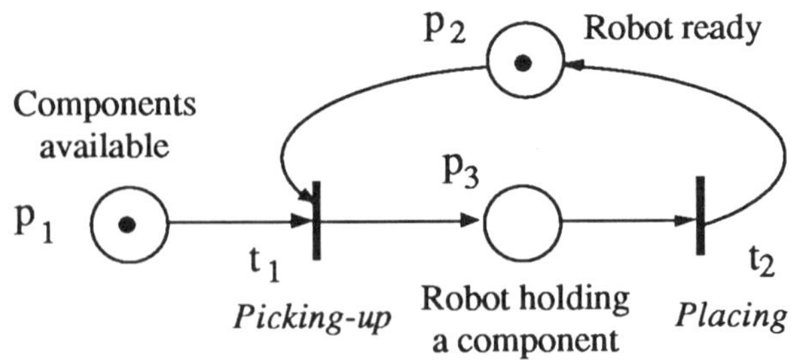

Figure 1: A simple PN example [69]

places. Pictorially, places are depicted by circles, transitions by bars or boxes. A place is an input place to a transition if there exists a directed arc connecting this place to the transition. A place is an output place of a transition if there exists a directed arc connecting the transition to the place. In its simplest form, a Petri net may be represented by a transition together with its input and output places. This elementary net may be used to represent various aspects of the modeled systems. For instance, input (output) places may represent preconditions (postconditions), the transition an event. Input places may represent the availability of resources, the transition their utilization, output places the release of the resources. An example of a Petri net is shown in Fig. 1. This net consists of three places, represented by circles, two transitions, depicted by bars, and directed arcs connecting places to transitions and transitions to places. In this net, places p_1 and p_2 are two input places of transition t_1. Place p_3 is an output place of transition t_1.

In order to study dynamic behavior of the modeled system, in terms of its states and their changes, each place may potentially hold either none or a positive number of tokens, pictured by small solid dots, as shown in Fig. 1. The presence or absence of a token in a place can indicate whether a condition associated with this place is true or false, for instance. For a place representing the availability of resources, the number of tokens in this place indicates the number of available resources. For example, a token in p_1 and p_2 means a component and a robot available, respectively. At any given time instance, the distribution of tokens on places, called Petri net marking, defines the current state of the modeled system. A marking of a Petri net with n places is represented by an $(n \times 1)$ vector m, elements of which, denoted as $m(p)$, are nonnegative integers representing the number of tokens in the corresponding places. A Petri net containing tokens is called marked Petri net. For example, in the Petri net model shown in Fig. 1, $m = (1,1,0)^T$.

Formally, a Petri net can be defined as follows:

PN $= (P, T, I, O, m_0)$; where

1. $P = \{p_1, p_2, ..., p_n\}$ is a finite set of places,

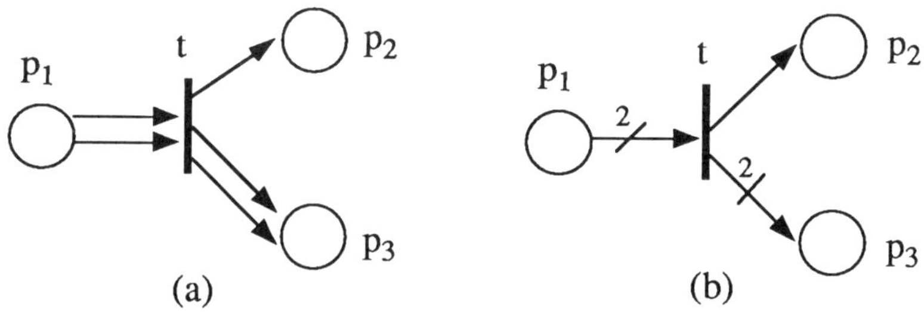

Figure 2: (a) Multiple arcs and (b) Compact representation of multiple arcs

2. $T = \{t_1, t_2, ..., t_s\}$ is a finite set of transitions, $P \cup T \neq \emptyset$, and $P \cap T = \emptyset$,
3. $I : (P \times T) \mapsto N$ is an input function that defines directed arcs from places to transitions, where N is a set of nonnegative integers,
4. $O : (P \times T) \mapsto N$ is an output function which defines directed arcs from transitions to places,
5. $m_0 : P \mapsto N$ is the initial marking.

If $I(p,t) = k$ ($O(p,t) = k$), then there exist k directed (parallel) arcs connecting place p to transition t (transition t to place p). If $I(p,t) = 0$ ($O(p,t) = 0$), then there exist no directed arcs connecting p to t (t to p). Frequently, in the graphical representation, parallel arcs connecting a place (transition) to a transition (place) are represented by a single directed arc labeled with its multiplicity, or weight k. This compact representation of multiple arcs is shown in Fig. 2.

By changing distribution of tokens on places, which may reflect the occurrence of events or execution of operations, for instance, one can study dynamic behavior of the modeled system. The following rules are used to govern the flow of tokens:

1. *Enabling Rule:* A transition t is said to be enabled if each input place p of t contains at least the number of tokens equal to the weight of the directed arc connecting p to t, i.e., $m(p) \geq I(p,t)$ for any place p in P.
2. *Firing Rule:*
   (a) An enabled transition t may or may not fire depending on the additional interpretation, and
   (b) A firing of an enabled transition t removes from each input place p the number of tokens equal to the weight of the directed arc connecting p to t. It also deposits in each output place p the

number of tokens equal to the weight of the directed arc connecting t to p. Mathematically, firing t at m yields a new marking $m'(p) = m(p) - I(p,t) + O(p,t)$ for any p in P.

The enabling and firing rules are illustrated in Fig. 3. In Fig. 3(a), number 3 in place p_1 means there are three tokens (a compact representation for more than one token in a place). Transition t is enabled as its input place p_1 contains three tokens, and $I(p_1,t) = 2$. The firing of the enabled transition t removes from the input place p_1 two tokens as $I(p_1,t) = 2$, and deposits one token in the output place p_2, $O(p_2,t) = 1$, and two tokens in the output place p_3, $O(p_3,t) = 2$. This is shown in Fig. 3(b).

The net in Fig. 1 in fact models a pick-place robotic assembly cell. The robot picks up a component and then places it in a desirable position. Two events are *picking-up* and *placing*, represented by two transitions t_1 and t_2, respectively. The first one can take place if a component is available and the robot is ready. These two conditions are modeled and labeled by two places p_1 and p_2. Putting at least one dot, a token, into each circle represents that each condition is true. Two arcs link p_1 and p_2 to t_1. After the robot picks up a component, a new condition results, i.e., the robot holds the component, represented by another place p_3. Then event *placing* represented by t_2 can occur. Arcs from t_1 to p_3 and from p_3 to t_2 show the relationships. Once the robot has placed the component, it is ready for the next pick-up operation. This is modeled by the arc from t_2 to p_2. Since the component is already inserted, there is no arc from t_2 to p_1.

Tokens (dots) in a PN appear identical although they may carry different physical interpretations in different places. For example, a token in p_1 means one available component and one in p_2 the ready robot. The number of tokens in a place defines a local condition. The system operation can now be described by the execution rule defined above. The initial system state for the above system is $(1,1,0)^T$ given the token distribution in Fig. 1. Only transition t_1 is enabled. The state changes to $(0,0,1)^T$ when t_1 fires. At this new marking, only t_2 is enabled. Firing it yields $(0,1,0)^T$. At this marking, no more transitions are enabled. This clearly shows the evolution of the robotic assembly cell when there is only one component to insert.

The modeling power of Petri nets can be increased by adding the zero testing ability, i.e., the ability to test whether a place has no token. This is achieved by introducing an inhibitor arc. The inhibitor arc connects an input place to a transition, and is pictorially represented by an arc terminated with a small circle. A Petri net with an inhibitor arc is shown in Fig. 4. The presence of an inhibitor arc connecting an input place to a transition changes the transition enabling conditions. In the presence of the inhibitor arc, a transition is regarded as enabled if each input place, connected to the transition by a normal arc (an arc terminated with an arrow), contains at least the number of tokens equal to the weight of the arc, and no tokens are

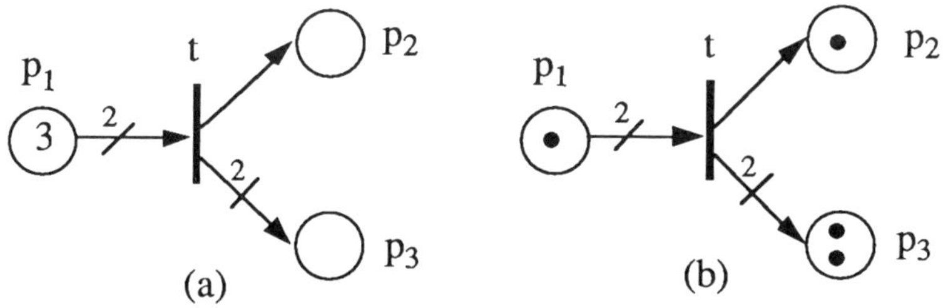

Figure 3: (a) Transition t enabled and (b) after t fires

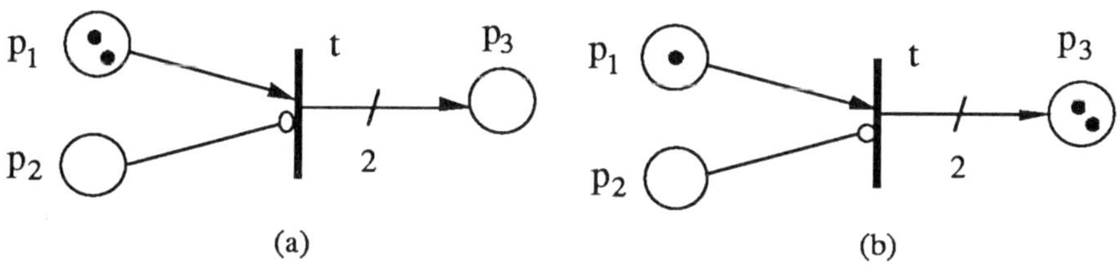

Figure 4: (a) Inhibitor arc and enabled transition (b) After firing t once

present on each input place connected to the transition by the inhibitor arc. The transition firing rules are the same for normally connected places. The firing, however, does not change the marking in the inhibitor arc connected places. For example, t in Fig. 4(a) is enabled at $(2, 0, 0)^T$ and firing it yields $(1, 0, 2)^T$ as shown in Fig. 4(b). The inhibitor arc can be generalized to one with multiplicity. For example, suppose that the inhibitor arc from p_2 to t has multiplicity k (> 0). Then if the number of tokens in p_2 is equal to or greater than k, its output transition t is disabled.

A Petri net is said to be pure or self-loop free if no place is an input place and output place of the same transition. A Petri net that contains self-loops can always be converted to a pure Petri net as shown in Fig. 5.

4. Modeling with Petri Nets

An application of Petri nets to industrial systems starts with their understanding and modeling. To design a classical continuous or discrete-time control systems, designers have to understand the operation of a physical plant. Their design methodologies require the understanding and modeling of a system to a certain degree. For example, optimal control of a plant requires very accurate mathematical descriptions and fuzzy logic control requires rule-based knowledge of the system's operations in terms of the input-output relations. To analyze and evaluate a control system, analysts again need the mathematical

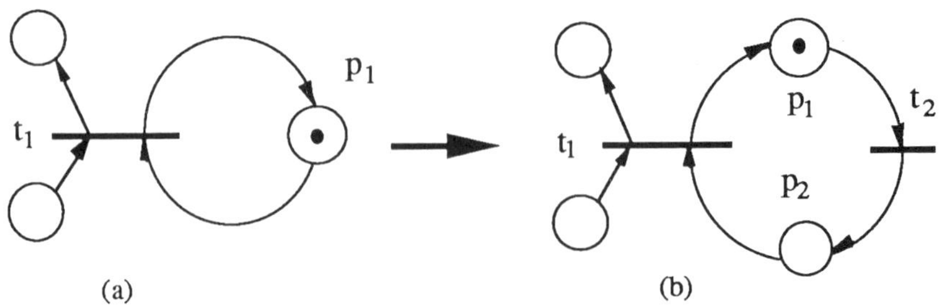

Figure 5: (a) A net with self loop and (b) Conversion to self-loop-free net

modeling of system components' characteristics and the relationships among them. Necessary assumptions have to be made to achieve the efficiency and simplicity in design and analysis. In the continuous and discrete-time systems, there exist physical laws governing the components and their interconnections which can be used to build a model with, e.g., differential or difference equations. Are there any such laws in discrete event system modeling? How do we build a model from the system description? This section tries to answer these questions. A adhoc modeling method will be introduced and two manufacturing examples are given.

An automated manufacturing system consists of a variety of components ranging from robots, machines, raw materials, sensors, actuators, computers and accessories related to a specific process. At the modeling stages, one needs to focus on the major operations and their sequential or precedent, concurrent, or conflicting relationships. The basic relations among these processes or operations can be classified as follows:

Sequential: If one operation follows the other, then the places or transitions representing them should form a cascade or sequential relation in Petri nets.

Concurrent: If two operations are initiated by an event, they form a parallel structure starting with a transition, i.e., two places are two outputs of the same transition. The pipeline concurrent operations can be represented with a sequentially-connected series of places/transitions in which multiple places can be marked simultaneously or multiple transitions are enabled at certain markings.

Conflicting: If either of two operations can follow an operation, then two transitions will be two outputs from a same place.

Cyclic: If a sequence of operations follow one after another and completion of the last one initiates the first one, then the cyclic structure is formed among these operations.

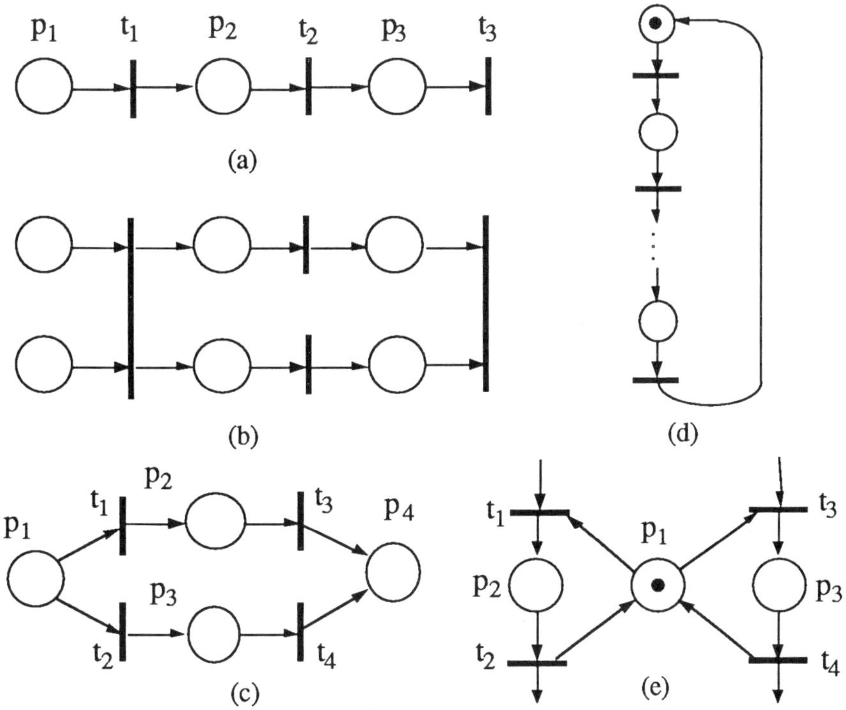

Figure 6: Examples (a) Sequential (b) Concurrent (c) Conflicting (d) Cyclic (e) Mutual exclusive

Mutual Exclusive: Two processes are mutually exclusive if they cannot be performed at the same time due to constraints on the usage of shared resources. A structure to realize this is through a common place marked with one token plus multiple output and input arcs to activate these processes. For example, a robot may be shared by two machines for loading and unloading. Two such structures are parallel mutual exclusion and sequential mutual exclusion discussed in [63].

Their typical Petri net examples are given in Fig. 6.

The resources can be classified into dedicated and shared ones. The dedicated one features with a place with single input and single output arcs; the shared one a place with multiple-input and/or multiple-output arcs. A condition representing the status of a sensor or an actuator is also modeled with a place whose holding a token implies the truth of the modeled condition. The same kind of resources may be represented by a place with the number of tokens corresponding to the number of resources. Initiation of an operation requires often several kinds of conditions and resources available - modeled as a transition with several input places. Completion of an operation may release some resources and change the status of the conditions - modeled as a transition with several output places.

A modeling method can be summarized into the following steps:

1. Identification of Operations and Resources: Given an industrial automated system description, identify the major events, operations, processes, resources, conditions, routing information, etc.

2. Identification of Relations: The relationships among the above identified events, operations, and/or processes have to be determined. The resources should be divided into different classes and major allocation policies for shared ones and routing information on products should be determined. Then the initial net structure can be roughly decided.

3. Petri Net Design: Design and label the places and/or transitions which represent events, operations, and/or processes. Arrange them according to the relationship identified in Step 2. Designate and label places which model status of resource(s) and the condition(s). Insert necessary places and transitions such that no two places can link to each other and neither two transitions. For each transition, draw an input arc to it from a place if enabling it requires the resource(s), truth of the condition, or the completion of the operations represented in the place; draw an output arc from it to a place if firing it releases resource(s), changes the condition(s), or signals the initiation of the operation in the place. The number of input arcs from a place to the transition should equal the required quantity of the tokens (often implying resources) in the place to enable the transition. The number of output arcs from the transition to a place should equal the quantity of the tokens to the place to be produced due to the transition's firing. Determine the initial number of tokens over all places according to the system's initial state. Associate other characteristics, e.g., timing, with places and transitions if needed.

4. Petri Net Modification: Check the net model to see whether it reflects the system operation and modify the net until it models the system.

It should be noted that the above method generally cannot lead to a net model with known behavioral properties [19, 68]. In addition, some systems cannot be modeled using ordinary Petri nets unless some extensions such as inhibitor arcs or priority transitions are used.

A Two Machine Production Line

The production line consists of two machines (M1 and M2), two robot arms (R1 and R2), and two conveyors. Each machine is serviced by a dedicated robot

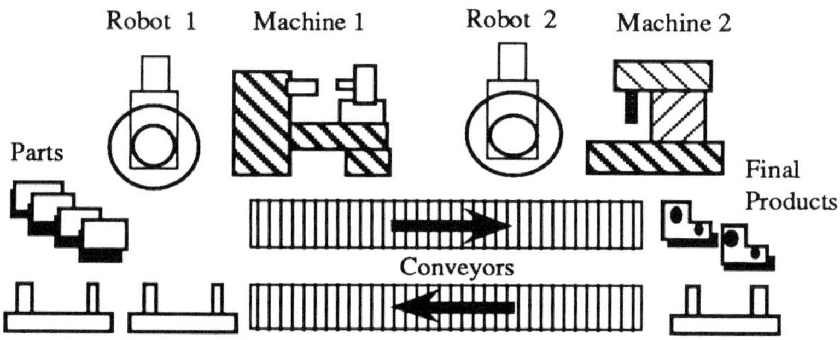

Figure 7: A two-machine production line [73]

which performs load and unload tasks. One conveyor is used to transport workpieces, a maximum of two at a time. The other conveyor is used to transport empty pallets. There are three pallets available in the system. The arrangement is shown in Fig. 7. Each workpiece is machined on M1, and M2, in this order. The machining on M1 (M2) takes 10 (16) time units. The load and unload tasks require 1 time unit. The conveyor's transportation times are negligible. The Petri net modeling steps follow:

1. Major activities are: *R1 loading, M1 processing, R1 unloading, R2 loading, M2 processing*, and *R2 unloading*. The resources are raw materials with pallets, Conveyors, M1, M2, R1 and R2.

2. The relations among these operations form a strict sequential sequence for products. All the resources are dedicated.

3. Design four transitions for four relatively short operations, i.e., *R1 loading, R1 unloading, R2 loading*, and *R2 unloading*, two places for M1 and M2 operations, and one place for intermediate parts are ready in conveyor for M2. Due to their strict sequential relations, they are connected with direct arcs from t_1 to p_2, p_2 to t_2, ..., p_3 to t_4 as shown in Fig. 8(a). Next, designate six places to represent six resources, pallets, conveyor to carry intermediate parts, M1, M2, R1 and R2. Note that the conveyor carrying empty pallets back to the raw material entry area is ignored. Add arcs based on, for each operation's start and completion, which resources are required and released.

   For example, the input arcs from p_1, p_5, and p_8 to transition t_1 are constructed due to the fact that starting *R1-loading* requires the availability of raw materials with pallets, M1, and R1. the output arc from t_1 to p_2 is formed because the precedence relation between *R1-loading* (t_1) and *M1-processing* (p_2). Other direct arcs are drawn as shown in Fig. 8(b).

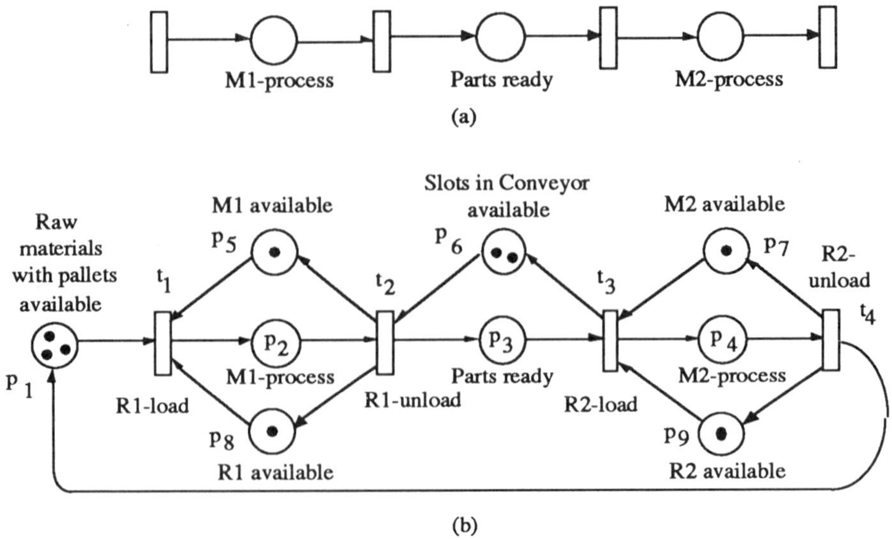

Figure 8: (a) Major Operations represented by 4 transitions and 3 places, (b) Complete Petri net model for the two-machine production line

Mark all the resource places with the tokens whose number is determined based on the number of resources in the specification. For example, there is only M1 and one token is thus added to p_5. There are three pallets, and three tokens are thus added to p_1.

Finally, associate the timing needed to perform operations with the corresponding places and transitions. The explanation and timing information are summarized in Table 2.

4. Check it and we find that the model represents the specifications if conveyor's transportation is ignored.

The following example uses the places for all operations while transitions represent start and/or completion of operations in a system.

An AT&T Flexible Workstation

The AT&T FWS-200 flexible workstation is developed for high-volume, low-cost production of printed circuit boards. The front view of FWS-200 is shown in Fig. 9. It consists of the following major components:

1. The frame structure supports the control cabinet, robot arms, platen, work table, etc.

Table 1: Interpretation and delays of places and transitions in Fig. 8(b)

Place	Interpretation	Time Delay
p_1	Workpieces and pallets available	0
p_2	M1 processing a workpiece	10
p_3	Workpiece available for processing at M2	0
p_4	M2 processing a workpiece	16
p_5	M1 available	0
p_6	Conveyor slots available	0
p_7	M2 available	0
p_8	R1 available	0
p_9	R2 available	0
Transition	Interpretation	Time Delay
t_1	R1 moves a workpiece to M1	1
t_2	R1 moves a workpiece from M1 to Conveyor	1
t_3	R2 moves a workpiece to M2	1
t_4	R2 removes a final workpiece from M2	1

2. The Control Cabinet houses hardware including an AT&T personal computer, Industrial I/O modules and STD bus, control electronics, display, pneumatic and ventilation systems, and power supplies.

3. The Robot Arms are able to accomplish pick-and-place operations, with their X, Y, Z, and θ motions in high accuracy (± 0.001 inch along X, Y, and Z axes and $\pm 0.15^o$ about Z axis).

4. The AT&T PC 6386 computer serves as the supervisory controller for the system.

5. The Drive Electronics contains the electronics for controlling the various axes, with two servo motor amplifiers for separate Z and θ motions and a stepper motor amplifier for both X and Y motions.

6. The Control Panel contains power buttons, touch display, pressure and vacuum gauges, and emergency stop buttons.

7. The Vision System includes IRI SV512 and ICOS M10000 modules.

8. The Peripheral Equipment includes part feeders and others.

The system can be used to accomplish tasks such as assembly of surface mount components, assembly of through-hole components, wire harnessing, etc. The common work area is of size 24 inch $\times$ 32 inch (60.96 cm $\times$ 81.28 cm); refer to [3] for details.

With focus on the activities of the two robot arms in the system, we design a Petri net model in the following steps:

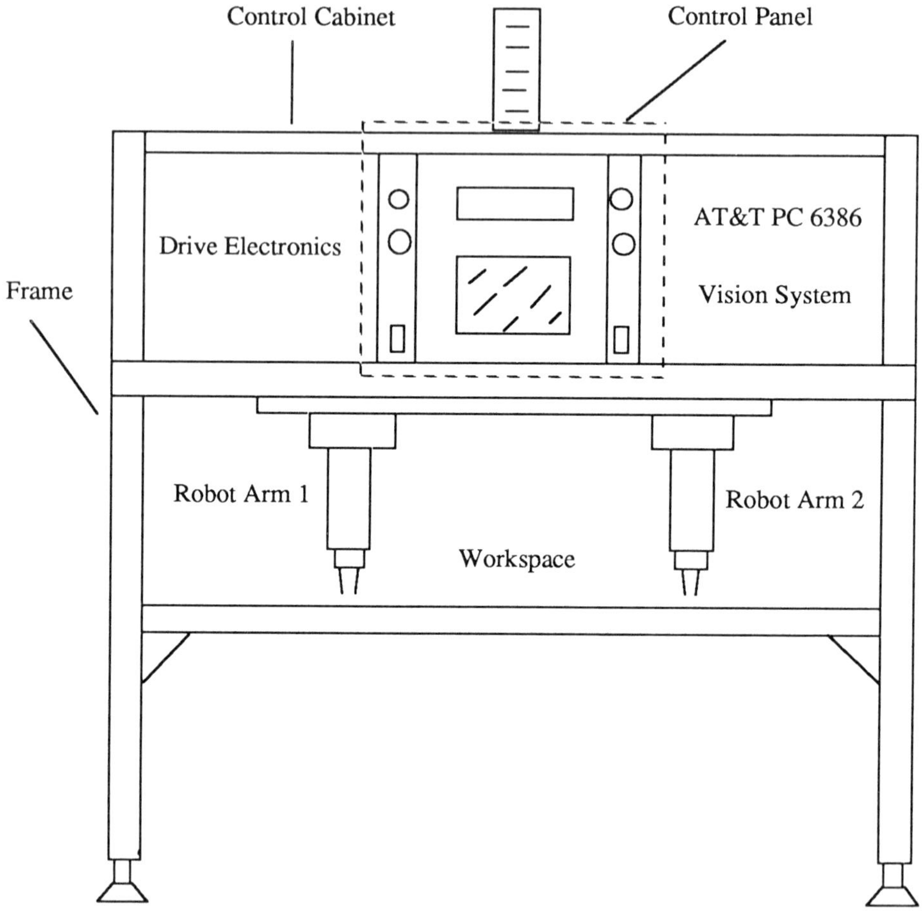

Figure 9: The front view of FWS-200 developed by AT&T [64]

1. Each robot performs repeatedly the following operations: *picking, moving, inserting*, and *moving*. To use the common workspace, each robot has to request first, then gets access to, and leaves the common workspace. The resources include the shared feeder area, shared PCB area, and the two robots.

   However, since the two arms work on the same circuit board and obtain components from the same feeder area, avoidance of arm collision has to be considered. Collision avoidance can be achieved by mutual exclusion structure implemented using semaphores in the Modular Manufacturing Language (M^2L) program [3]. Thus, when one robot is doing a job above the PCB area, the other is prohibited from moving into the same area. Possible collision may also occur in the feeder area. Based on similar reasoning, the use of mutual exclusion structure is also needed for this area.

2. The relations among these operations and events for each robot are sequential. R1 and R2's starting pick-up as well moving to PCB area for insertion are mutually exclusive, respectively and thus the robot collision can be avoided.

3. For each robot, design the four places for R1 and R2's operations, and five transitions for their start and/or end and arrange them in a sequential order, as shown in Fig. 10(a). Designate six places to represent the availability of two robots, two kinds of components, feeder area, and PCB area, respectively. Label these places and transitions as shown in Fig. 10(a). Next, add the arcs. Since the precedence relation between picking, moving, inserting, and moving, we add the arcs from t_{i1} to p_{i2}, p_{i2} to t_{i2}, ..., t_{i4} to p_{i5}, and p_{i5} to t_{i5} for $i = 1, 2$. Since enabling t_{i1} requires the availability of the feeder area, robot arms Ri, and Ri's component, input arcs from p_3, p_i, and p_{i1} are added to t_{i1}. Firing t_{i2} makes feeder area available, and thus output arc from t_{i2} to p_3 is formed. Enabling t_{i2} requires the availability of the PCB area, and thus input arc from p_4 to t_{i2} is created. Continue this process till all the relations are represented. Finally, we add a token to all the resource places to signify the initial system state. The use of one token in p_{i1} and the arc from t_{i5} back to p_{i1} models actually the case that components to be inserted will be always available over the concerned period. The final model is shown in Fig. 10(b).

4. Check it and we find that the model represents the specifications.

When timing information, possible changes of tools for robots, and insertion errors are considered, the above model can be further extended [64].

Petri net modeling and synthesis is a very important research area which attracts much attention. Top-down, bottom-up, hybrid methods are proposed

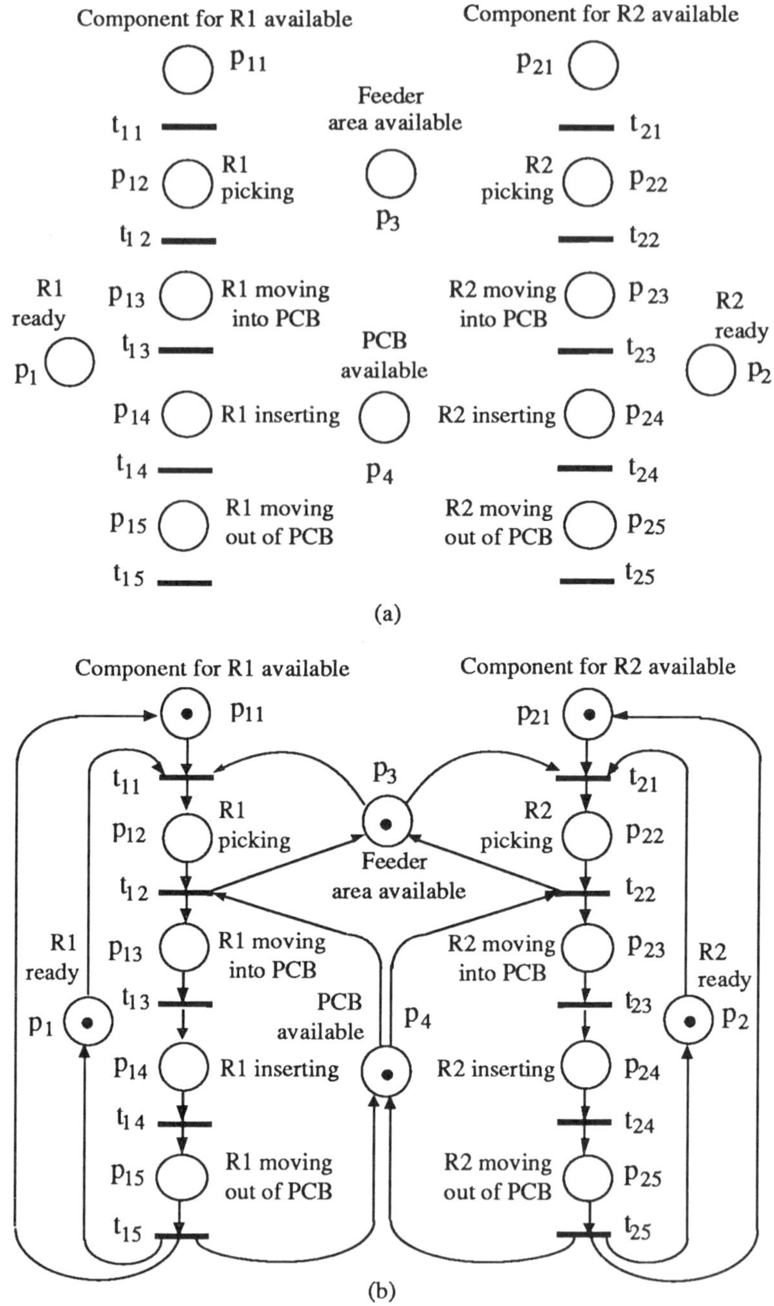

Figure 10: (a) Places and transitions and their arrangement, and (b) Complete net model after adding arcs

to handle Petri net modeling of flexible manufacturing systems. The current focus is on the synthesis of a Petri net model which satisfies both the system specifications and certain behavioral properties such as freedom from deadlock, no capacity overflow and cyclic behavior. An in-depth treatment of this topic can be seen in [68] and an overview of these techniques can be referred to [33].

5. Petri Net Properties

Petri nets as a mathematical tool possess a number of properties. These properties, when interpreted in the context of the modeled system, allow the system designer to identify the presence or absence of the application domain specific functional properties of the system under design. Two types of properties can be distinguished, behavioral and structural ones. The behavioral properties are these which depend on the initial state or marking of a Petri net. The structural properties, on the other hand, do not depend on the initial marking of a Petri net. They depend on the topology, or net structure, of a Petri net. In this section, we provide an overview of some of the most important, from the practical point of view, behavioral properties. They are reachability, boundedness, conservativeness, liveness, reversibility and home state. Other behavioral properties such as coverability, persistence, synchronic distance, and fairness, structural properties, and their analysis are discussed in [43, 48, 51].

Reachability

An important issue in designing distributed systems is whether a system can reach a specific state, or exhibit a particular functional behavior. In general, the question is whether the system modeled with Petri nets exhibits all desirable properties, as specified in the requirements specification, and no undesirable ones.

In order to find out whether the modeled system can reach a specific state as a result of a required functional behavior, it is necessary to find such a sequence of firings of transitions which would transform a marking m_0 to m_i, where m_i represents the specific state, and the sequence of firings represents the required functional behavior. It should be noted that real systems may reach a given state as a result of exhibiting different permissible patterns of functional behavior. In a Petri net model, this should be reflected in the existence of specific sequences of transition firings, representing the required functional behavior, which would transform m_0 to the required m_i. The existence in the Petri net model of additional sequences of transition firings which transform m_0 to m_i indicates that the Petri net model may not be exactly reflecting the structure and dynamics of the underlying system. This may also indicate the presence of unanticipated facets of the functional behavior of the real system, provided that the Petri net model accurately reflects the underlying system requirements specification. A marking m_i is said to be reachable from a marking m_0 if there exists a sequence of transitions firings which transforms a marking m_0 to m_i. A marking m_1 is said to be immediately reachable from m_0 if firing an enabled transition in m_0 results in m_1. For instance, in the Petri net model of the two-machine production line in Fig. 8(b), the state in which one pallet is ready to enter the line, Machine M1 is processing, Conveyor holds one intermediate part for loading, and M2 is idle, is represented by the marking $m_i = (1,1,1,0,0,1,1,0,1)^T$. m_i can be reached from the initial marking $m_0 = (3,0,0,0,1,2,1,1,1)^T$, by firing the following sequence of transitions - $t_1t_2t_1$. The marking $(1,1,0,1,0,2,0,0,0)^T$, which represents that one pallet is ready to enter the line, M1 and M2 are processing, and Conveyor is empty, is immediately reachable from the marking m_i when transition t_3 fires. The set of all possible markings reachable from m_0 is called the reachability set, and denoted by $R(m_0)$. This will be discussed in more detail in Section 6. The set of all possible firing sequences from m_0 is denoted by $L(m_0)$. Thus the problem of identifying the existence of a specific state m_i on which the system can take, can be redefined as the problem of finding if $m_i \in R(m_0)$.

Boundedness and Safeness

Places are frequently used to represent information storage areas in communication and computer systems, product and tool storage areas in manufacturing systems, etc. It is important to be able to determine whether proposed control strategies prevent from the overflows of these storage areas. The Petri net property which helps to identify in the modeled system the existence of overflows is the concept of boundedness. A place p is said to be k-bounded if the number of tokens in p is always less or equal to k (k is a nonnegative integer number) for every marking m reachable from the initial marking $m_0, m \in R(m_0)$. It is safe if it is 1-bounded. A Petri net is k-bounded (safe) if each place is k-bounded (safe). A Petri net shown in Figs. 1 and 10(b) are safe. In these two nets no place can contain more than one token at any marking. The net in Fig. 8(b) is 3-bounded where p_3 is 3-bounded, p_3 and p_6 are 2-bounded, and other places are safe. The examples of unbounded Petri nets can be seen in [43, 73].

Conservativeness

In real systems, the number of resources in use is typically restricted by the financial as well as other constraints. If tokens are used to represent resources, the number of which in a system is typically fixed, then the number of tokens in a Petri net model of this system should remain unchanged irrespective of the marking the net takes on. This follows from the fact that resources are neither created nor destroyed, unless there is a provision for this to happen.

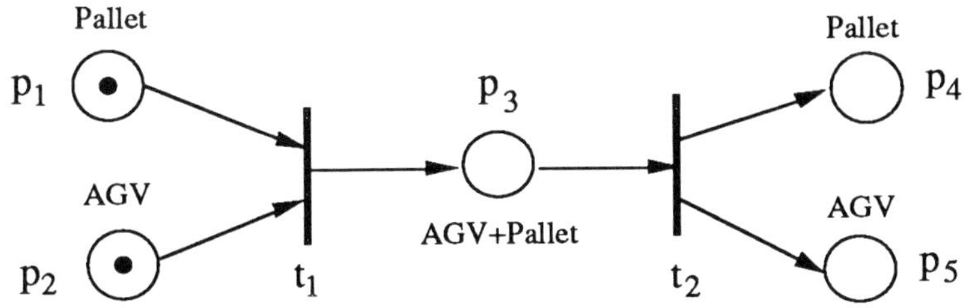

Figure 11: A conservative but not strictly conservative Petri net example

For instance, a broken tool may be removed from the manufacturing cell, thus reducing the number of tools available by one.

A Petri net is conservative if the number of tokens is conserved. From the net structural point of view, this can only happen if the number of input arcs to each transition is equal to the number of output arcs. However, in real systems resources are frequently combined together so that certain task can be executed. Then they are separated after the task is completed. For instance, in a flexible manufacturing system an automatic guided vehicle collects from a machining cell a pallet carrying products, and subsequently delivers it to the unload station where the vehicle and pallet are separated. This scenario is illustrated in Fig. 11. Transition t_1 models loading a pallet onto a vehicle; transition t_2 represents the pallet being delivered to the unload station and subsequently removed from the vehicle. Although the number of tokens in the net changes from two to one when t_1 fires, and then back to two tokens when t_2 fires, the number of resources in the system does not change. In order to overcome this problem, weights may be associated with places allowing for the weighted sum of tokens in a net to be constant. A Petri net is said to be conservative if there exists a vector $w = (w_1, w_2, ..., w_n)^T$, where n is the number of places, and $w(p) > 0$ for each $p \in P$, such that the weighted sum of tokens remains the same for each marking m reachable from the initial marking m_0. The Petri net shown in Fig. 11 is conservative with respect to vector $w = (1, 1, 2, 1, 1)^T$ as the weighted sum of tokens in each marking is two. An example of a Petri net which is not conservative is shown in Fig. 1 since there is no such a positive weighing vector such that $w^T m =$ constant. If a Petri net is conservative with respect to a vector with all elements equal to one, then the net is said to be strictly conservative. If each transition has the same number of input and output arcs, then the net will be strictly conservative.

Liveness

The concept of liveness is closely related to the deadlock situation, which has been studied extensively in the context of computer operating systems. The following four conditions must hold for a deadlock to occur [15]:

1. Mutual exclusion: a resource is either available or allocated to a process which has an exclusive access to this resource.

2. Hold and wait: a process is allowed to hold a resource(s) while requesting more resources.

3. No preemption: a resource(s) allocated to a process cannot be removed from the process, until it is released by the process itself.

4. Circular wait: two or more processes are arranged in a chain in which each process waits for resources held by the process next in the chain.

For instance, in a flexible manufacturing system, a deadlock occurs when the input/output buffer of a machining tool holds a pallet with already machined products, and another pallet with products to be machined has been delivered to the buffer. Assuming that the buffer can hold one pallet only at a time, and an automated guided vehicle (AGV), for instance, has a space for one pallet, a deadlock occurs. The pallet with machined parts cannot be moved from the buffer to the AGV. The pallet with parts to be machined cannot be moved from the AGV to the buffer. In this example, all four conditions hold, with the buffer and AGV space for pallets regarded as resources. Unless there is a provision in the control software for deadlock detection and recovery, a deadlock situation, initially confined to a small subsystem, may propagate to affect a large portion of a system. This frequently results in a complete standstill of a system. Additional examples of deadlock can be referred to [4, 18, 31, 68]. A Petri net modeling a deadlock-free system must be live. This implies that for all markings m, which are reachable from the initial marking m_0, it is ultimately possible to fire any transition in the net by progressing through some firing sequence. This requirement, however, might be too strict to represent some real systems or scenarios which exhibit deadlock-free behavior. For instance, the initialization of a system can be modeled by a transition (or transitions) which fires a finite number of times. After initialization, the system may exhibit a deadlock-free behavior, although the Petri net representing this system is no longer live as specified above. For this reason, different levels of liveness for transition t, and marking m_0, were introduced. A transition t in a Petri net is said to be:

- $L0$-live (or dead) if there is no firing sequence in $L(m_0)$ in which t can fire.
- $L1$-live (potentially firable) if t can be fired at least once in some firing sequence in $L(m_0)$.
- $L2$-live if t can be fired at least k times in some firing sequence in $L(m_0)$ given any positive integer k.
- $L3$-live if t can be fired infinitely often in some firing sequence in $L(m_0)$.

- $L4$-live (or live) if t is $L1$-live (potentially firable) in every marking in $R(m_0)$.

Following this classification, a Petri net is said to be Li-live, for marking m_0, if every transition in the net is Li-live. In Fig. 1, t_1 and t_2 are $L1$-live since they can fire only once from the initial marking m_0. All transitions in Figs. 8(b) and 10(b) are live ($L4$-live) since for each transition under any marking, one can find a sequence of transitions whose firing enables it.

Reversibility and Home State

An important issue in the operation of manufacturing systems, process control systems, etc., is the ability of these systems for an error recovery [18-19, 68]. These systems are required to return from the failure states to the preceding correct states. This requirement is closely related to the reversibility and home state properties of a Petri net. A Petri net, for the initial marking m_0, is said to be reversible if for each marking m in $R(m_0), m_0$ is reachable from m. The home state property is less restrictive, and more practical, then the reversibility property of a Petri net. A Petri net state m_i is said to be a home state if for each marking m in $R(m_0), m_i$ is reachable from m. The Petri net shown in Fig. 8(b) and 10(b) are reversible. The Petri net shown in Fig. 1 is not reversible.

6. Analysis Methods

The defined properties of Petri nets are useful to analyze modeled systems. An important issue to be considered during analysis is whether there exists one-to-one functional correspondence between the Petri net model and the original requirements specification; typically expressed in an informal way [73]. The construction of Petri net models from informal requirements specifications is not a trivial task, which requires a great deal of modeling experience, as well as the knowledge of the techniques assisting in the model construction. As a result, a Petri net model may differ considerably from its original specification. This is especially true when large Petri net models of complex systems are involved. The existence of the one-to-one functional correspondence between an original requirements specification and its Petri net representation allows to project the analysis results, obtained for the Petri net model, onto the original description. This provides a feedback to the customers which can, in many instances, help themselves clarify their perception of the system. Another important issue to be addressed during the analysis stage is the completeness of the requirements specification. In most cases, the specification defines the external functional behavior of a system. This is typically expressed in terms of the system input-output relationships. Inputs are generated by the environment of the system. Outputs are responses of the system to these inputs.

If some inputs, generated by the environment of the system, are not included in the requirements specification, then the system will be unable to respond to these inputs properly when they occur during the system normal operation. The completeness of the requirements is especially important in the case of safety-critical systems. In these systems, the incompleteness of the requirements specification may lead to catastrophic events to occur in the environment of the system. The consistency of the requirements specification is another issue to be considered during analysis. The inconsistency occurs when for a given permissible, temporal combination of inputs, a requirements specification allows for two or more different permissible temporal combinations of outputs. It is mainly due to a vague, incomplete, and frequently incorrect perception of the system functionality. This chapter discusses briefly four methods, i.e., Reachability Tree, Invariant Analysis, reduction, and simulation, for analyzing these properties.

A. Reachability Tree

This fundamental method is based on the enumeration of all possible markings reachable from the initial marking m_0. Starting with an initial marking m_0, one can construct the reachability set by firing all possible transitions enabled in all possible markings reachable from the initial marking m_0. In the reachability tree, each node is labeled with a marking; arcs are labeled with transitions. For unbounded Petri nets, symbol ω meaning the infinite needs to be used to generate a finite tree and it satisfies, $\omega > k$, $\omega \geq \omega$, and $\omega \pm k = \omega$ for any finite integer k. The tree can be generated as follows:

1.0) Let the initial marking m_0 be the root of the tree and tag it "new"
2.0) While "new"markings exist do the following:
3.0) Select a "new" marking m
3.1) If m is identical to another marking in the tree, then tag m "old", and go to another "new" marking
3.2) If no transitions are enabled in m, tag m "deadend"
4.0) .For every transition t enabled in marking m do the following:
4.1) Obtain the marking m' which results from firing t in m
4.2) If on the path from the root to m, there exists a marking m'' such that $m'(p) \geq m''(p)$ for each place p, and $m' \neq m''$, then replace $m'(p)$ by ω for each p wherever $m'(p) > m''(p)$
4.3) Introduce m' as a node, draw an arc from m to m' labeled t, and tag m'' "new".

The resulting tree is called coverability tree. The net is bounded if and only if ω does not appear in any node. For this case, the tree is called reachability tree, and

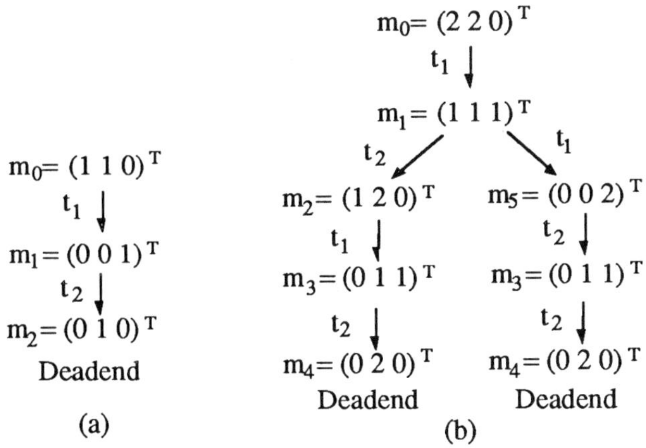

Figure 12: Reachability trees for the net in Fig. 1 with (a) $m_0 = (1,1,0)^T$ (b) $m_0 = (2,2,0)^T$

1. We can find the maximum number of tokens in a place over all markings, say, k, then the place is k-bounded. Similarly, if K is the maximum component over all markings, then the net is K-bounded.
2. If each node of the tree contains only zeros and ones, then the net is safe.
3. The number of different deadends in the tree is the number of deadlock states of the net.
4. A transition is dead if it does not appear as an arc label in the tree.
5. If given any two nodes in the reachability tree, there is a direct path and all transitions are present, then the net is live.
6. If there is a direct path from any node to the initial marking in the tree, then the net is reversible.

The limitation of this method is the combinatorial state explosion problem.

The reachability trees for the net in Fig. 1 with $m_0 = (1,1,0)^T$ and $m_0 = (2,2,0)^T$ are shown in Fig. 12. It can be observed that different initial markings result in different reachability trees. Other examples can be seen in [18-19, 43, 73].

B. Invariant Analysis

Arcs describe the relationships among places and transitions and can be represented by two matrices. By studying linear equations based on the execution rule and the matrices, one can find subsets of places over which the sum of the tokens remains unchanged. One may also find that a transition firing sequence brings the marking back to the same one.

Mathematically, denote $C = O - I$ incidence matrix. The positive integer solution x of $A^T x = 0$ is called a P-invariant. The P-invariants can be explained intuitively in the following way. The non-zero entries in a P-invariant represent weights associated with the corresponding places so that the weighted sum of tokens on these places is constant for all markings reachable from an initial marking. These places are said to be covered by a P-invariant.

The integer solution y of $Cy = 0$ is called a T-invariant. The non-zero entries in a T-invariant represent the firing counts of the corresponding transitions which belong to a firing sequence transforming a marking m_0 back to m_0. Although a T-invariant states the transitions comprising the firing sequence transforming a marking m_0 into m_0, and the number of times these transitions appear in this sequence, it does not specify the order of transition firings.

Invariant findings may aid in net analysis for some PN properties. For example, if each place in a net is covered by a P-invariant, then it is bounded. However, this approach is of limited use since invariant analysis does not include all the information of a general PN. Examples and applications of invariant analysis can be seen in [18, 43-44, 73].

C. Reduction

The reduction approach applies the decomposition principle to analyze a net. By simplifying a subnet or structure while preserving the concerned properties, one is able to derive the properties of a complex net. The limitation lies in that reducible subnets may not exist or are difficult to find. Some reductions rules could be very hard to apply. A set of easy-to-use reductions are given in Fig. 13, including fusion of series places/transitions, fusion of parallel places/transitions, and elimination of self-loop places/transitions. It can be proved that these reduction rules preserve the boundedness, liveness, and reversibility. An example to use these rules are given in Fig. 14. By fusing two transitions t_1 and t_2 into t_{12}, and then eliminating the self-loop place p_2, the most left net is obtained. The net is bounded, nonlive and nonreversible. Thus the original net is bounded, nonlive and nonreversible. A reader can also try to simplify Petri nets in Figs. 8(b) and 10(b). An in-depth treatment of the topic can also be seen in [51].

D. Simulation

For complex Petri net models, discrete-event simulation is another way to check the system properties. The idea is simple, i.e., using the execution algorithm to run the net. Simulation is an expensive and time-consuming technique. It can show the presence of undesirable properties but cannot prove the correctness of the model in general cases. Despite this, Petri net simulation is indeed a convenient and straightforward yet effective approach for engineers

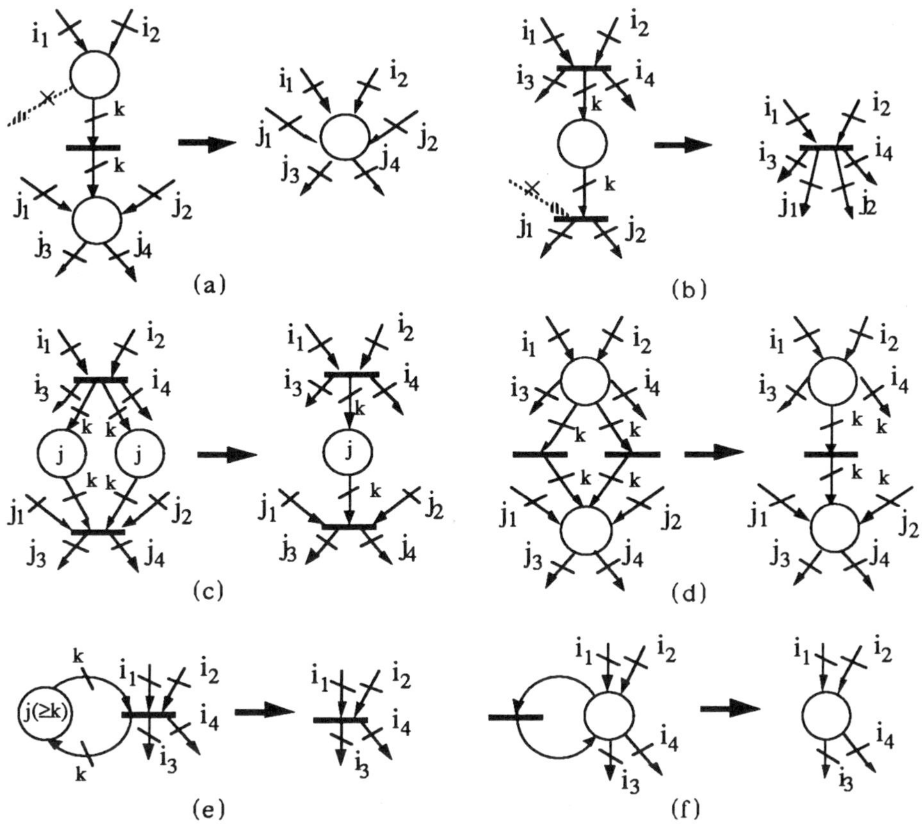

Figure 13: A set of reduction rules which preserve boundedness, liveness, and reversibility where k, j, i_j and j_i are non-negative integers and dotted arcs with a cross are prohibited in Reduction Rules (a) and (b)

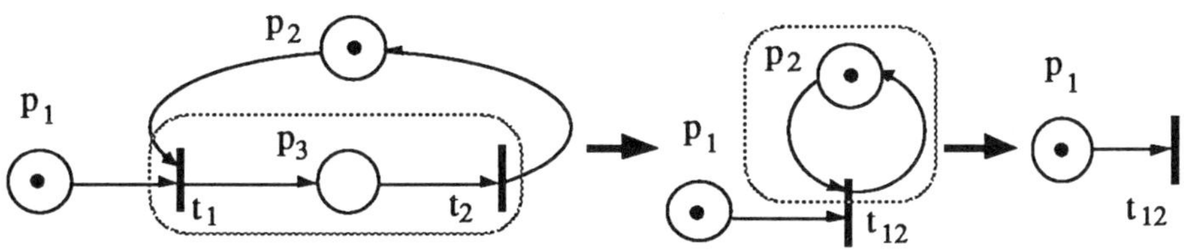

Figure 14: A reduction example

to validate the desired properties of a manufacturing system to a certain degree [40, 53, 55, 68]. The algorithm is given as follows:

1. Initialization: decide the initial marking and the set of all enabled transitions at the marking;

2. If the number of preset simulation steps or certain stopping criteria are met, stop. Otherwise, if there is no transition enabled, report a deadlock marking and either stop or go to Step 1.

3. Randomly pick a transition to fire. Remove the same number of tokens from each of its input places as the number of arcs from that place to the transition and deposit the same number of tokens to each of its output places as the number of arcs from the transition to that place.

4. Remove enabled transitions which are modified at the new marking by checking the output transitions of the input places used in Step 3. If the output transitions of the output places in Step 3 become enabled, add those enabled ones. Go to Step 2.

The above algorithm can be modified to simulate extended Petri nets such as timed ones. The advantage of the simulation method is to allow one to derive the temporal performance for a system under very realistic assumptions.

7. Performance Analysis

The development of technological systems requires that both functional and performance requirements are met [73]. Depending on the development stage of a system, the knowledge of either an approximate or exact (or both) performance may be required. For example, at the design stage, the approximate performance of the alternative design models is required in order to eliminate those proposals which are highly unlikely to meet the performance requirements when fully developed and implemented. Analytical techniques play an important role at this stage. They allow the designer to obtain the required performance measures, involving a relatively small time investment needed for the model construction and its solution. The selected design alternatives are then refined by increasing the level of details in order to include in the model the actual operational policies and time characteristics. As a result, the model complexity, or the presence of heuristic algorithms may prohibit the use of the analytical techniques. The discrete-event simulation is, then, the only viable alternative for the performance evaluation although it is an expensive and time consuming technique.

The ordinary Petri nets do not include any concept of time. With this class of nets, it is possible only to describe the logical structure of the modeled system, but not its time evolution. Responding to the need for the temporal performance analysis of discrete-event systems, time has been introduced into Petri nets in a variety of ways. In this section two fundamental types of timed Petri nets are discussed in the context of the performance evaluation. These are deterministic timed Petri nets, and stochastic ones. The previous two-machine production line is used to illustrate the basic solution techniques for these two classes of nets.

Deterministic Timed Petri Nets

Several approaches are possible for the introduction of temporal specifications into Petri net models. Time may be associated with places (timed place Petri net - TPPN). A token deposited in a place becomes available only after a certain period of time (time delay). Only available tokens can enable transitions. Time may be also associated with transitions (timed transition Petri net - TTPN). TTPN can be classified into three-phase firing TTPN, and atomic firing TTPN. In the former class, tokens are removed from the input places when a transition becomes enabled. The transition fires after a certain period of time (time delay), depositing tokens on the output places. In the later class, tokens remain on the input places of a transition which is enabled. After a certain period of time (time delay), the transition fires by removing tokens from the input places, and deposits tokens on the output places.

The presence of the conflict structures (a structure involving a place having two, or more output transitions) in a Petri net, execution of which leads to conflicts (in the conflict structure only one transition can fire, others become disabled), requires a conflict resolution mechanism to be introduced. Since this mechanism is, typically, based on a probabilistic function, the net becomes stochastic. For this very reason, the use of the deterministic timed Petri nets for the performance evaluation, as reported to date, has been restricted to the choice-free or conflict-free nets, which can be further modeled as marked graphs, or event-graphs [29, 43, 47].

A marked graph is a PN (P, T, I, O, m_0) such that $\forall p \in P, t \in T$, $I(p,t) \leq 1$, $O(p,t) \leq 1$, and given any $p \in P$, $|\{t \in T : O(p,t) = 1\}| = 1$, and $|\{t \in T : I(p,t) = 1\}| = 1$.

In a marked graph, each place has exactly one input transition and exactly one output transition. Thus no conflict is possible. The marked graph models are suitable to represent and study performance of the asynchronous concurrent systems which do not involve conflicts [47]. This class of nets has been extensively used to model and analyze performance of the industrial automated systems such as job-shop production systems, robotic assembly cells, and flexible manufacturing cells [8, 26, 29, 67].

A timed marked graph is a marked graph $(P, T, I, O, m_0, \gamma, \tau)$, where $\gamma : P \mapsto R^+$ is the place delay function, R^+ is the set of nonnegative real numbers, and $\tau : T \mapsto R^+$ is the transition firing time function.

In a timed marked graph, deterministic time delays are associated with places and transitions. A token in a place, with an associated delay time, can be available or not-available to enable its output transition. The previous firing rules for ordinary PNs are expanded to include the following:

1. At any time instance, a transition t becomes enabled if each input place p of t contains at least one available token, and
2. Once transition t is enabled, it starts firing by removing one token from each input place. It completes firing after the time delay $\tau(t)$, and deposits one token into each output place p. These tokens become available after time $\gamma(p)$.

A marked graph is strongly connected if there is a directed path from any node to any other node, where a node can be either a transition or place.

An elementary path is a sequence of nodes: $x_1x_2...x_n$, $n > 1$, such that there is an arc from x_i to x_{i+1}, where $1 \leq i < n$, and $x_i = x_j$ implies that $i = j, 1 \leq i, j \leq n$. An elementary loop, or circuit, is a sequence of nodes: $x_1x_2...x_n$, $n > 1$, such that there is an arc from x_i to x_{i+1}, where $1 \leq i < n$, and $x_i = x_j$, where $1 \leq i < j \leq n$, implies that $i = 1$ and $j = n$.

The fundamental approach to studying timed marked graphs is based on two concepts; the total time delay in a loop, and the total number of tokens in a loop. A strongly-connected marked graph consists of a limited number of elementary loops. The total time delay in a loop is obtained as a sum of delays introduced by all transition and places comprising the loop. The total number of tokens in a loop is obtained as a sum of tokens present in the places which belong to the loop. Then, the minimum cycle time of the marked graph is [47]:

$$\pi = \max_i \{D_i/N_i\}$$

where D_i is the time delay associated with the i-th loop, N_i is the total number of tokens present in the places of the i-th loop. The ratio D_i/N_i is called the cycle time of loop i.

The upper and lower cycle times can be derived for the timed (deterministic or random) marked graphs using the methods proposed in [8]. Linear programming was demonstrated to be a computationally efficient method for computing the cycle times [8, 42].

The previous production line is used to demonstrate the method for obtaining cycle times of a marked graph. The deterministic timed Petri net model of this system is shown in Fig. 8 and Table 1. The initial marking of the net is $(3\ 0\ 0\ 0\ 1\ 2\ 1\ 1\ 1)^T$. The Petri net model contains four loops. The time delays associated with these loops, as well as their token content are shown in Table 2. Then, the minimum cycle time is 18 time units, as determined by the

Table 2: Loops and their delays, token sums, and cycle times for Fig. 8

Loop	Loop Delay	Token Sum	Cycle Time
$t_1p_2t_2p_3t_3p_4t_4p_1t_1$	30	3	10
$t_1p_2t_2p_5$(or p_8)t_1	12	1	12
$t_2p_3t_3p_6t_2$	2	2	1
$t_3p_4t_4p_7$(or p_9)t_3	18	1	18

loops $t_3p_4t_4p_7$(or p_9)t_3. This means that it takes a minimum of 18 time units to transform a raw workpiece into a final product.

It should be noted that the increase in the number of available pallets will not make the minimum cycle time of the system any shorter; and the decrease in the number by one will not increase the cycle time either. This is because of the time involved in machining at M2 which is a bottleneck machine in this system. Thus, Petri net analysis can help decide the appropriate number of pallets and/or fixtures and bottleneck machines. More application examples can be found in [28-29, 67].

Stochastic Timed Petri Nets

When time delays are modeled as random variables, or probabilistic distributions are added to the deterministic timed Petri net models for the conflict resolution, stochastic timed Petri net models are yielded. In such models, it has become a convention to associate time delays with the transitions only. When the random variables are of general distribution or both deterministic and random variables are involved, the resulting net models cannot be solved analytically for general cases. Thus simulation or approximation methods are required. The stochastic timed Petri nets in which the time delay for each transition is assumed to be stochastic and exponentially distributed are called stochastic Petri nets (SPN) [25, 41]. The SPN models which allow for immediate transitions, i.e., with zero time delay, are called generalized SPN (GSPN) [12]. Both models, including extensions such as priority transitions, inhibitor arcs, and probabilistic arcs can be converted into their equivalent Markov process representations, and analyzed analytically.

To demonstrate the methodology, we consider the example of the system shown in Fig. 7. Suppose that the machining tool M1 is faster than M2, however subject to failures. M2 and the two robots are failure-free. Given the processing rates of the machining tools and robots, as well as the failure and repair rates of M1, we are interested in obtaining the average utilization of M1, and the production rate of the system (throughput), assuming that only one pallet is available [73]. The SPN model of the system, is shown in Fig. 15(a). A number of modeling assumptions were made for the convenience of the presentation:

Table 3: Interpretation and firing rates of places and transitions in Fig. 15(a)

Place	Interpretation	–
p_1	Workpieces and pallets available	–
p_2	M1 processing a workpiece	–
p_3	Workpiece ready for processing at M2	–
p_4	M1 in repair	–
p_5	M1 available	–
p_6	Conveyor slots available	–
Trans.	Interpretation	Firing Rate
t_1	Loading involving R1, and M1	$\lambda_1 = 40$
t_2	Processing and unloading with M1 and R1	$\lambda_2 = 5$
t_3	Loading, processing, and unloading with R2 and M2	$\lambda_3 = 4$
t_4	M1 breaks down	$\lambda_4 = 0.5$
t_5	M1 is repaired	$\lambda_5 = 4$

1. The tasks performed by M2 and R2 were aggregated into a single transition t_3 representing *loading, processing and unloading at M2.*

2. The time associated with the unloading tasks performed by R1 is incorporated into transition t_2 representing *processing at and unloading M1.*

3. The failure-repair loop is added to M1.

4. The time delays are associated with transitions only.

In the model, λ_i is used to represent the firing rate of transition t_i. The firing rate of a transition equals the reciprocal of the average firing delay time of the corresponding event or operation. The description of places and transitions is given in Table 3, together with the adopted values of the firing rates.

The reachability graph of the model is shown in Fig. 15(b). The graphical representation of the corresponding embedded continuous time Markov chain is shown in Fig. 15(c). In this representation, the links between states are labeled with the firing rates of the corresponding transitions.

The transition rate matrix (a_{ij}) $(i, j = 0, 1, \dots, q)$ for an embedded continuous time Markov chain with the number of states q can be derived as follows. For $i = 0, 1, \dots, q$, the ith row elements, i.e. a_{ij}, $j = 0, 1, \dots, q$ are determined:

1. if $j \neq i$, a_{ij} is the sum of the rates of all the outgoing arcs from state i to j.

2. Since any row of elements in A satisfies $\sum_{j=0}^{q} a_{ij} = 0$, after Step 1,

$$a_{ii} = -\sum_{j \neq i}^{q} a_{ij}.$$

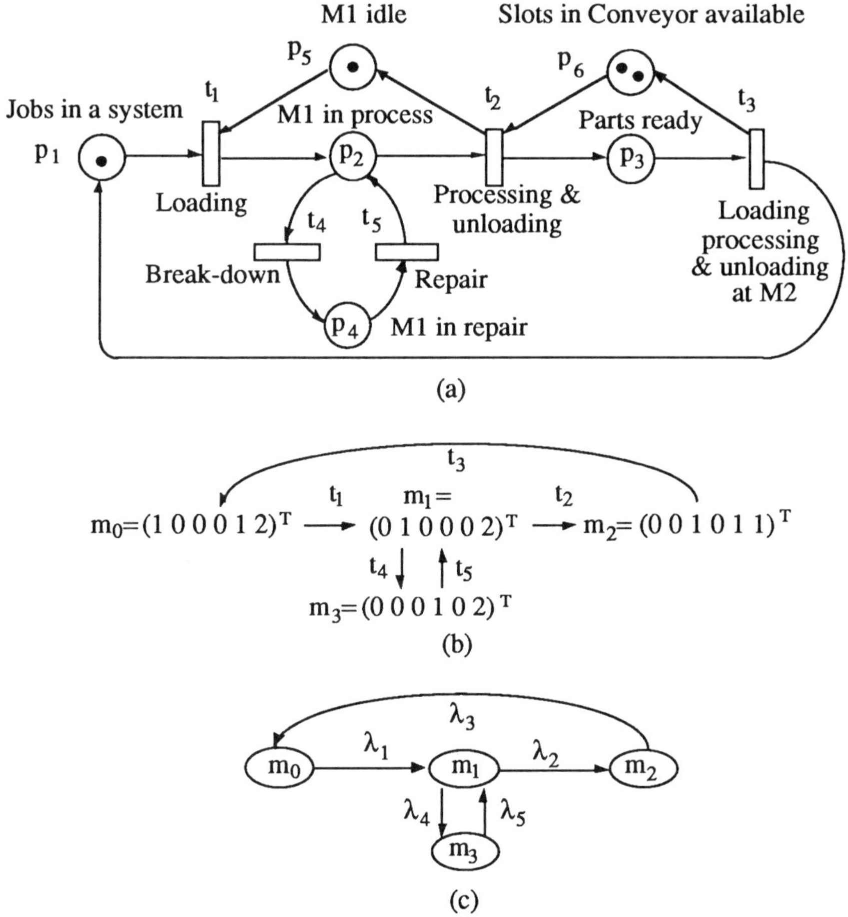

Figure 15: (a) SPN model (b) Reachability graph (c) Markov chain [73]

For example, when $i = 0$, a_{0j}, $j = 1, 2, 3$ are first obtained:

$$a_{01} = \lambda_1, \ a_{02} = a_{03} = 0.$$

Then $a_{00} = -\lambda_1$.

By the same token, we can find all other elements in the transition rate matrix which is:

$$A = \begin{bmatrix} -\lambda_1 & \lambda_1 & 0 & 0 \\ 0 & -\lambda_2 - \lambda_4 & \lambda_2 & \lambda_4 \\ \lambda_3 & 0 & -\lambda_3 & 0 \\ 0 & \lambda_5 & 0 & -\lambda_5 \end{bmatrix}$$

The steady-state probabilities, denoted as π_i, can be obtained by solving the following equations:

$$(\pi_0 \ \pi_1 \ \pi_2 \ \pi_3)A = 0$$
$$\pi_0 + \pi_1 + \pi_2 + \pi_3 = 1$$

This yields
$\pi_0 = \lambda_2\lambda_3\lambda_5/\lambda$, $\pi_1 = \lambda_1\lambda_3\lambda_5/\lambda$, $\pi_2 = \lambda_1\lambda_2\lambda_5/\lambda$, $\pi_3 = \lambda_1\lambda_3\lambda_4/\lambda$
where $\lambda = \lambda_2\lambda_3\lambda_5 + \lambda_1\lambda_3\lambda_5 + \lambda_1\lambda_2\lambda_5 + \lambda_1\lambda_3\lambda_4$.

When using the firing rates from Table 3, and the derived formulas for the steady-state probabilities, the following values are obtained $\pi_0 = 0.05$, $\pi_1 = 0.40$, $\pi_2 = 0.50$, and $\pi_3 = 0.05$.

1. The M1 executing a task is modeled by the presence of a token on place p_2. This corresponds to the marking m_1, and state probability π_1. The expected utilization of M1 is then 40%.

2. The completion of a product is modeled by firing transition t_3. This transition is enabled in marking m_3. Thus, the system production average rate is given by the product of the probability that t_3 is enabled and the firing rate of t_3, i.e., $\pi_2\lambda_3 = 2$ workpieces/time unit.

The analysis of a Markov model may involve the steady-state (as shown in the example) or transient analysis (the study of the system evolution from its initial state to the steady state), or both depending on the requirements and the Markov process itself. For example, if the net is not deadlock-free, i.e., the underlying Markov process has an absorbing state, then the steady-state analysis makes no sense. General-purpose software packages are available for solving the above stochastic net models, including GreatSPN [12] and SPNP [14], and more example applications on stochastic timed Petri nets can be referred to [8, 13, 18-19, 37, 41, 43, 64].

8. Petri Nets: Application Prospects

The development of Petri nets has been, to a large extent, motivated by the need to model the industrial systems. The introduced Petri nets are not always sufficient to represent and analyze complex industrial production systems. This prompted the development and research of new classes of nets. For example, high-level Petri nets are proposed to handle tokens with identity [34]. Thus they can offer a compact model for a system with many similar subsystems or components. Meanwhile, it should be noted that the model may lose the feature to visualize the properties. Object-oriented Petri nets allow for the representation and manipulation of an object class, in which tokens are considered as instances or tuples of instances of object classes [1, 5]. Fuzzy Petri nets are developed to deal with approximate and uncertain information in robotic assembly systems [9, 54].

Although attempts to combine Petri nets with other techniques, such as neural networks, fuzzy logic, artificial intelligence, etc., seem to be on the increase [54], it appears that the use of Petri nets is still restricted to research laboratories and academic institutions. This situation, to a large extent, results from the lack of widely available inexpensive software tools suitable for the development of industrial type of systems. This type of tools would be required to provide facilities for dealing with the application domain specific problems at a relatively skill-free level which would not require the knowledge of Petri nets, and the analysis methods. The facility for translating Petri net models to the executable code will also be essential. They will allow for rapid prototyping of the developed systems in the operational environment. In the past few years, a large number of tools have been reported in the Petri net literature. However, majority of these tools are used mostly for research and educational purposes [73]. An overview of some of the Petri net tools can be found in [22].

Another reason why the use of Petri nets is largely confined to academic and research institutions is the difficulty involved in constructing Petri net models. Constructing Petri net models of systems, especially large scale systems, is not a trivial task. It requires a great deal of experience to guarantee one-to-one correspondence between a net model and its requirements specification that is often informal or semi-formal. No methodology is available, yet, which would allow for a fully automatic construction of Petri net models. From our observations, in most cases, Petri net models are constructed in an ad hoc manner [19]. However, attempts have been recently made to make this particular approach more systematic. These were reported in [10, 33, 65, 68, 70-71], and the work in this area still continues. These approaches, using the terms of software engineering, can be broadly classified into bottom-up, top-down, and hybrid approaches whose comprehensive discussion can be found in [33, 68]. The reuse of Petri net models is also restricted. This is mainly due to fact that

Petri net models are, typically, constructed on one-off basis. The development is, in most cases, not supported by a proper documentation. It is clear that if Petri nets were to be widely used, especially by the industry people, methods, and the supporting tools, allowing for an automatic or semi-automatic construction of Petri net models from requirements specifications would have to be developed. Studies on the automatic construction of restricted classes of Petri net models from requirements specifications expressed using production rules, flow diagrams, state machines, temporal logic, application domain dependent semi-formal languages, etc. have been reported and remain of great interest to both researchers and industrial engineers [73].

To establish Petri nets in discrete-event control areas as a powerful and easy-to-use tool, there is a great need to perform benchmark studies among Petri nets, sequential function charts, ladder logic diagrams, high-level programming languages, etc. More industrial engineers have to join the Petri net community and make this tool and technology really useful for the development of modern manufacturing systems.

A graphical and reusable model assumes an important role in development, implementation and maintenance of flexible and agile manufacturing systems. In a Petri net framework, the algorithms for deadlock-avoidance, planning and scheduling, and discrete-event control can often be constructed more easily and efficiently than other methods do. With CAD tools, rapid-prototyping of manufacturing control systems can be accomplished. The research and development of Petri net approaches in automation shall bring a greater possibility to fulfill the system flexibility and agility, i.e., to reconfigure quickly manufacturing systems to the changing markets. Therefore, the appropriate use of these approaches by industrial firms will help them maintain and improve their competitiveness in today's world market.

9. Acknowledgment

The authors would acknowledge the partial financial support by the Center for Manufacturing Systems at the New Jersey Institute of Technology and the National Science Foundation under Grant No. DMI-9410863.

10. References

1. Ahmed Ben, S., M. Moalla, M. Courvoisier, and R. Valette(1991). "Flexible manufacturing production system modelling using object Petri nets and their analysis," *IMACS Symposium MCTS*, Lille, France, pp. 553-560.

2. Amar, S., E. Craye, and J-C. Gentina (1992). "A method for hierarchical specification and prototyping of flexible manufacturing systems", in

Proc. IEEE Workshop on Emerging Technologies and Factory Automation, Melbourne, Australia, pp. 44-59.

3. AT&T (1989). *Flexible Work Station: User Guide and Manual,* Version 1.01.

4. Banaszak, Z. A. and B. H. Krogh (1990). "Deadlock avoidance in flexible manufacturing systems with concurrency competing process flows," *IEEE Trans. Robotics and Automation,* 6(6), pp. 724-734.

5. Bastide, R., C. Silbertin-Blanc (1991). "Modelling of flexible manufacturing system by means of cooperative objects," in *Computer Applications in Production and Engineering,* Doumeingts G., et. el. (Eds.), Elsevier Science, pp. 593-600.

6. Berge, N., M. Samaan, G. Juanole, and J. Atamna (1994). "Methodology for LAN modelling and analysis using Petri net based models," in *Proc. Int. Workshop on Modelling, Analysis and Simulation in Telecommunication Systems,* Durham, NC, pp. 269-275.

7. Bruno, G. and G. Marchetto (1986). "Process-translatable Petri nets for the rapid prototyping of process control systems," *IEEE Trans. on Software Engineering,* 12(2), pp. 346-357.

8. Campos, J., G. Chiola, J. M. Colom and M. Silva (1992). "Properties and Performance Bounds for Timed Marked Graphs," *IEEE Trans. on Circuits and Systems - I: Fundamental Theory and Applications,* 39(5), pp. 386-401.

9. Cao, T. and A. C. Sanderson(1992). "Task decomposition and analysis of assembly sequence plans using Petri nets," in *Proc. 3rd Int. Conf. on Computer Integrated Manufacturing,* Troy, NY, pp. 138-147.

10. Chao, D. Y., M. C. Zhou, and D. T. Wang (1994). "Extending knitting technique to Petri net synthesis of automated manufacturing systems," *The Computer Journal,* Oxford University Press, 37(1), pp. 67-76.

11. Chen, Q. and J. Y. Luh (1992). "Operational scheduling using truncated Petri net technique," in *Proc. IEEE Workshop on Emerging Technologies and Factory Automation,* Melbourne, Australia, pp. 230-235.

12. Chiola, G. (1987). "A graphic Petri net tool for performance analysis," in *Proc. of Int. Workshop on Modeling Techniques and Performance Evaluation,* France, pp. 323-333.

13. Christodoulou, S. and M. C. Zhou (1994). "A Petri net approach to modeling and performance analysis of fiber distributed data interface (FDDI) network," in *Proc. of IEEE Workshop on Emerging Technologies and Factory Automation,* Tokyo, Japan, pp. 373-380.

14. Ciardo, G. (1989) *Manual for the SPNP Package*, Duke University.

15. Coffman E. G., Elphick M. J., and Shoshani A. (1971). "System Deadlocks," *Computing Surveys*, 3, pp. 67-78.

16. Crockett, D., A. Desrochers, F. DiCesare, and T. Ward (1987). "Implementation of a Petri net controller for a machining workstation," in *Proc. of IEEE Int. Conf. Robotics and Automation,* Raleigh, NC, pp. 1861-1867.

17. David, R. and H. Alla (1992). *Petri Nets and Grafcet*, Prentice-Hall, New York.

18. Desrochers, A. A. and R. Y. Al-Jaar (1994). *Applications of Petri nets in Manufacturing Systems*, IEEE Press.

19. DiCesare, F. and A. A. Desrochers (1991). "Modeling, control, and performance analysis of automated manufacturing systems using Petri nets," *Control and Dynamic Systems*, C. T. Leondes (Ed.), Vol. 47, pp. 121-172, Academic Press.

20. Di Mascolo, M. and Y. Frein, Y. Dallery and R. David (1992). "A unified modeling of Kanban systems using Petri nets," *Int. J. Flexible Manufacturing Systems*, 3, pp. 275-307.

21. DiStefano, A. and O. Mirabella (1991). "Evaluating the Fieldbus data link layer by a Petri net based simulation," *IEEE Trans. on Industrial Electronics*, 38(4), pp. 288-297.

22. Feldbrugge, F. (1990). "Petri Net Tools Overview 1989," in *Advances in Petri Nets 1989*, G. Rozenberg (Ed.), Vol. 424, Springer-Verlag, pp. 151-178.

23. Ferrarini, L. (1992). "An incremental approach to logic controller design with Petri nets," *IEEE Trans. on System, Man, and Cybernetics*, 22(3), pp. 461-473.

24. Ferrarini, L., M. Narduzzi, and M. Tassan-Solet (1994). "A new approach to modular liveness analysis conceived for large logic controllers' design," *IEEE Trans. on Robotics and Automation* 10(2), pp. 169-184.

25. Florin, G. and S. Natkin (1982). "Evaluation based upon stochastic Petri nets of the maximum throughput of a full duplex protocol," in *Application and Theory of Petri Nets,* C. Girault and W. Reisig (Eds.),Springer-Verlag, NY, 208-288.

26. Freedman, P. (1991) "Time, Petri nets, and robotics," *IEEE Trans. on Robotics and Automation,* 7(4), pp. 417-433.

27. Genrich, H. J. and K. Lautenbach (1991). "System modelling with high-level Petri nets," *Theoret. Comp. Sci.*, 13, pp. 109-136.

28. Hanisch, H.-M. (1993). "Analysis of place/transition nets with timed arcs and its application to batch process control," in *Applications and Theory of Petri Nets 1993, Lecture Notes in Computer Science,* Vol. 691, M. A. Marsan (Ed.), Springer Verlag, pp. 282-299.

29. Hillion, H. P. and J. M. Proth (1989). "Performance evaluation of job-shop systems using timed-event-graphs," *IEEE Trans. on Automatic Control,* 34(1), pp. 3-9.

30. Holloway, L. E. and B. H. Krogh (1990). "Synthesis of feedback control logic for a class of controlled Petri nets," *IEEE Trans. on Automatic Control,* 35(5), pp. 514-523.

31. Hsieh, D.-Y. and S.-C. Chang (1994). "Dispatching-driven deadlock avoidance controller synthesis for flexible manufacturing systems," *IEEE Trans. Robotics and Automation,* 10(2), pp. 196-209.

32. Jafari, M. A. (1992). "An architecture for a shop-floor controller using colored Petri nets," *Int. J. of Flexible Manufacturing Systems,* 4, pp. 159-181.

33. Jeng, M. D. and F. DiCesare (1993). "A review of synthesis techniques for Petri nets with applications to automated manufacturing systems," *IEEE Trans. on Systems, Man and Cybernetics,* 23(1), 301-312.

34. Jensen, K. (1990). "Coloured Petri nets: a high level language for system design and analysis," in *Advances in Petri Nets 1990,* G. Rozenberg (Ed.), Springer-Verlag, pp. 342-416.

35. Lee, D. Y. and F. DiCesare (1994). "Experimental study of a heuristic function for FMS scheduling," *IEEE Trans. Robotics and Automation,* 10(2), pp. 123-132.

36. Merabet, A. (1986). "Synchronization of operations in a flexible manufacturing cell: the Petri net approach," *J. of Manufacturing Systems,* 5(3), pp. 161-169.

37. Marsan A. M., and V. Signore (1987). "Timed Petri net performance of fibre optics LAN architectures", *in Proc. IEEE Int. Workshop on Petri Nets and Performance Models,* Madison, USA, pp. 66-74.

38. Martinez, J., P. Muro, and M. Silva (1987). "Modeling, validation and software implementation of production systems using high level Petri nets," in *Proc. of IEEE Int. Conf. on Robotics and Automation,* Raleigh, NC, pp. 1180-1185.

39. McCarragher B. (1993). "Robotic assembly and trajectory planning using discrete-event modelling," in *Proc. IEEE Int. Workshop on Emerging Technologies and Factory Automation,* Palm-Cove, Cairns, Australia, pp. 187-196.

40. Mitchell, Jr. F. H. (1991). *CIM Systems: An Introduction to Computer-Integrated Manufacturing,* Prentice Hall, Englewood, NJ.

41. Molloy, M. K. (1982) "Performance analysis using stochastic Petri nets," *IEEE Trans. on Computers,* 3(9), pp. 913-917.

42. Morioka, S. and T. Yamada (1991). "Performance evaluation of marked graphs by linear programming," *Int. J. of Systems Science,* 22(9), 1541-1552.

43. Murata, T. (1989). "Petri nets: properties, analysis, and applications," *Proceedings of the IEEE,* 77(4), pp. 541-580.

44. Murata, T., B. Shenker, and S. Shatz (1989). "Detection of ADA static deadlocks using Petri net invariants," *IEEE Trans. on Software Engineering,* 15(3), pp. 314-326.

45. Murata, T., N. Komoda, K. Matsumoto, and K. Haruna (1986). "A Petri net based controller for for flexible and maintainable sequence control and its application in factory automation," *IEEE Trans. on Industrial Electronics,* 33(1), pp. 1-8.

46. Nketsa, A. and M. Courvoisier (1990). "A Petri net based single chip programmable controller for distributed local controls," in *Proc. of IECON'90,* Vol. 1, 542-547.

47. Ramamoorthy, C. V. and G. S. Ho (1980). "Performance evaluation of asynchronous concurrent systems using Petri nets," *IEEE Trans. on Software Engineering,* 6(5), pp. 440-449.

48. Reisig, W. (1985). *Petri Nets: An Introduction,* Spring-Verlag, New York.

49. Rembold, U., B. O. Nnaji, and A. Storr (1993). *Computer Integrated Manufacturing and Engineering,* Addison-Wesley, Wokingham, England.

50. Shatz S., K. Mai, C Black., S. Tu (1990). "Design and implementation of a Petri net based toolkit for Ada tasking analysis," *IEEE Trans. on Pararell and Distributed Systems,* 1(4), pp. 424-441.

51. Silva, M. (1985). *Las redes de Petri en la Automatica y la Informatica,* Editorial AC, Madrid.

52. Silva, M., and R. Valette (1990) "Petri nets and flexible manufacturing," in *Advances in Petri Nets 1989,* G. Rozenberg (ed.), Springer-Verlag, pp. 374-417.

53. Valavanis, K. S. (1990). "On the hierarchical analysis and simulation of flexible manufacturing systems with extended Petri nets," *IEEE Trans. on Systems, Man, and Cybernetics,* 20(1), pp. 94-100.

54. Valette, R. and M. Courvoisier (1993). "Petri nets and artificial intelligence," *Modern Tools for Manufacturing Systems,* R. Zurawski and T. Dillon, (Eds.), Elsevier, pp. 385-405.

55. Venkatesh, K., K. Mehdi, M. C. Zhou, and R. J. Caudill (1994). "Augmented timed Petri nets for modeling of robotic systems with breakdowns," *J. of Manufacturing System,* 13(4), pp. 289-301.

56. Venkatesh, K., M. C. Zhou, R. J. Caudill, and E. B. Fernandez (1994). "A control software design method for CIM systems," in *Proc. of Conf. on Computer Integrated Manufacturing in Process Industry,* New Brunswick, NJ, pp. 565-579.

57. Viswanadham, N., Y. Narahari, and L. Johnson(1990). "Deadlock prevention and avoidance in flexible manufacturing systems using Petri net models," *IEEE Trans. on Robotics and Automation,* 6(6), pp. 713-723.

58. Wang, F.-Y. and Saridis, G. (1990). "A coordination theory for intelligent machines," *Automatica,* 26(5), 833-844.

59. Willson, R. G. and B. H. Krogh (1990). "Petri net tools for the specification and analysis of discrete controllers," *IEEE Trans. on Software Engineering,* 16(1), pp. 39-50.

60. Xiong, H. H. and M. C. Zhou (1994). "Computer communication networks for automated manufacturing," in *Proc. of 1994 Int. Conf. on Electronics and Information Technology,* Beijing, China, pp. 178-183.

61. Yamalidou, E. C. and J. Kantor (1990). "Modelling and optimal control of discrete-event chemical processes using Petri nets," *Computers and Chemical Engineering,* 15, pp. 503-519.

62. Zhang, D. (1992). "Planning using timed Pr/T Nets," in *Proc. Japan-U.S.A. Symp. on Flexible Automation,* San Francisco, CA, pp. 1179-1184.

63. Zhou, M. C. and F. DiCesare (1991). "Parallel and sequential mutual exclusions for Petri net modeling for manufacturing systems," *IEEE Trans. on Robotics and Automation,* 7(4), pp. 515-527.

64. Zhou, M. C. and M. C. Leu (1991). "Modeling and performance analysis of a flexible PCB assembly station using Petri nets," *Trans. of the ASME, Journal of Electronic Packaging,* 113(4), pp. 410-416.

65. Zhou, M. C., F. DiCesare, and A. A. Desrochers (1992). "A hybrid methodology for synthesis of Petri nets for manufacturing systems," *IEEE Trans. on Robotics and Automation,* 8(3), pp. 350-361.

66. Zhou, M. C., F. DiCesare, and D. Rudolph (1992). "Design and implementation of a Petri net supervisor for a flexible manufacturing system," *Automatica*, 28(6), pp. 1199-1208.

67. Zhou, M. C., K. McDermott, and P. A. Patel (1993). "Petri net synthesis and analysis of an FMS cell," *IEEE Trans. on Systems, Man, and Cybernetics*, 23(2), pp. 523-531.

68. Zhou, M. C. and F. DiCesare (1993). *Petri Net Synthesis for Discrete Event Control of Manufacturing Systems,* Kluwer Academic Publishers, Boston, MA.

69. Zhou, M. C. and A. D. Robbi (1994). "Application of Petri net methodology to manufacturing systems," *Computer Control of Flexible Manufacturing Systems*, S. Joshi and G. Smith (eds.), Chapman and Hall, pp. 307-330.

70. Zurawski, R., T. Dillon (1991). "Systematic construction of functional abstractions of Petri net models of typical components of flexible manufacturing systems," in *Proc. Int. Workshop on Petri Nets & Performance Models*, Melbourne, pp. 248-257.

71. Zurawski, R. (1992). "A class of Petri net models used to represent primary components of flexible manufacturing systems," in *Proc. Japan-U.S.A. Symposium on Flexible Automation*, San Francisco, CA, pp. 1695-1702.

72. Zurawski, R. and T. Dillon (1993). "Modelling and verification of flexible manufacturing systems using Petri nets", in *Modern Tools for Manufacturing Systems*, R. Zurawski, T. Dillon (Eds.), Elesevier Scientific Publishers, pp. 237-261.

73. Zurawski, R. and M. C. Zhou (1994). "Petri nets and industrial applications: a survey", *IEEE Trans. on Industrial Electronics*, to appear in Dec. 1994.

2

Application of Petri Nets to Sequence Control Programming

Tomohiro Murata
Systems Development Laboratory, Hitachi, Ltd.
1099 Ohzenji, Asao-ku, Kawasaki 215, JAPAN

ABSTRACT

Petri net-based sequence control software, called Station Controller (SCR) can improve the flexibility and maintainability of factory automation application software. Sequence control programs for Flexible Manufacturing Systems (FMS) can be described intuitively using Control net (C-net), which is a high-level sequence control language based on a Predicate/Event net (P/E-net). C-net can be directly executed by an interpreter installed in a 16-bit microcomputer system. This chapter briefly introduces several conventional models for specifying sequence control and describes some features and successful applications of SCR.

1. HIERARCHICAL CONTROL STRUCTURE OF FLEXIBLE MANUFACTURING SYSTEMS

Figure 1 shows the hierarchy of the various levels in FMS configurations [5]. From the view point of functional classification, the factory level has off-line scheduling functions and usually main frame computers are used. The shop level has functions gathering data on production results and for real-time scheduling, and middle class control computers or engineering workstations are used. The cell level has sequence control functions of the production line and microcomputer-based small class control computers are used. Procedural languages are conventionally used for control programming at shop and cell levels. The production line level has direct control functions of automatic equipments, and PLCs have been conventionally used to process i/o sequencing where relay ladder diagrams are conventionally used as programming means. The unit level consists of several automatic FA facilities such as NC machines, robots, and other peripheral equipments.

As the life cycles of industrial products have become shorter, it is becoming necessary to rapidly develop or modify control software, especially that in the cell level. It is difficult, however, even for a trained programmer to develop or modify control software using only conventional languages because it is not easy to derive control specifications from programs described by such languages.

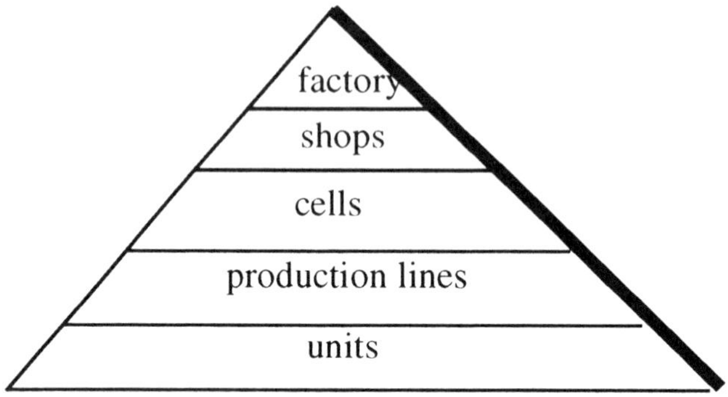

Figure 1. Functional hierarchy of FMS

To maintain sequence control programs flexibly, we therefore need new problem-oriented languages that can hierarchically and intuitively describe asynchronous and concurrent control logic necessary in the cell level control.

2. SEQUENCE CONTROL SPECIFICATION MODELS

Figure 2 shows a conceptual model of sequence control. The functional structure of sequence control software consists of several local autonomous objects each of which behaves as a state machine and has a supervising object which behaves as a global state machine to manage the concurrent execution of the local objects: sequencing, parallel execution, synchronization and exclusion control. This section describes several models that are convenient for describing these specifications.

2.1 State-based Model

2.1.1. State-transition diagram (STD)

This simple and well known model is based on finite-state-machine model [Figure 3]. Each circle of the STD corresponds to a state, and an arc connecting two circles corresponds to a state-transition triggered by an event associated with the arc. STD is a suitable model for describing a local state machine's behavior with no concurrency, but it is not powerful enough to describe the behavior of a global statemachine controlling the sequence with concurrent actions of local objects.

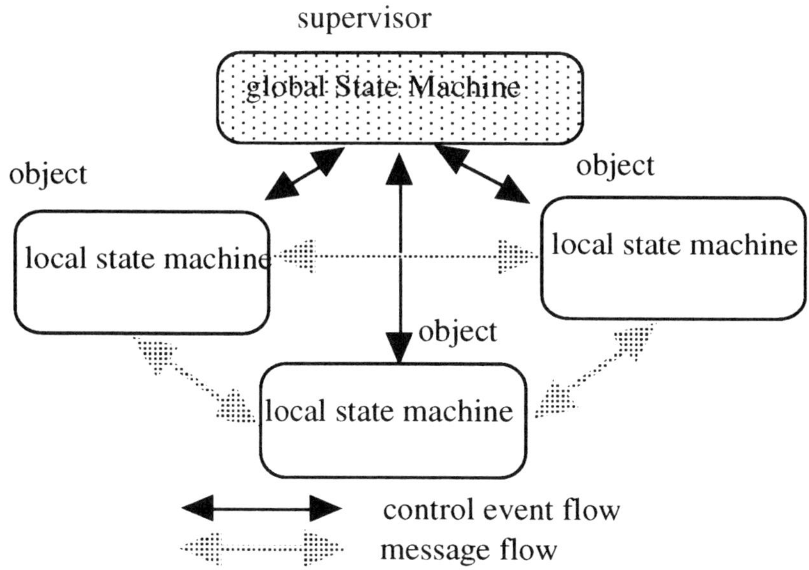

Figure 2. Concurrent object-based sequence control model

2.1.2 Statechart

Statechart [4] is an extended STD including AND/OR decomposition of states so that it could be used for hierarchical STD description: interlevel transitions and implicit intercomponent event broadcasting. A statechart specification can be regarded as a tree of states, where the leaf states correspond to the states in finite-state-machines and where all other states have relations of superstate-substate property corresponding to the tree structure. In the statechart of Figure 4, state **s1** is a root-state and **s3**, **s6**, **s7**, **s8**, **s9** are leaf-states. The state **s1** is an OR-state which consists of substates **s2** and **s3** which can exist exclusively. The state **s2**, on the other hand, is an AND-state consisting of substate **s4** and **s5** which behave concurrently. Entering the **s2**, thus results in entering both **s4** (an initial state is **s6**) and **s5** (an initial state is **s8**).

A statechart is useful for describing complex system behaviors including concurrency. One of the limitations of statechart however, is that global states compounded from AND-states cannot be comprehended intuitively. Hierarchical State Diagram Matrices [11] which can explicitly represent hierarchical compound states of concurrent state-transitions using a matrix frame, has been presented to avoid this problem.

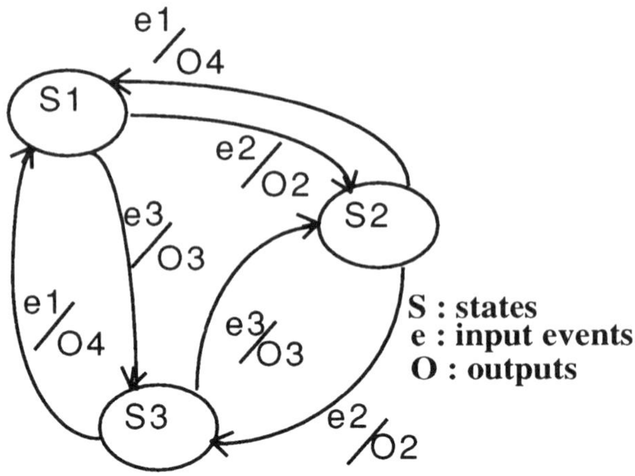

Figure 3. State-based model

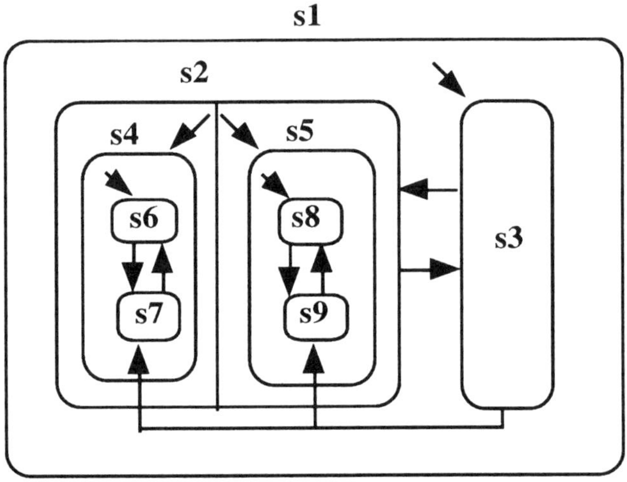

Figure 4. Statechart

2.2 Rule-based Model

Ladder logic diagrams (LLDs) [8] are traditionally used to specify the input and output procedures of the PLCs using relay ladder circuit descriptions and can be regarded as a Boolean expression or a set of production rules [Figure 5].

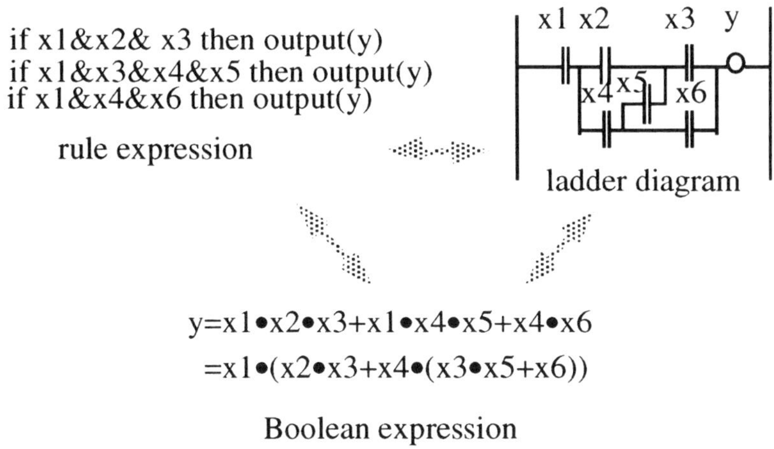

Figure 5. Rule-based description

The circuit of a LLD can be logically divided two parts; a condition part and an action part, connected in series. The condition part consists of switches or relays for holding some existing states, and the action part has solenoids that output control signals when the Boolean logic of the condition part is satisfied. LLDs are useful for easy programming of on/off-based sequence control which drives repetitive operations of automatic equipments, but it is difficult to derive the order of operations from a large LLD description.

2.3 Petri- Net-based Model

Models based on Petri nets [13] can describe both sequential and concurrent control logic explicitly, and are useful for describing sequence control specifications.

2.3.1 Condition/Event Net

In a Condition/Event net (C/E-net), which is a subclass of Petri net, each place can contain at most one token (i.e. marking is Boolean). This constraint is called safeness.

A C/E-net R is defined formally: $R = (P,T,I,O,m)$, where P is a set of places $\{p_1, p_2,\ldots p_n\}$ and T is the set of transitions $\{t_1, t_2, \ldots t_m\}$. The functions I: $P \times T \rightarrow N$, and O: $P \times T \rightarrow N$ are respectively called input and output functions, and m: $P \rightarrow N$,where $N = \{0, 1\}$, is called a marking. A place p_i such that $I(p_i,t_j)=1$ (or $O(p_i,t_j)=1$) is called an input place (or an output place) of t_j. Conversely, a transition t_j

such that $O(p_i,t_j)=1$ (or $I(p_i,t_j)=1$) is called an input transition (or an output transition) of p_i. The notation $m(p_i) = 1$ (or 0) means place p_i has a token (or no token).
A transition t_j becomes "enabled" if each of the input places of t_j contains a token and each of the output places of t_j contains no token. This condition is called "firing condition". Firing of a transition t_j consists in withdrawing a token from each of the input places of transition t_j and in adding a token to each of the output places of transition t_j. Figure 6 shows the well-known dining philosophers problem (most simple case: two philosophers) described by a graphical representation of C/E-net. Circles (corresponding to places) represent some conditions, and squares (corresponding to transitions) represent some event occurrences. A directed arc from

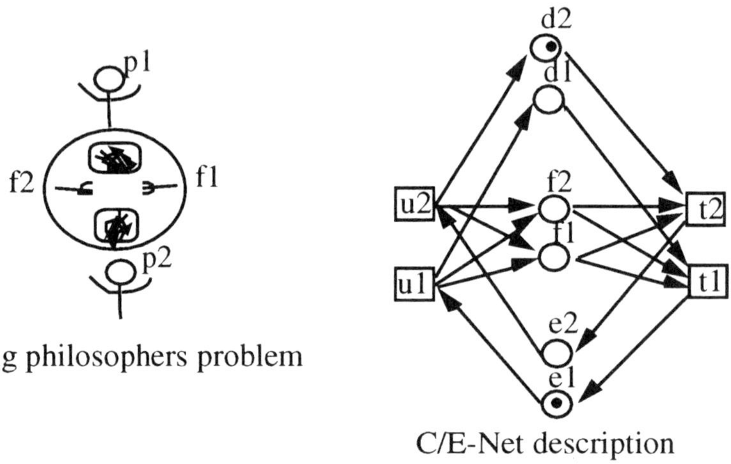

Figure 6. C/E-net description of the "dining philosophers" problem

a place to a transition represents a precondition, and one from a transition to a place represents a postcondition. In Figure 6, places **d1** and **e1** respectively mean that philosopher-1 is meditating and eating, place **f1** means that fork-1 is not being used, and transitions **t1** and **u1** respectively mean that philosopher-1 takes and releases two forks on both sides at the same time. A black dot (corresponding to a token) indicates that the condition corresponding to the place that includes the token is satisfied, and token movement from some places (preconditions) to other places (postconditions) via a transition represents occurrence of the event corresponding to that transition.

2.3.2 Grafcet

Grafcet [2] is a C/E-net based model that was developed in order to describe logic controller specifications. An example of Grafcet specification for a vessel production line control is shown in Figure 7. A Grafcet is made of steps (represented by rectangles and looking like a place) and transitions (represented by bars and looking like a transition) connected by direct links, and it has only one initial token. Actions (which are the control outputs) are associated with each step and receptivities (which are functions of the input process state) are associated with each transitions. Each step can have two states (active or inactive) and the active state is

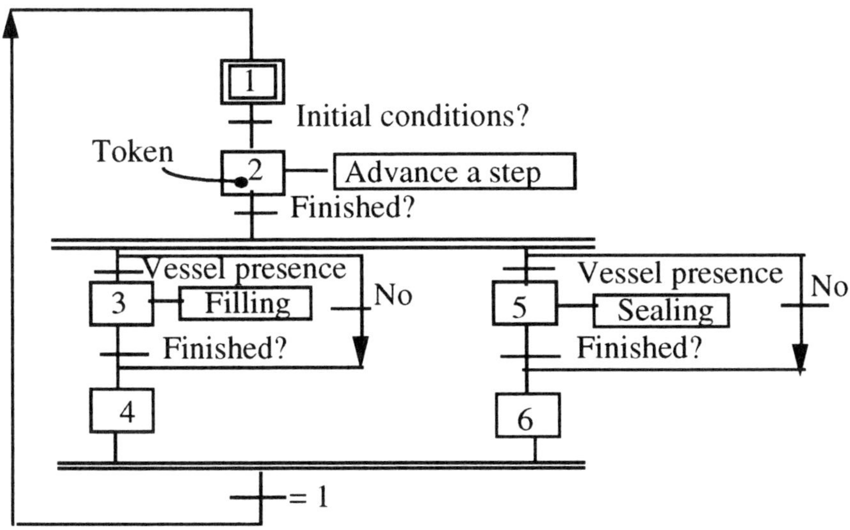

Figure 7. Grafcet description

represented by a token in the step. A transition firing condition is same as the condition for a C/E-net except that, additionally, the receptivity condition of the transition is true. We can also regard Grafcet as an extended STD that introduces parallel state transitions (corresponding to double bars in Figure 7) to describe parallel activation and synchronization of plural steps. PLC that can directly execute Grafcet-based control program has been developed [7].

2.3.3. Predicate/Event Net

The Predicate/Event net (P/E-net) is an extended C/E-net that introduces color tokens (token-IDs) to identify each token. P/E-net S is defined formally, $S = (P,T,I,O,\underline{D},\lambda,m)$. Here, P,T,I, and O are the same as that of C/E-net [14]. The notation $\underline{D} = (D, \Phi)$ is an algebra where D is a set of token-IDs and Φ is a set of mappings $M \rightarrow D$, ($M \subseteq D^n$, $n=\{1,2,...\}$). The symbol λ is a set of arc-label: set of arcs$\rightarrow$ power set of $\Psi(\underline{D}, X)$) except for $\{\phi\}$, where X is a set of variables, and $\Psi(\underline{D}, X)$ is a minimal set of expressions X and f, satisfying the following conditions,

(a) $X \in \Psi(\underline{D}, X)$,

(b) if $X \in \Psi(\underline{D}, X)$ and $f \in \Phi$, then $f(X) \in \Psi(\underline{D}, X)$.

The marking m: denotes a function $P \rightarrow$ power set of D, namely, $m(p_i)$ represents a set of token-IDs in a place p_i. A transition t_j becomes enabled, if and only if, both of the following conditions (*firing condition*) are satisfied.

(i) Marking of any input place p_i of t_j includes token-IDs that equals $\beta(\lambda(a(p_i, t_j)))$ where $a(p_i, t_j)$ is an arc connecting p_i to t_j, $\lambda(a(p_i, t_j)) \in \Psi(\underline{D}, X)$ is an arc label of $a(p_i, t_j)$ and $\beta(\lambda(a(p_i, t_j)))$ is a function $\lambda(a(p_i, t_j)) \rightarrow D$.

(ii) Marking of any output place p_k does not include token-ID that equals $\beta(\lambda(a(t_j, p_k)))$ where $a(t_j, p_k)$ is an arc connecting t_j to p_k, $\lambda(a(t_j, p_k)) \in \Psi(\underline{D}, X)$ is an arc label of $a(t_j, p_k)$ and $\beta(\lambda(a(t_j, p_k)))$ is a function $\lambda(a(t_j, p_k)) \rightarrow D$.

A new marking of input place p_i is obtained by $m'(p_i)=m(p_i) - \beta(\lambda(a(p_i, t_j)))$, and new marking of output place p_k is obtained by $m'(p_k)=m(p_k)+\beta(\lambda(a(t_j, p_k)))$ is obtained after firing of t_j. Figure 8 shows the same dining philosophers problem in Figure 6(a) described by graphical representation of a P/E-net. The modeling power

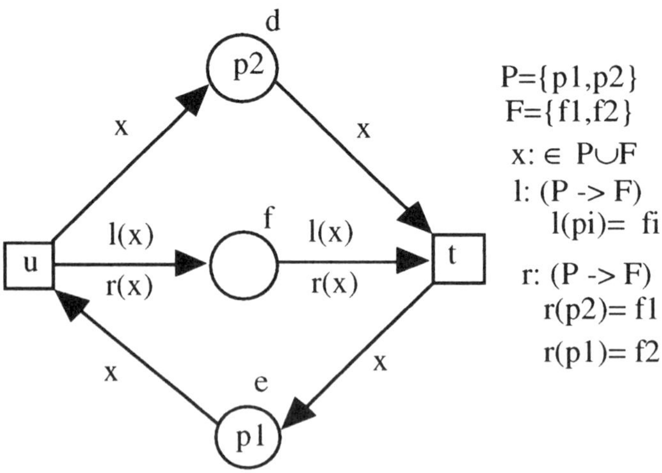

Figure 8. P/E-net description of the "dining philosophers" problem

of P/E-net is the same as of C/E-net because P/E-net can be described equivalently by using C/E-net, a P/E-net is, nevertheless, useful to describe complex sequence control specifications because its representation is very compact.

3. EXTENSION OF P/E-net FOR SEQUENCE CONTROL PROGRAMMING

3.1 Control-net (C-net)

C-net [9] is a P/E-net-based high-level sequence control programming language directly executable on a real-time microcomputer system. In a C-net program, finite capacity is associated to each places, and an input transition of a place can be fired only if firing of the transition does not result in a number of tokens in the place which exceeds this capacity. Concrete transaction processing functions can be also assigned to each places, and the content of each transaction can be represented as a colored token.

C-net can be considered as a programming language consisting of a 2-level hierarchy. The higher level is a P/E-net-based visual program that manages the execution sequence of autonomous control objects corresponding to places of C-net. The lower level is a procedural program (place program) assigned to each control object and it performs concrete control actions. Elements of C-net program description are as follows.

3.1.1 Individual token with data

Each token can have data-slots for storing values. Data structure of a token can be defined according to a token type [Figure 9].

```
TOKEN:  type name;
   Slot name-1,  array (i) of data type (int, real, char)
   Slot name-2,  array (j) of data type (int, real, char)
                  :
   Slot name-n,  array (k) of data type (int, real, char)
END_TOKEN;
```

Figure 9. Token type definition

Each slot has a slot-name and data type of an array to hold plural field data. Instance values representing some real transaction data are given to each slot when a new token is generated. To distinguish instances, an identifier (token-ID) is given to each token. The token-ID is a combination of token-type and serial token-number.

3.1.2 C-net description and Place Elements

C-nets are directed graphs described by connecting several functional places via arcs. Symbols of these functional places are shown in Figure 10.

symbols	kinds of places	contents
	start place	*Detect input events and generate new tokens.*
	action place	*Hold at most one token, execute assigned place programs when a token arrives.*
	queue place	*Hold several tokens.*
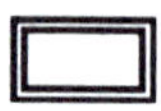	dummy place	*Hold at most one token to exclude plurarity of control place execution.*
	branch place	*Branch a token by firing an output transitions depending on the result of executing an assigned place program.*
	hierarchy place	*Describe subnet hierarchically.*

Figure 10. Place functions of C-net

Arcs in C-net have no arrows for simple description, because of the tacit rules that arcs connecting to the left side of a place mean inputs of the place and arcs connecting to the right side of a place mean outputs of the place. A simple C-net description is shown in Fig. 11.

Firing conditions of transitions with colored tokens (corresponding to predicates of P/E-net) are assigned to each transitions as production rule-type descriptions as follows:

if token(<input place name><token-id necessary to exist in the place>)
then fire(<output place name><new token -id to be output to the place after firing>).

A firing condition, in Fig. 11, "transition g1 requires its input places token-id 100 and 204 to fire and output token-id 100 to its output place" is described as follows:

if (token<place1><100> token<place4><204>)
then (fire<place2><100>).

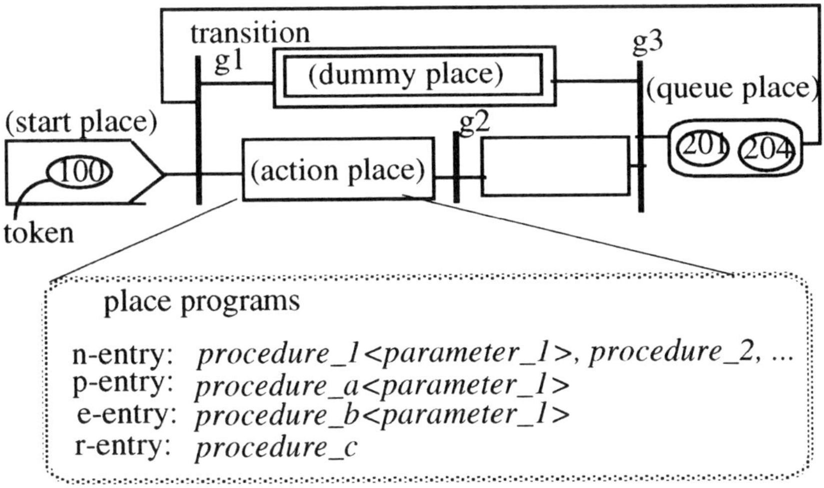

Figure 11. C-net description

The contents of each functional places are as follows:

(1) ***Action-place and place-programs.*** A place-program consisting several subroutines can be assigned to each place. A place in which place-programs are assigned is called an action-place. In a place-program, as shown in Figure 11, subroutines are assigned to four kinds of entry: **n**, **p** **e**, and **r** entries. Execution sequence of these subroutines are controlled according to the inner state transition of the action place shown in Figure 12 .

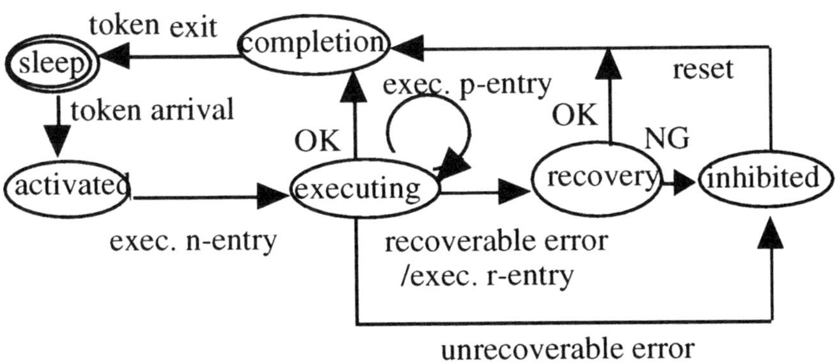

Figure 12. Innerplace state-machine configuration

Assigning the start address of a subroutine to n-entry in a place causes to start execution when a token enters the place. Slot data of an entered token can be referred to or changed by using the variable name <TOKEN> as a subroutine parameter to indicate the memory address storing the token's slot data. The action-place offers two types of wait mechanisms: p-entry (periodical execution entry) and e-entry (event-driven execution entry). A place-program assigned to p-entry is executed periodically until it satisfies some condition, and a place-program assigned to e-entry is executed when the assigned event has been detected. The status of a normal place is *executing* while the return code value of a place program assigned to these entry is non-zero, and status of the place becomes *completed* when the value becomes zero. If these place-program's return codes are negative (which means some error occurred in a subroutine), the place program assigned to r-entry is automatically executed.

(2) ***Start-place.*** A start-place is a place that, to provide autonomous receptivity for input signals, has a special function to execute its place program periodically with no token. An initial token is generated in a start-place when its place-program satisfies the receptivity condition assigned to that place.

(3) ***Branch-place.*** A branch-place has more than one output transition, and a place-program assigned in the place can choose one of its output transitions to be fired according to its execution result.

(4) ***Dummy-place.*** A dummy-place has no place program and is simply a place in which only one token can be held in order to prevent simultaneous activation of several action places. Figure 13 shows an example of a dummy place used to ensure that only one of the action-places (action-place 1, 2, 3, 4) has a token and, therefore, the one action-place having the token can execute its place program. A token generated in a dummy-place is called a dummy-token and has no token data.

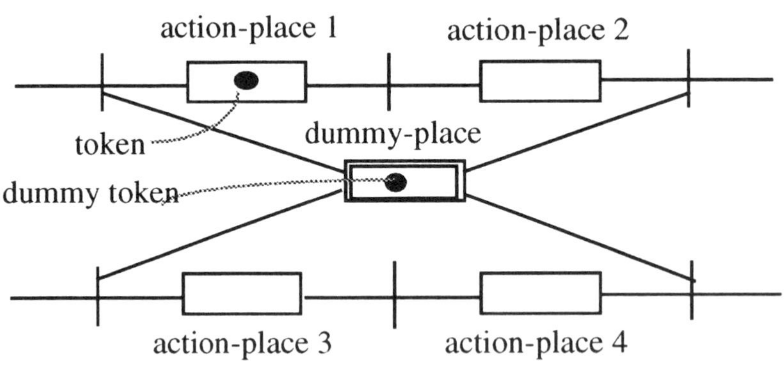

Figure 13. Dummy-place ensuring exclusive execution of action-places

(5) ***Queue-place.*** This place is for queuing a finite number of tokens in it. Any token in the place can be put out according to the assigned predicate functions of the place. If no predicate functions are assigned to the place, queued tokens are put out in the FIFO manner.

(6) ***Hierarchy-place*** .A hierarchy place can be developed into a more detailed subnet. Subnets are automatically combined into an upper-level net during net execution.

3.1.3. Transition firing rules

As shown in Figure 14, transition t_k has three inner states and becomes an enabled state when it satisfies the following conditions:

(i) All input places of t_k have tokens and the execution of their assigned place procedures has been completed.

(ii) A number of tokens in any output place of t_k does not exceed the capacity associated to the place.

(iii) t_k satisfies the same *firing condition* as that of P/E-net .

(iv) A logical function g (gate function) assigned to t_k outputs a *true* value. This function is used to synchronize the firing of t_k with the inputs of external signals/events.

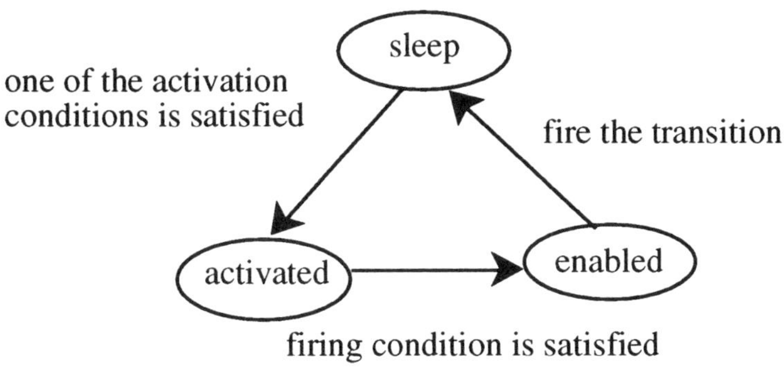

Figure 14. Innertransition state-machine configuration

An enabled transition can be fired at any time and a new marking is obtained according to the same *firing operation* as of P/E-net. Arc-labels of P/E-net for handling individual tokens are assigned to each transition t_k as a set of production rules represented as follows:

if (*an input-box ID, arc-label of the arc connecting the input-box and* t_k)

then (*an output-box ID, arc-label of the arc connecting* t_k *and the output-box*).

In new marking, a token that has the same token ID as before firing has the same slot values as before the firing, and a token that has a new token ID has new slot values. A new token output in a dummy-place is a dummy-token that has no slot data.

3.2. C-net DESCRIPTION EXAMPLE

Figure 15 shows an example of an assembly station. At this station, robot R1 gets two assembly parts from a part feeder and assembles them on the assembly table. Robot R2 gets four screws from the screw feeder and inserts them at the specified points on the assembled parts. The control specifications for these two robots can be described with C-net as shown in Figure 15.

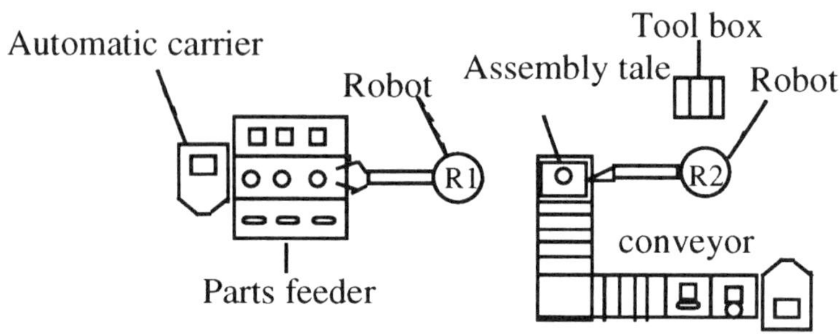

(a)

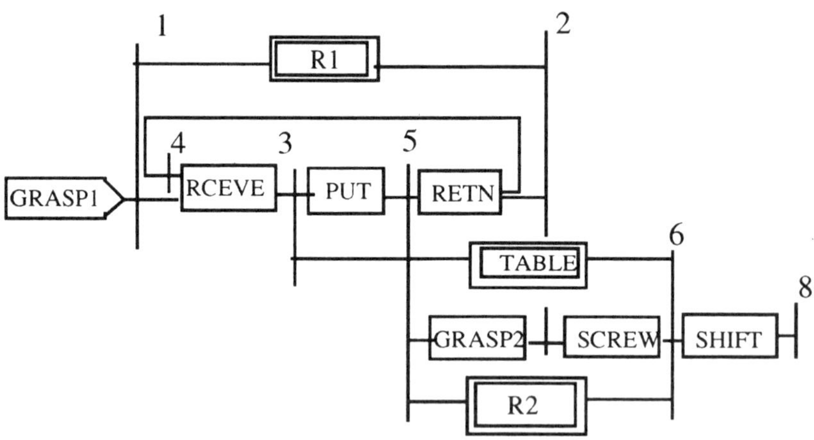

(b)

Figure 15. (a) An assembly station, (b) A C-net program for the station control

Each place (rectangle) and transition (bar) has a unique place ID and transition number. The places GRASP1, PUT, and RETN include place program which perform robot R1's actions and the places GRASP2 and SCREW include place programs that perform robot R2's actions. Place SHIFT includes a place program to trigger conveyor actions. When a job start request is detected by the place program in RCEVE, a numbered token with data slots is generated there according to required job type. This token moves through the boxes in the C-net program according to the firing rules, and the action sequence of the two robots is controlled.

3.3 A COMPARATIVE STUDY : C-net v.s. GRAFCET

Grafcet and C-net are both based on Petri net but Grafcet was originally developed as a specification standard for describing sequence control logic and does not suppose any specific programming language for implementing the specifications. C-net, on the other hand, was originally developed as a problem-oriented programming language for sequence control that can be executed directly on a control computer system. Although Grafcet and C-net cannot be compared directly, because their original objectives are different, their modeling concepts can be compared.

In Grafcet, a place (step) corresponds to a state and a transition corresponds to action and receptivity. This is because grafcet specifies an individual state as a place in order to provide explicit and strict state descriptions. In C-net, on the other hand, a place corresponds to both actions and states, and transition corresponds only to receptivity. This is because C-net describes an elementary action behavior as a place object that capusulizes both inner states and methods (programs) in order to perform control input/output functions. This improves modularity and reusability of elementary control functions.

A token in Grafcet is Boolean and describes only that some of conditions are currently satisfied. A token in C-net, on the other hand, has data type, data slots and numerical values, and represents not only activation state of places but also messages sent/received between places. Grafcet can specify states and its transitions clearly, and can easily be translated to conventional LLD expressions. C-net, on the other hand, provides rapid application development features for a sequence programming environment where sequence control program can be graphically programmed by networking place-objects each of which performs elementary control functions.

4. REAL-TIME INTERPRETATION OF C- NET

To use C-net for real-time control programming, a C-net interpreter for a 16 bits real-time microcomputer system has been developed. This section describes an effective interpretation scheme, and theoretically evaluates its execution overhead.

4.1 Selective scanning of enable transitions

The simplest method of net execution is to test all transitions and its input/output places to determine whether or not they are enabled. This enable test, however, requires a large overhead if the number of transitions is large. This overhead can be considerably reduced by using C-net structure and marking information, and this in turn is done by introducing the concept of transition activation. A transition t_i is called an activated transition when one of the following conditions (transition activation conditions) is satisfied:

(i) A new token enters one of the input places of t_i.

(ii) The status of an input place changes from executing to completed.

(iii) A token exits from one of the output places of t_i.

The C-net interpreter executes C-net at a practical speed by selecting only activated transitions as candidates for the enable test because only activated transitions have the possibility of being fired. This idea is called the selective scanning method, and it reduces the total execution time (response time) of Petri net interpretation.

4.2 Performance Evaluation

The selective scanning method shown in Figure 16 consists of four execution steps (S1, S2, S3, and S4) and four stacks (H1, H2, H3 and H4). Parameter ε ($0 < \varepsilon < 1$) is the average number of input and output arrows per place, and σ ($0 < \sigma < 1$) is the average number of input and output places per transition. The parameters ε and σ are called structural property parameters of C-net. Parameter μ ($0 < \mu < 1$), the completion rate of place programs is given as τ / υ ($\tau < \upsilon$), where τ is the iteration interval of C-net interpretation, and υ is the average execution time of place programs. Parameter μ' ($\mu' = h3 / h4$) is called a firing rate. The contents of each stack are

H1: Places that contain tokens and whose place-programs are uncompleted.
H2: Places whose place-programs have been completed.
H3: Activated transitions which satisfy one of the activation conditions.
H4: Transitions which satisfy the firing conditions.

The average number of places/transitions in Hi is indicated as *hi* . The contents of each execution step are

S1: Judge completion of executing places.
S2: Select activated transitions.
S3: Perform enable testing on all of the selected activated transitions.
S4: Fire enabled transitions and stack the output places of the transition on H1.

By firing a transition in S4, average $\sigma/2$ places (input places) have lost tokens (an

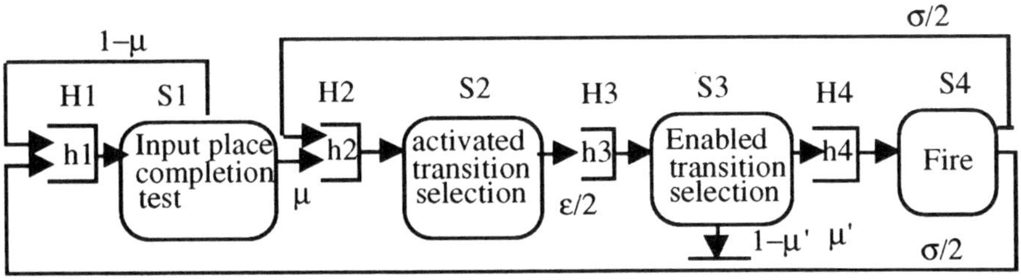

parameters μ : completion rate of executing place per iteration cycle
μ' : firing rate of transition per iteration cycle
ε : average number of input and output arcs per place
σ : average number of input and output places per transition

Figure 16. C-net interpretation scheme (Selective scanning)

arc from S4 to H2) and average $\sigma/2$ places (output tokens) have got tokens (an arc from S4 to H1). On average, *h1, h2, h3*, and *h4* must satisfy the following equations:

$$h1 = h1 \bullet (1-\mu) + h4 \bullet (\sigma/2) \quad (1)$$

$$h2 = h1 \bullet \mu \quad + h4 \bullet (\sigma/2) \quad (2)$$

$$h3 = h2 \bullet (\varepsilon/2) \quad (3)$$

$$h4 = h3 \bullet \mu' \quad (4)$$

If these simultaneous equations have solutions other than $h1 = h2 = h3 = h4 = 0$, the following matrix equation must be completed :

$$\begin{vmatrix} -\mu & 0 & 0 & \sigma/2 \\ \mu & 0 & 0 & \sigma/2 \\ 0 & \varepsilon/2 & -1 & 0 \\ 0 & 0 & \mu' & -1 \end{vmatrix} = 0, \quad (5)$$

From which, $\mu(1-\varepsilon\sigma\mu'/2) = 0$ is obtained. The firing rate μ must be not 0, so $\mu' = 2/(\varepsilon \bullet \sigma)$. This equation gives the balance condition, and we can use this condition to obtain the total execution time Z of S1 through S4 as

$$Z = h1 \bullet a_j + h2 \bullet a_i + h3 \bullet a_e + h4 \bullet a_f$$
$$= (a_j + (2a_i + \varepsilon \bullet a_e + 2a_f / \sigma)\mu)\, h1 . \quad (6)$$

Here a_j , a_i , a_e , and a_f are the average run times for S1, S2, S3, and S4. These values were measured on a C-net interpreter implemented on a 16-bit microcomputer and were used to calculate Z as

$$Z=(0.91+3.12\mu)hl . \qquad (7)$$

Z values experimentally measured for various values of μ and hl coincide with the value calculated from equation (7) (see [Figure 17]).

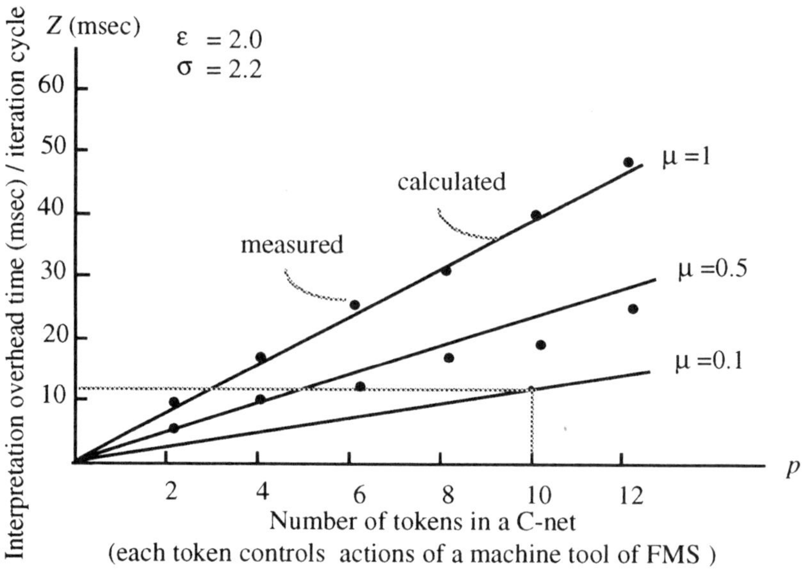

Figure 17. Evaluation of C-net interpretation overhead

When μ is 0.1, the number of arcs is 1100 and the number of places is 550, the response time Z is expected to be about 12 ms for the concurrent processing of hl=10 transactions. This should be shorter enough for ordinary transaction processing control. Because the C-net interpreter evaluates only transitions and places that are activated, overhead time of the interpretation is not directly proportional to the number of total places / transitions of C-net, but to the number of tokens in C-net. Therefore, C-net can be executed efficiently even if the nets become much more complicated with more places and transitions.

5. IMPLEMENTATION OF C-NET BASED SEQUENCE CONTROLLER

The proposed C-net interpretation function and other necessary functions for real-time control have been implemented on a real-time microcomputer system using C-language. This program product, called station controller (SCR), consists of the C-net editor, translator, monitor and interpreter, and its configuration is shown in Figure 18.

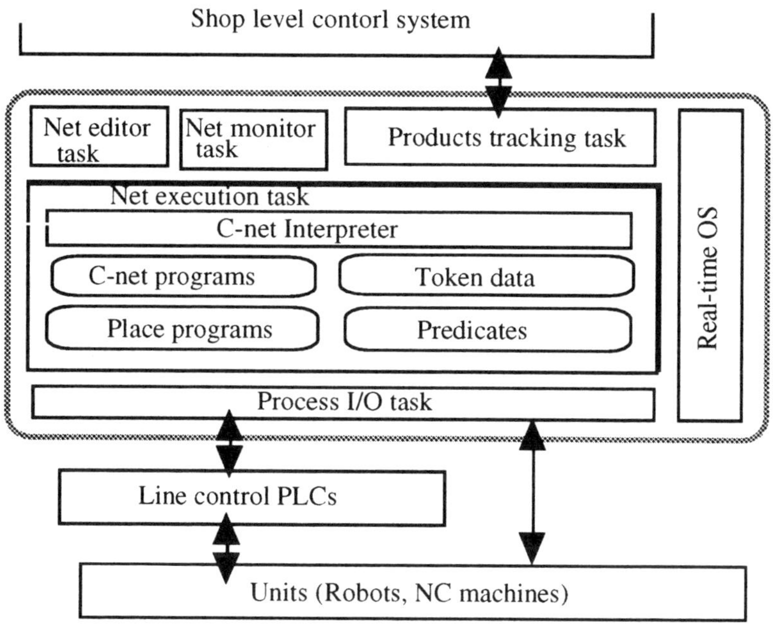

Figure 18. C-net based sequence control software: SCR

Several features of SCR are the following:

1) ***Net editor/translator***: Net editor helps a programmer interactively draw a C-net program on a graphical screen by using a pointing device. The translator makes the C-nets conform to the executable data format and inputs the format to control tables.

2) ***Interpreter:*** SCR can execute several C-net programs concurrently. Each C-net program starts its execution when a trigger event assigned to the net occurs. If several nets become executable at the same time, the SCR interpreter dispatches one of them according to an assigned priority. Each net can control other nets' start/stop by using internet event-post macros.

x3) ***monitor and test/debugging support functions***: The interpreter executes C-net programs according to the generated control tables. A monitor displays token locations and place program execution statuses in each action-place of the C-net program. The SCR monitor has also test and debugging supports in off-line mode.

6. APPLICATIONS OF SCR

The SCR has been successfully applied to several real systems.
1) ***Automatic assembly sequence control:*** A parts assembly station consisting of one industrial robot and six machine tools was controlled by the SCR. [Figure 19].

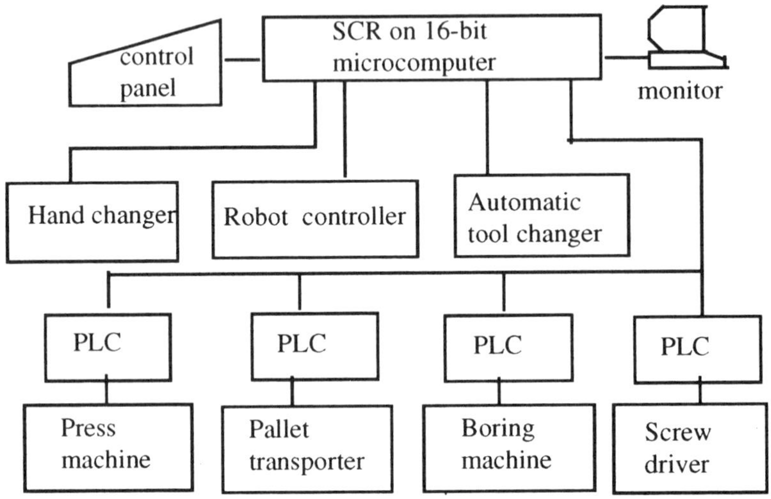

Figure 19. Configuration of an automatic assembly station controlled by SCR

The SCR controls the robot and machine tools through digital i/o interfaces, and the execution cycle time is about 10 ms. In this application, the total scale of the C-nets is about 1000 places and 1000 transitions. The total C-nets are divided to ten subnets and each subnet is developed correspondingly to each works to be assembled.
2) ***Material flow control and tracking:*** Figure 20 shows a simple example of a C-net program for pallet tracking in an automatic conveyor line. Each pallet's tracking information is represented as a colored token's slot data, and the pallet tracking procedure is implemented as a place program in a place corresponding to each tracking zone. For example, the place program updates the pallet tracking file according to the information about the token when it enters the place. The branch places execute branching control of a pallet on the conveyor depending on the contents of the token corresponding to the pallet.
3) ***Information processing control:*** The SCR was also applied to an automatic train arrival/departure guidance system. This system [Figure 21] consists of the SCR on a microcomputer-based controller which controls display/announce equipment on platforms, a centralized traffic controller (CTC) that detects trains approaching the station and sends train ID data to the SCR.

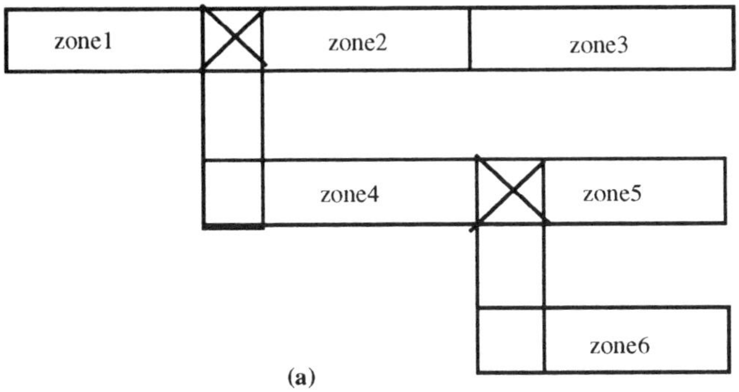

(a)

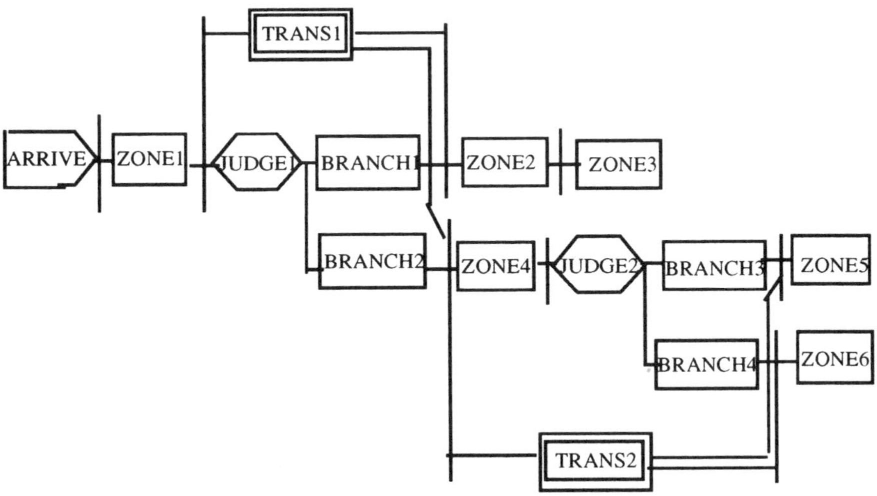

(b)

Figure 20. (a) Conveyor layout, (b) A C-net program for conveyor control

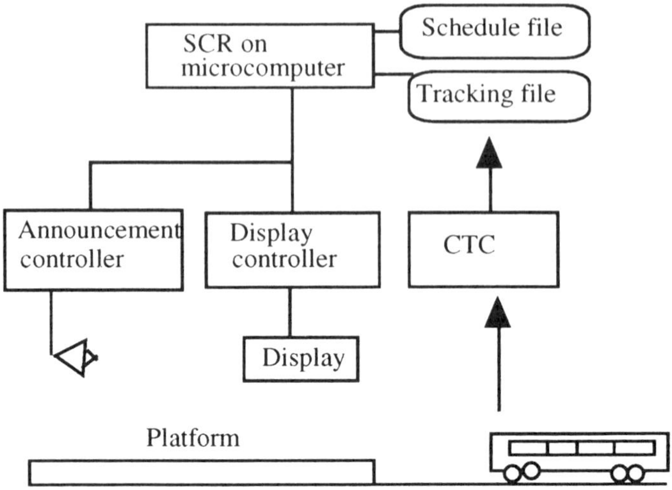

Figure 21. (a) Conveyor system, (b) A C-net program for conveyor control

The control logic for train arrival/departure guidance on each line are described by C-net programs and tracking information of each train (train ID, arrival time, departure time, destination, etc.) is represented as colored tokens. The C-nets program consists of an event analysis net, a train identifying/tracking net, and a broadcast control net. The SCR generates a new token in an event-analysis net corresponding to the detected train ID when CTC sends a train detection signal to the SCR. The event-analysis net then starts processing for the token. Part of an event analysis net is shown in Figure 22. It reads the train data from a train arrival schedule file, performs data rationality checks, stores the train data in the colored token's slots, and sends the token to the corresponding train identifying net.

The train identifying/tracking net transfers the arriving train ID to a train tracking file. A broadcast control net reads guidance message data corresponding to the approaching train from the scheduling file and sends the data to a automatic broadcast/display equipments on platforms. The total number of places in the C-net programs was about 500 and the total number of place programs (subroutines) was about 50. All place programs were implemented in C language.

In these applications, initial software development man-hours were reduced to about half of what would be needed if the conventional programming method using only a procedural language were used. Furthermore, several changes of control specifications were carried out easily because C-net-level changes and place-program-level changes can be performed independently.

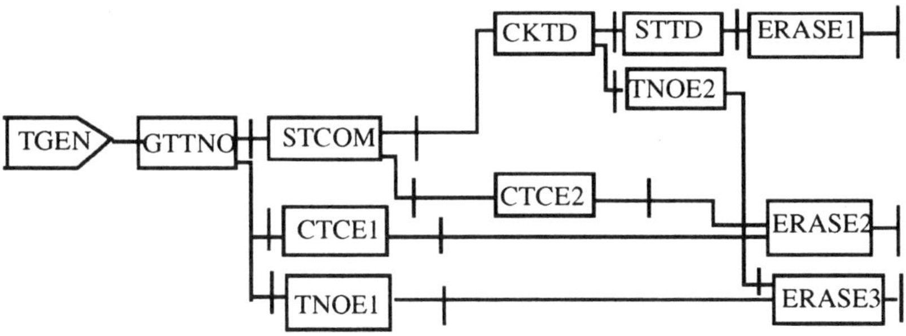

meanig of places

TGEN	Token generation	STTD	Train ID setting
GTTNO	Train scheduling file read	CTCE1, 2, 3	CTC error
STCOM	approaching signal input	TNOE 1, 2	Train no. mismatch
CKTD	Train ID checking	ERASE1,2, 3	Tain data delete

Figure 22. Part of a C-net program for arriving train identification

7. VERIFICATION OF SEQUENCE CONTROL LOGIC

A mishap is an unplanned event or series of events that results in damage or loss of property of a component. Verification of sequence control specifications is important to show that system operation under the specifications will not result in a mishap. A hazard is a set of conditions within a state from which there is a path in the reachability graph to a mishap [6]. To generate the reachability graph, we must develop a combined model as shown in Figure 23 consisting of the control target system behavioral model (environment model) operated under the the controller specification model (controller model).

Figure 24 is a simple example of a combined environment model (Petri net representation of two robots) with a controller model (black box model). The controller model can be verified as a black box, and any specification description can be incorporated into the environment model via the interface places of current-state-input (places s1, s2,..., s6) and control-event-output (places c1, c2,..., c6). The scheme to translate the environment model described as a Petri net to Prolog clauses has been developed. Figure 25 shows the Prolog descriptions transformed from the Petri net model of robot 1 and parts loader 1 in Figure 24 and expression of

[e1]>[s1]>>[s2] means that the state **s1** transits to the state **s2** when control event **e1** has been input. The reachability state analyzer shown in in Figure 23 is implemented as a Prolog program and creates the reachability state space of the environment model by iterating the following procedures.

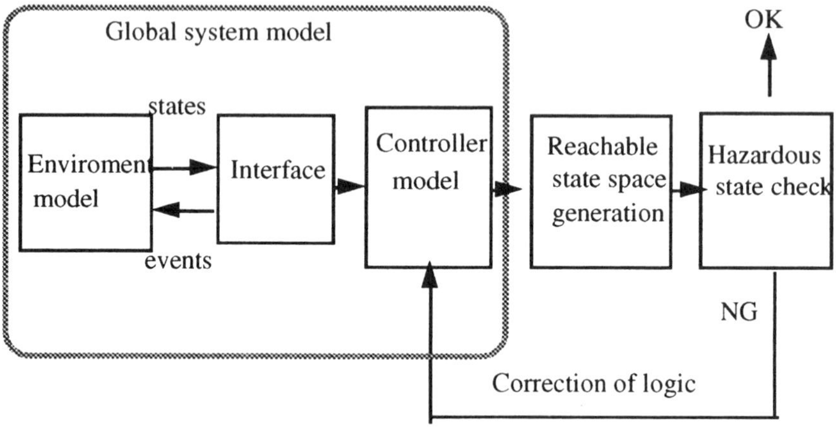

Figure 23. Configuration of hazardous analysis of control specifications

1) Input current states of the environmental model and execute the controller model which generates control events for the environment model depending on the current states.
2) Input the control events and, by applying the Petri net reachability analysis scheme, execute the Prolog clauses of the environmental model and create all of the next states of the environment model depending on the control events.

This method handles a controller model as a black box model and an environmental model to be controlled as Petri nets which has usually smaller set of states than that of controller models, therefore, complexity of a control logic is nothing to the size of the Petri nets. Designers can ensure that the control logic satisfies safety requirement by verifying that the reachability tree of the Petri net of the environmental model include no hazardous state from which a mishap is reachable.

8. FUTURE STUDY

A variety of models based on Petri nets have been studied in real-time control specification modeling, controller design and control programming [12]. Petri net is a basic model which enables designers to model a variety of discrete event systems' behaviors and its enhanced models have their own specific characters and application

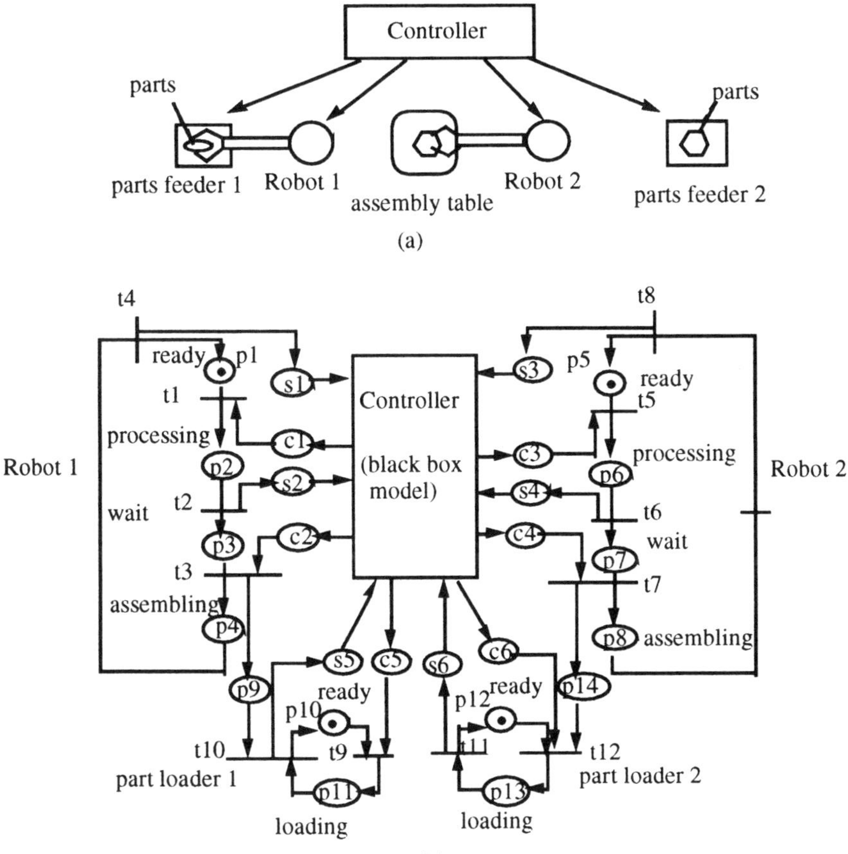

Figure 24. Combined environmental model with a controller model (Petri net model)

```
                    control event list
Robot 1                   previous state list
                                  next state list
    t1:   [c1]>[p1]>>[p2].
    t2:   [  ] >[p2]>>[p3].
    t3:   [c2]>[p3]>>[p4].
    t4:   [  ] > [p4]>>[p1].

Parts loader 1
    t9 :  [c5]>[p10]>>[p11].
    t10:  [  ] >[p11]>>[p10].
```

Figure 25. Prolog descriptions of behavioral model of robot 1 and parts loader 1

fields [10]. The main weak point is that the Petri net itself has no well-defined structured description ability, as can have a conventional structured programming language [3], and, therefore, it is difficult to develop a large size system based on Petri nets. To solve this problem, some synthesis methodology is essential [15]. Object-oriented Petri nets is being studied in software engineering field for concurrent programming [1].

Petri nets and their enhanced models are attracting attentions and expected to be increasingly applied to real-time or concurrent system control because of their advantages in terms of improving understandability, modifiability, and verifiability of specifications.

9. ACKNOWLEDGEMENT

I wish to thank Prof. Norihisa Komoda of Osaka University for encouraging me to submit this chapter and I thank Dr. Koichi Haruna, general manager of Systems Development Laboratory, Hitachi, Ltd., for his valuable suggestions and supports to this research. I also thank related members for their cooperation in SCR implementation and application.

REFERENCES

[1] Bastide, R., P. Palanque and C. Sibertin-Blank, 1993, Cooperative Objects: A Concurrent, Petri-Net Based, Object-Oriented Language, *Proc. IEEE International Conference on Systems, Man and Cybernetics,* Vol. III, 286-291

[2] David, R. and H. Alla, 1992. *Petri Nets and Grafcet:Tools for Modeling Discrete*

Event Systems, Prentice Hall, London.
[3] David. R and H. Alla, 1994, Petri Nets for Modeling of Dynamic Systems - A Survey, *Automatica,* Vol. 30, No. 2, 175-202.
tems, Man and Cybernetics, Vol. III, 286-291.
[4] Harel, D., 1987. Statecharts: A Visual Approach to Complex Systems, *Science of Computer Programming,* 8(3), 231-274.
[5] Komoda, N., K. Kera, and T. Kubo, 1984. An Autonomous Decentralized Control Systems for Factory Automation, *Computer, IEEE,* 1984, Vol. 17 No. 12 , 73-83.
[6] Leverson, N.G. et al., 1987, Safety Analysis Using Petri Nets, *Trans. Software Engineering,* IEEE, SE-13-3, 386-397.
[7] Matuzaki, K., S. Hata, et al.,1985. Petri-Net Structured Sequence Control Language with GRAFCET-like Graphical Expression for Programmable Controllers, *Proc. IECON'85*, 433-438.
[8] Michel, G., 1990. Programmable Logic Controllers: Architectures and Application, John Wiley and Sons, England.
[9] Murata, T., N. Komoda and K. Haruna,1986. A Petri Net- Based Controller for Flexible and maintainable Sequence Control and its Applications in Factory Automation, *Trans. on Industrial Electronics,* IEEE, IE-33, 1-8.
[10] Murata. T, 1989, Petri Nets: Properties, Analysis and Applications, *Proc. IEEE,* 77, 541-580.
[11] Murata, T., 1993. Hierarchical Specification of Reactive Systems using State Diagram Matrix, *Proc. IECON '93,IEEE,* 1993, 138-143.
[12] Peterson, J. L., 1983. *Petri net Theory and the Modeling of Systems*, Prentice Hall,Inc.
[13] Petri, C.A., 1962. Kommunication mit Automaten, Schriften des IIM Nr.3, *Bonn : Inst. fur Instrumentelle Mathematik.* Also, English translation, Communication with Automata, New York: Griffiss Air Force Base, *Tech. Report RADC-TR-65-337, Vol. 1, Suppl. 1.*
[14] Reisig, W. 1985. *Petri Nets,* Spring-Verlag.
[15] Zhou, M. C., and DiCesare, F., 1993. *Petri Net Synthesis for Discrete Event Control of Manufacturing Systems*, Kluwer Academic Publishers, Boston, MA.

3

Computer Aided Design of Logic Controllers with Petri Nets

Luca Ferrarini
Dipartimento di Elettronica e Informazione
Politecnico di Milano
Piazza L. da Vinci, 32
20133 Milano - ITALY
E-mail. ferrarin@elet.polimi.it

Abstract

In the present chapter, the application of Computer Aided Control System Design principles to the analysis and design of logic control systems is discussed.

1. Introduction

The potential for real applications of a Computer Aided Control System Design (CACSD) tool relies on two key issues: system representation and data management [1]. The first issue is relevant to the user interface, which should be characterized by a few standardized paradigms, as close as possible to the intuitive models the user builds in his mind [2, 18]. The second issue is relevant to the definition of a data environment, ensuring that the produced data are consistent, stored in a unified database, retrievable and usable by all the developed tools, available in various formats for the user. Further important elements are the simplicity of the processing tools, the possibility to customize the environment, the openness with respect to hardware platforms and software tools, the autodocumentation and the extendibility.

These requirements lead to a CACSD environment where diverse tools and functions are integrated to assist the control engineer in all the phases of the development, from specification, to functional design, to validation, to implementation and documentation. More specifically, the CACSD environment should support a *design methodology* to guide, and thus necessarily limit, the designer's operations.

Here the attention is focused on the design of logic controllers. Logic systems may be considered as a particular class of discrete-time systems, specialized to manipulate signals that may assume only a limited number of values or symbolic values and that may change in correspondence to the abrupt occurrence, at possibly unknown irregular intervals, of physical events, which also may be described in nonnumerical terms. The control of logic systems is usually performed by programmable logic controllers or CPU-based control devices. In particular, an integrated system to design logic controllers will be described, which is based on:

- a non-ambiguous model based on Petri nets formalism, conceived to use a few standard structures;
- a design methodology, which refers to the two fundamental principles applied in engineering practice to deal with complex designs: the bottom-up and the top-down approaches;
- a CAD package which incorporates the model, the design methodology, and a number of analysis and data manipulation functions which definitely reduce the designer's effort and ensure consistent development of controller design.

A very peculiar aspect of such a CACSD environment is the ability to account both for algorithmic properties of control systems and for hardware/software architectural properties, like multitasking operating mode or data exchange among different computing units. Thus, it can account for the *real* system response.

A second distinctive feature is the multi-level representation of logic control systems, which meets two important objectives: *a*) the possibility to employ abstraction mechanism, which are very useful in the preliminary phase of system specification; *b*) the contextual design of high level control system functions (supervisory functions) and of low level commands deeply related to field signals.

It is finally worth noting that the adopted design scheme has been conceived to be used along the whole project development, with the aim of bridging the gap between specification, design and documentation, thus keeping consistency between what a control system is required to do and what it actually does. The developments illustrated here are particularly effective in the perspective of design.

This chapter is organized as follows. In Section 2 the problem of logic control design is addressed. In Section 3 the definition of the Petri net model is given. In Section 4 the two used modeling approaches are described. Section 5 provides criteria for the analysis of the model. Section 6 and 7 illustrates the features of an integrated CACSD environment suitably developed, while in Section 8 a modeling example is given. Section 9 provides some concluding remarks.

2. The design of logic controllers

Because of the variety of specification formalisms, there is not a generally accepted and structured standard design method yet, which is capable of supporting unambiguous specification of a logic controller, its programming and its communication with the process and with other devices [12]. Actually, standardization efforts have been made in some countries. Examples of these are the well-known Ladder Diagrams formalism in the U.S.A. market and GRAFCET in France. Since there is no clear and effective methodology to deal with the complexity of the design procedure (from specification to implementation), the development of logic controller programs is generally performed through many successive steps (a step consisting of the interpretation and the translation of the scheme obtained at the preceding step). In this way, many errors may be introduced at any step, so that the final design may easily not fulfill the system requirements. In addition, when dealing with complex automated systems, a design involves inevitably a number of people and phases, so that its development and maintenance are often a source of trouble.

The development of our representation formalism followed two distinct lines of thought. The first one is to give the designer the possibility to separately design different parts of a control system, say *tasks*, and then to combine them through suitable connections. Here, the main idea is to avoid the introduction of limitations in the aggregation of tasks, thus enlarging the modeling power of the formalism, which is very important in real applications. In order to facilitate the understanding and analysis of the result of the aggregation, the same language (i.e., the same vocabulary symbols and syntax rules) should be used to describe both the single module and the aggregation of tasks.

The second line of thought is to allow stepwise refinement, but preventing translation errors from one step to another. Again, typically this can be obtained when the same language is used at each step. This means that such a language should be apt for abstract representation, each step of the synthesis process being one level of abstraction. At the lowest level of abstraction one can imagine to have the actual control program. In addition, when dealing with large systems, it would be desirable to have a language that admits the decomposition of the system into subsystems by possibly keeping the same formalism at any decomposition level. The problem arising in this case is how to define subsystems without altering the structure and the meaning of the formalism itself.

3. Definition of the Petri net model

A graphical formalism able to solve the conceptual and practical problems involved in the specification process of logic controllers, so as to improve productivity and automatize the logic controller design, is illustrated. As a base language, the Petri Nets (PNs) formalism has been chosen. Although, when dealing with logic controllers, PNs may degenerate into finite state machines, one major advantage in using PNs is that reference to the total state at any instant is not necessary

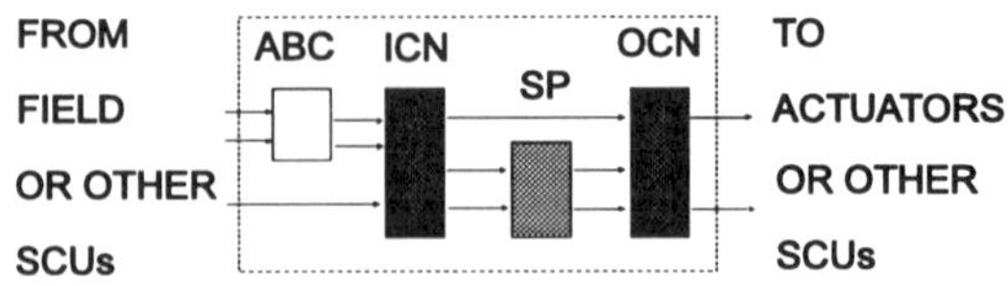

Figure 1: The Standard Control Unit

when designing a control sequence [22]. Advantage of this can be taken in both the synthesis and analysis phases.

The model developed for logic controllers is called Standard Control Unit (SCU). An SCU is the simplest, but complete, control unit which can make decisions on its own on the basis of the values measured from the field and send them to actuators or to other controllers. Each SCU is planned to solve a specific problem, like the control sequence of a conveyor with multiple feed lines. In a generic SCU it is possible to find four separate parts (Fig. 1) [7, 16]:

- an Analog-to-Binary Converter (ABC), where threshold criteria are applied to analogs to get binary signals;
- an Input Combinatorial Net (ICN), where ABC output and external binary inputs are combined to get signals affecting the Sequential Part behavior or directly the outputs;
- a Sequential Part (SP), implementing the memory of the logic controller;
- an Output Combinatorial Net (OCN), where output variables are obtained via combinatorial operations on ICN and SP outputs.

The SP has been substantially modeled as a PN. In order to let the SP be influenced by external signals coming from the ICN, the ICN outputs are used to modify the enabling conditions of transitions connected to them: if a signal to a transition is high, then the transition is allowed to fire according to the classical PN rules, otherwise it cannot be enabled. The analysis and synthesis of the sequential part will be discussed in next sections. As for combinatorial nets, their analysis reduces to check simple requirements due to the coupling of the sequential part and combinatorial nets, such as the fact that the ICN should not allow cyclic operations in the sequential part, when the SCU inputs are not varying, and that the net should not be prevented from evolving any longer for any external input combination. These concepts are dealt with in [15, 16].

The SCU representation is definitely *modular*. Actually, even if conceived as a model of just one control program, the SCU can be used to model also subprograms and, according to simple transformations exposed in [4], the aggregation of SCUs can result in a larger "equivalent" SCU. This fact guarantees that the analysis techniques and tools developed for a single PN scheme and a single SCU can be extended to any aggregation of SCUs. The important characteristic is that the unification of more SCUs into one can be made automatically and transparently by the primitive functions of a CAD environment.

4. Modeling approach

Bottom-up and top-down approaches can be incorporated, with care, into the PN model. In our approach the bottom-up approach is based on the possibility given to the designer to define elementary tasks, suitably modeled, and then to connect these tasks together through a few types of generic connections. The top-down approach is based on a multi-level (hierarchical) representation, which is extremely useful to deal with complex systems.

4.1. Bottom-up approach

In the real bottom-up approach considered here [4, 8, 10] the designer is given the possibility to define tasks (Elementary Control Tasks, ECTs), modeled as strongly connected state machines with just one token, and then to connect these tasks together through three kinds of connections, namely self-loops, inhibitor arcs and (generalized) synchronizations. ECTs may be used to model resources or machines involved in cyclic operations, like robots, machining workstations, AGVs. They can also be used to represent just process states. The meaning of connections is quite straightforward. A self-loop from an ECT, say E_1, to another ECT, say E_2, allows the behavior of E_2 to be influenced by the presence of a token in a place of E_1 without modifying the behavior of E_1; an inhibitor arc from an E_1 to E_2 allows the behavior of E_2 be influenced by the absence of a token in a place of E_1 without modifying the behavior of E_1; a (generalized) synchronization represents a rendez-vous between two ECTs that need to share resources or operations. Thus, a self-loop will be used, for example, to prevent a certain operation being executed until another one is terminated, or until a certain situation (empty buffer, full conveyor) is reached; an inhibitor arc will be used, for example, to execute a certain operation unless another one is in execution, or unless a certain situation is reached; synchronizations will be namely used to force simultaneous executions of more operations.

Observe that, since no particular restrictive rule is assumed for the connections, a net which is not live or not reversible can "easily" result from connection operations. On the other hand, the complete freedom in connection composition allows a great modeling power, which may prove very important in applications.

Finally, notice that, even if the above connections could be formally converted into just one connection, namely the synchronization [4, 7, 16, 19], both for the user and for the particularly useful analysis method here exploited it is convenient

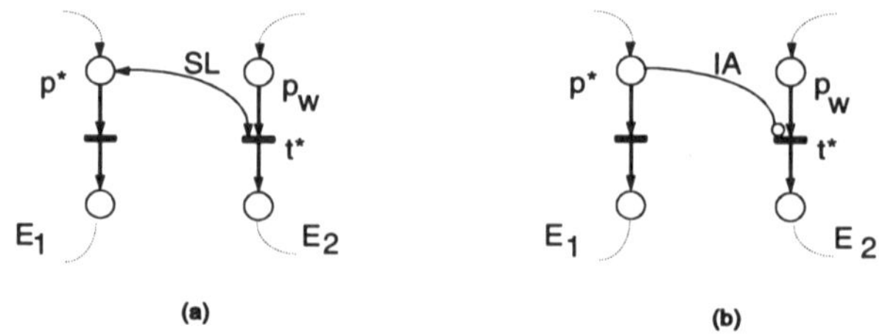

Figure 2: (a) Self loop - (b) Inhibitor arc

to keep three different connections instead of just one.

Recalling that: *a)* a state machine [14] is a PN where transitions are unbranched, i.e. there is just one place in their preset and postset, *b)* that a strongly connected state machine is a state machine in which there exists a path from any two nodes (places and transitions), *c)* a unitary marking is a marking under which one and only one place is marked, the following definition is given.

Definition 1 (Elementary Control Task). A strongly connected state machine E(P,T,I,O,m_0) with unitary initial marking m_0 is said to be an ECT.

Strongly connected state machines with an unitary initial marking are typically used to model cyclic sequential tasks. They will always have a unitary marking, are live and reversible, and they are covered by p-invariants and t-invariants. Such properties can be easily proved, as shown in [8, 21]. The connections between different ECTs are now introduced.

Definition 2a (Connections). Let p* be a place of an ECT E_1, and t* a transition of another ECT E_2. The following connections between E_1 and E_2 are defined:

1. self-loop (SL), starting from place p* and with arrival transition t* (see Fig. 2a); transition t* may be enabled only if the place p* is marked, while the usual enabling conditions for the places connected to t* with normal arcs still hold;
2. inhibitor arc (IA), starting from place p* and with arrival transition t* (see Fig. 2b); transition t* may be enabled only if the place p is unmarked, while the usual enabling conditions for the places connected to t* with normal arcs still hold;
3. (simple) synchronization (S) is the merging of the two or more ECTs through a common transition t*, as shown in Fig. 3a; the graphic symbol for a synchronization is depicted in Fig. 3b and consists of a double arrow line between any pair of graphically consecutive transitions involved by this connection.

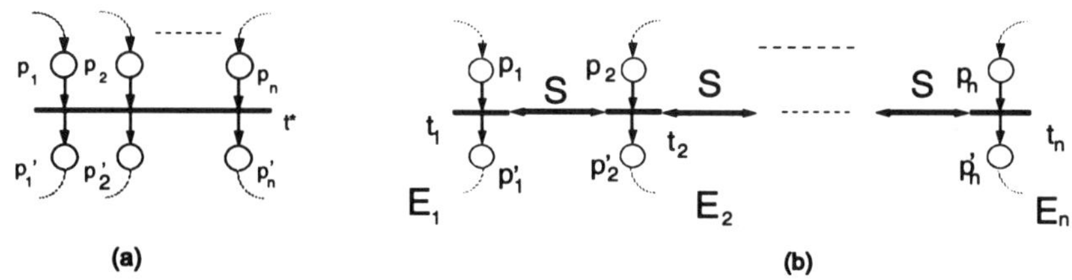

Figure 3: (a) Synchronization - (b) Its representation

Transitions receiving at least one connecting arc (self-loop, inhibitor arc, synchronization) are called ***conditioned transitions***. Given an ECT E_1 with a conditioned transition t_1, the place in the preset of t_1 is called *wait place* (called p_w in fig. 2(a) and 2(b)). It is apparent that possible liveness problems are necessarily associated with conditioned transitions. Note that, henceforth, when referring to a net N without any specification, it is meant that N is made, according to the model defined so far, of ECTs connected as in Def. 2a. Of course, the net obtained connecting two or more ECTs with the defined connections is not an ECT.

Now, the definition of the Generalized Synchronization is given. First of all, it is necessary to define the portion of an ECT that can be used in this new type of connection.

Definition 2b (Transition-included subnet). Let t_s and t_e be two transitions belonging to a given ECT $E(P,T,I,O,m_0)$ and let $EP(t_s,t_e)=t_s,...,t_e$ be a generic non-closed path of E which starts from t_s and ends in t_e. A transition-included subnet between t_s and t_e is a PN $Q(P_{in},T_{in},I_{in},O_{in})$ such that :

1. $T_{in}=\{t \in T \mid t$ is in at least one $EP(t_s,t_e)\}$;
2. $P_{in}=\{p \in P \mid p$ is in at least one $EP(t_s,t_e)\}$;
3. $\forall\ p \in P_{in}$, $(post(p) \in T_{in} \setminus \{t_s\}) \wedge (pre(p) \in T_{in} \setminus \{t_e\})$ (*locality condition*);
4. $\forall\ p \in P_{in}$, $\forall\ t \in T_{in}$, $I_{in}(p,t)=I(p,t)$ and $O_{in}(p,t)=O(p,t)$.

Remark. The locality condition is necessary to ensure that the token of an ECT accesses to (exits from) a transition-included subnet between t_s and t_e only through the firing of t_s (t_e). As it will be clear in the sequel, this particular requirement is fundamental for the consistency of the connection called generalized synchronization. Observe that when t_s coincides with t_e the transition-included subnet always exists and it is composed of the single transition t_s.

Finally, no marking has been defined for a transition-included subnet since such a subnet will always be considered within the corresponding ECT for which the marking is already given.

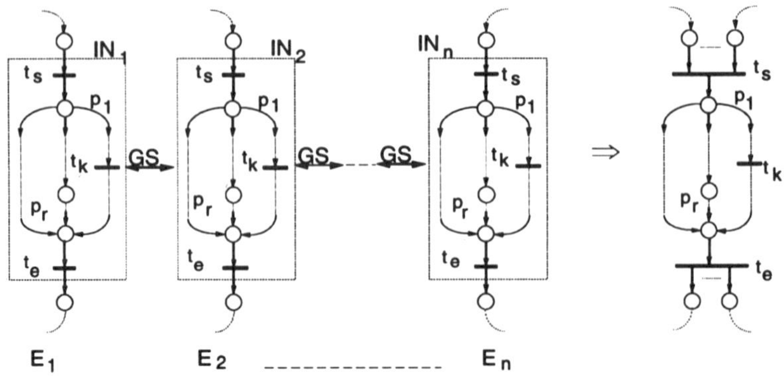

Figure 3(c): Generalized synchronization and its representation

Now, recalling the definition of isomorphic nets [21], the generalized synchronization definition is given:

Definition 2c (Generalized Synchronization). Let $IN_1, \ldots, IN_n$, $n \geq 2$, be n isomorphic transition-included subnets belonging to n different ECT. A generalized synchronization among $IN_1, \ldots, IN_n$ is the connection in which the n transition-included subnets are merged into one.

In Fig. 3c an example of generalized synchronization is given. According to Def. 2c, the GS connection implies the merging of isomorphic corresponding transitions and places. Actually, the transition merging can be simply obtained connecting each set of isomorphic corresponding transition with a simple synchronization.

However, the number of simple synchronizations, used to describe a GS, is proportional to the number of transitions belonging to a generic included transition subnet, which basically means that one single connection has been substituted for many (simpler) connections. Since the analysis complexity increases with the number of connections, analysis methods and tools would be practically unusable, even if they could always be in principle applicable.

In [10], this problem is addressed. There it is proved that, if an initial marking *consistent* with a GS in a net is chosen (that is *all* isomorphically corresponding places for a GS are either marked or unmarked), then that GS can be substituted by just *one* simple synchronization among the first transition (t_s) of that GS.

4.2. Top-down approach

Unfortunately, in their original formulation PNs do not support hierarchical modeling, thus making the design not manageable in most real cases. Valette [25] developed the concept of transformation of a transition into subnets and vice versa, giving conditions under which a subnet can be substituted for a single transition while preserving properties like liveness and boundedness. Hierarchical modeling with different types of places (and, possibly, of transitions and arcs) was treated by Crockett and Desrochers [3] and Valavanis [24]. However, their rich language does not necessarily facilitate the modeling procedure and the formal analysis; in addition, a systematic approach for the design of PN models of logic controllers remains undeveloped. In this direction, it is worth citing the works of Zhou et al. [27, 28], clearly pointing out basic properties and basic useful structures of PN models. However, the underlying synthesis method may still be hard to use in complex designs and the result of successive refinements is a "flat" large net, difficult to handle.

In the approach proposed here, a design is developed through repeated refinements, where any refinement operation associates a subnet to an element of the net (say, *root* element). The refinement is to be thought of as a substitution, according to sensible and obvious rules, of the root element with a given subnet. This concept is similar to that of the mentioned authors. On the other hand, however, in order to allow truly hierarchical design, this substitution does not really take place, but the root element coexists with the subnet detailing it. In this case, suitable rules must be given in order to clarify essential concepts like the marking of root places, the synchronization of a subnet with the net containing the corresponding root element, the multiple activation of a subnet. Thus, in this approach, provided that unambiguous rules are established, a net may be used for simulation or analysis either when the root elements it contains are connected with subnets or not; it can also be used when some root elements are refined and some other not yet. In the case when all the root elements are associated to their subnets, the whole net really corresponds to the actual final program to be loaded onto the PLC (with all the I/O signals and all the timings completely specified), so that it can be tested in a real-time environment. In the other cases, the net corresponds to an abstraction of that program.

As conceived here, subnets refinement is a strictly *local operation*, that is the refinements of two elements of a given (sub)net are not a priori related to each other. As a consequence, they do not necessarily express the same level of detail. Thus, there is not the concept of level of abstraction for the whole net, but of levels of abstraction for a single net element. Observe that in this (local) refinement, not only PN elements are involved, but also I/O signals, which represent one of the most troublesome problem for the designer.

To meet these requirements an important issue must be addressed: synchronization and management of the signal exchanges between two parts of the net must be dealt with at the level those parts belong to, in case they belong to a unique net level, or at higher levels, if they are contained in different subnets. As a consequence, all the subnets may exchange I/O signals only with the immediately higher level net and/or with the process. They must not exchange information among themselves whatever level they belong to, in order not to modify unexpect-

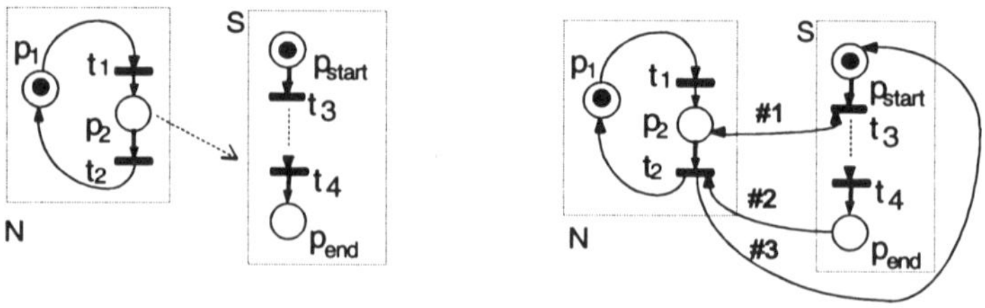

Figure 4: (a) Subnet for the refinement of p_2 - (b) The net after the refinement

edly and dangerously the synchronization structure of the net. These limitations are to be overcome in future developments. Finally, observe that hierarchical decomposition of the logic controller design in subsystems may be effectively dealt with only within a suitable CAD environment, especially developed to this aim, supported by a convenient structured data-base.

Let's now consider place refinement. Suppose that you want to detail the control program represented, at higher level, by the place p_2 of the net N of Fig. 4a. Let the net describing the control program be the net S in the same figure. Observe that in the subnet S of Fig. 4a only one place can be seen as starting place (p_{start}) and one place can be seen as ending place (p_{end}). The idea is to synchronize the execution of S with that of N. This can be done for the two-level net of Fig. 4a as represented in Fig. 4b, where just one "flat" net is used exploiting the connections defined in the previous section. Such connections have been labeled with a ♯ sign plus a number. The execution of this net is now straightforward. Due to arc ♯1 (self-loop), the subnet S with one and only one token in p_{start} can not be executed until the root place p_2 is marked. Once S begins its execution (i.e. when transition t_3 fires), net N cannot evolve (none of its transitions can fire) until the place p_{end} of S is marked. The cause of that is arc ♯2. Once t_2 fires, N can evolve free, while the initial marking in S is re-established through arc ♯3. Notice that t_3 can fire only after p_{end} has been marked. As a consequence, with the structure of Fig. 4b simultaneous multiple activation of the subnet is prevented. Anyway, this semantics allows the modeling of *while-do* cycles [15].

The formalization of these concepts follows the really effective line tracked by Valette for transition refinement [25] used as a design methodology.

Definition 3a (P-block). A net with one and only one initial place (a place with empty preset) p_{start} and one and only one final place (a place with empty postset) p_{end} is called a P-block.

Definition 3b (Adjoint P-block). The net obtained adding a transition t_a to a P-block PB such that pre(t_a)=$\{p_{end}\}$ and post(t_a)=$\{p_{start}\}$ and with initial marking M_{beg} such that $M_{beg}(p)=1$ for $p=p_{start}$ and 0 otherwise, is called adjoint P-block of PB.

Definition 3c (Well-defined P-Block). A P-block PB is called a well-defined P-block iff its adjoint P-block APB is live and safe.

Definition 4 (P-connection). The P-connection of a net, through one of its place p^* (root place), with a P-block is performed in the following way:

1. insert one self-loop from p^* to each transition in $post(p_{start})$;
2. insert one arc from p_{end} to each transition in $post(p^*)$;
3. insert one arc from each transition in $post(p^*)$ to p_{start}.

The refinement of transitions is a little subtler problem. As already exposed, we do not use step-wise refinement only as a design methodology to arrive at the final complete net. In that case, in fact, the various nets corresponding to each step may also not have a precise meaning and may be temporary in the design process; they may be used only once in the design development [27, 29].

Now, if you want the behavior of a given net to be the same *before* and *after* the refinement of some transition, as it is supposed to be in the place refinement, it is necessary to remove one of the basic assumption of PNs [20] - the instantaneous firing of transitions - in order to avoid the absurd situation of an instantaneous firing of a root transition while the execution of the associated subnet may not be necessarily instantaneous. Thus, either one accepts that the behavior of a net can not be the same before and after the refinement (so that modification of the standard PN definition is not necessary, but one has to deal with two different nets, one before and one after the refinement) or a new type of transition is to be introduced (and, consequently, new firing conditions and suitable interaction between transitions of different type). While the first possibility does not allow really hierarchical (multi-level) modeling and design, the second makes the design not easily readable, the "tracking" of tokens almost impossible and the redefinition of basic concepts and properties necessary.

In our choice, non-primitive transitions have been implemented not as new basic elements, but as a compound element, structured as shown in Fig. 5, where t_1 relates to the start-of-the-subnet-execution event, t_2 to the end-of-the-subnet-execution event and the place p_1 can be suitably refined with a subnet (similar examples can be found in [23]). While forcing the explicit re-formulation of non-primitive transitions, the problem of transition refinement has been transformed into a problem of place refinement, thus maintaining the full compatibility with the philosophy of P-connections and keeping the number of "concepts to learn" about the model limited for the user.

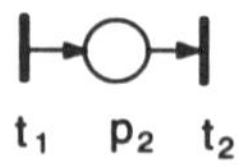

Figure 5: Place refinement

5. Analysis and validation

Through a set of specifically developed theoretical analysis results, it is possible to check a given model in order to find out conceptual errors and inconsistencies. In some cases, remedies are "automatically" suggested. The investigated properties are boundedness, liveness and reversibility.

Bottom-up approach

As it can be easily seen from the definition of ECT and of connections, the model here illustrated is safe, thus bounded. As for liveness, the analysis of a net can be performed constructing additional structured information, drawn from the PN model. First, exclude the presence of the trivial ***dead connection bundles***, that is of groups of connections going from one ECT, say E_1, to one transition, say t_1, of another ECT such that the token in E_1 structurally can never enable t_1, whatever transition t_1 is considered. As listed in [8], dead connection bundles represent only a few cases; they are due to some conceptual mistake, as in the case of a self-loop and an inhibitor arc both starting from just one place. In [11] it is proved that two fundamental causes for which a net not containing a dead connection bundle is not live are the presence of particular structures, called ***essentially locked B-nets*** and the presence of ***locked transitions***. An essentially locked B-net represents a subset of the ECTs and connections of a net, in which each conditioned transition cannot be enabled if the initial marking is a *wait marking*, i.e. a marking where all the marked places are wait places (see sect. 4.1). Thus, an essentially locked B-net defines a mutual wait (locking condition) among many ECTs, variably connected, thus extending the classical concept of circular wait.

Basically, it can be proved that the wait marking M_{wa} of an essentially locked B-net ELB_1 and any marking $M \in [M_{wa}>$ are non-live markings for a net N containing all the ECTs and all the connections that contribute to ELB_1. In addition, call Σ the set of the markings M above discussed, and N* a generic net containing ELB_1. Consider now a marking M* of N* from which a marking M' is reachable whose restriction to the ECTs of ELB_1 is a marking belonging to Σ. If M* is the initial marking of N*, then N* is non-live and ELB_1 is said to be *active* with M*. This means that the presence of that essentially locked B-net may prevent some transitions from firing.

The second cause of liveness loss is due to the presence of a locked transition.

This situation is due to the impossibility to mark at the same time an enabling place relative to each CB arriving at such a transition, despite that both its wait place and its enabling places can be marked in some marking M reachable from the initial one.

One of the most important results in the liveness analysis is the proof that the two mentioned causes of loss of liveness are independent of one another; another crucial property is that if a net is not live, then at least an active essentially locked B-net or a locked transition is present in the net. The illustrated analysis is especially effective if a CAD tool is available, since analysis can be performed each time a new connection has been established. As a matter of fact, in this case, only newly formed B-nets should be analyzed, since the presence and the effect of previously analyzed B-nets can not be modified while adding new connections. Please also refer to [8] for possible suggestions of automatic solutions in case of detected faults.

As for reversibility and reachability analysis, it can be proved that the marking of a net N can be partitioned into two sets, say A and B, according to the topological structure of the given (unmarked) PN model. Let M_A and M_B be two markings of N belonging respectively to A and B. Then M_A does not belong to $[M_B>$. In addition, if the net is live, M_A is not reversible. These results can be directly exploited to derive criteria to decide if an essentially locked B-net is active and to choose the initial marking of the net. Further conditions to decide the reversibility of a given marking are in development.

Top-down approach

In this case, a very simple and effective result can be obtained. First observe that the marking M_{beg} of the adjoint of a well-defined P-block is reversible. In addition, let N be a net composed of ECTs, with initial marking M_N and p* the place to which a well-defined P-block PB is to be P-connected. Let F be the net obtained after the P-connection of N with PB through p*. Finally, let M_F be the initial marking of F such that $M_F(p)=M_{beg}(p)$ for any place p of PB and $M_F(p)=M_N(p)$ for any place p of N. Then, it can be proved [15] that:

- M_F is reversible iff M_N is reversible;
- F is live iff N is live;
- F is safe iff N is safe.

Stemming from what exposed in the preceding sections, a design procedure is easily established which is able, with the minimum effort of the designer, to guarantee formally good results in the synthesis phase. For any of the identified SCUs, the designer can decide what ECTs and connections he may need; then he may design the ECTs, connect them, refine some places, if necessary. If some other ECTs are necessary, they can be defined at any moment of the design procedure, and then refined and connected, since the bottom-up and top-down approach here exposed can be alternatively exercised.

6. Real system representation

The mere validation of the functional design and the pure algorithmic checking are, in general, not sufficient to guarantee the correctness of the control system, especially when non trivial architectures are used, e.g. distributed control systems, and when non trivial hardware/software implementation are needed. Thus, PN models of logic control tasks are allowed to communicate among themselves via a few standard signal exchange modes, which emulate the most widely-spread communication mechanism employed in real multi-tasking or multi-unit architectures. In this way, the performance of the target hardware/software system can be realistically predicted within the same CACSD environment.

6.1. Simulation of a single PLC

In order to avoid the non-determinism typical of PN and to give a simple way to solve conflict situations, firing priorities are introduced, labeling transitions with ordering numbers. Similar problems arise when simulating PN models [13]. However, the semantics of the model is not complete if the way of using such priorities is not clearly provided [17, 26]. As discussed in [5], in order to prepare the simulation code, the hierarchical structure of the net is "flattened". The result of this operation is a list of elements, each one related to a transition and containing all the information needed to fire it (preset, postset, Boolean functions of external signals, etc.). The elements of this list are ordered according to decreasing priorities. The execution *cycle* is composed of the following phases. 1st: reading of input signals (R); 2nd: evolution of PN (EXE); 3rd: writing of output signals (W). A random execution time is associated to each phase with a probability distribution selected by the user.

Note that the actions associated to the marking of places are executed only if a place is marked at the end of the 2nd phase (PN evolution). On the contrary, intermediately reached markings of the PN have no effect on the external world.

Two approaches to the execution policy of the model have been considered:

- "static" approach: PN execution consists of a sequential scan of all the transitions; within a cycle, a transition is evaluated only once;
- "dynamic" approach: the execution policy consists of repeated scan of the list of transitions until a steady state (that is, a state in which no transition can fire) is reached; output signals are updated only when a steady state has been reached.

In the latter approach two sub-policies may be defined according to how a single scan is performed. Actually, *each* scan may consist of the sequential evaluation of the transitions until either *all* the transitions have been analyzed (once a transition is enabled, it fires; thus a steady state is reached when during a scan no transition fires) or *one* fireable transition has been found (in this case that transition fires and a new scan begins from the *first* transition of the list, so that a steady state is reached when the *last* transition of the list is evaluated and it is not fireable).

With the static approach, there is a direct mapping from the topologic structure of the PN into the structure of the PLC internal code. This is what it is actually done for a PLC programmed with the Ladder Diagrams graphical language.

On the other hand, dynamic approach is suitable for wider applications. This has been done in the literature, with the realization, for example, of a PNC (Petri Net-based Controller), in which the PN is *interpreted* producing an algorithm running on a general-purpose computer [3].

6.2. Communication mechanisms

Particular care must also be paid to the communication mechanisms among single-processor devices, like PLCs, composing non trivial control system architectures like a distributed control system. The devices can fundamentally be:

- **tightly coupled**, when processors share memory, and there is a system clock synchronizing the different machines; usually communication between machines is achieved by global variables in the shared memory;
- **weakly coupled**, when each processor has its local clock and memory and communication is achieved by high-speed buses; in this case every machine evolves in a totally asynchronous way.

The first kind is usually applied to logic control based on PLC's networks, while the other one is often used in more complex applications, such as real-time control systems.

As for shared-memory-based systems, when two tasks share a global variable, they can use it in any way, provided that they access it in *mutual exclusion*. The same variable can be accessed by a task both in reading and writing mode. This is a *competitive* mechanism because tasks, though logically independent, compete to access the shared resource (memory). In this context, sequences of operations like "*..., task1 writes, task1 writes, task2 reads, ...*" are possible on the same shared variable and there is no means of guaranteeing that all data sent from the producer be consumed by the receiver. Solving conflicts due to simultaneous writings or writing/reading operations is done by giving tasks user-defined priorities in accessing the shared memory.

As for message-based systems, a *buffer* handled as a FIFO queue where messages are stored is inserted for each message link between two tasks. Note that now communication is *unidirectional*, since each message is characterized by a sender and a receiver. When two tasks are linked through a buffer, only one of them can write into the buffer, while the other can only read. This can be seen as a *co-operative* mechanism, since tasks logically depend on each other and it is necessary that *all produced* data be *consumed*. The use of buffers guarantees this kind of communication, absorbing the tasks' different speeds in producing and consuming data. To simplify the handling of all these buffers (one for each message link) circular buffers have been implemented.

7. Implementation

A prototype CACSD environment has been implemented to support all the phases involved in the specification process (specification, functional design, coding, formal testing, simulation, validation and automatic documentation), integrating complex data structures (textual, graphical and semantic) with the established design procedure.

This environment has been developed for the modeling, evaluation and simulation of *hybrid control systems*, which are generally composed of three basic subsystems: namely, continuous-time (essentially for industrial processes), discrete-time (essentially for generic micro-processor based systems) and logic systems. Also, particular care has been paid to the simulation of real-time systems, that can be customized according to the most wide-spread software architecture.

The CACSD environment has the capability to allow customization of representation formalisms, through a suitable meta-editor package. This package is conceived to guide the user towards a complete definition of models, starting from the definition of the symbols vocabulary and semantics and of the syntactic rules of the model. The representable formalisms are both block diagram-like ones and formalisms based on the well-known concept of *physical metaphor*.

Peculiar of the environment are the concepts of *machine* and *bus*, integrated with a control-oriented system description. Thus, a unified framework is provided which combines the algorithmic (functional) description of control systems with that of their actual implementation, that is with the relevant computer-dependent features. The clear separation between algorithms and activation/communication policies allows control algorithms to be debugged by themselves and the suitability of a specific target architecture to be checked for their integration in a global strategy. Please refer to [6, 9] for an explanation of the details of the CACSD environment.

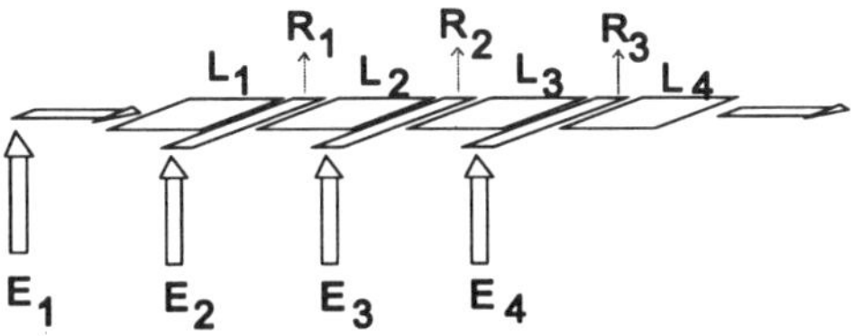

Figure 6: The super-elevated conveyor

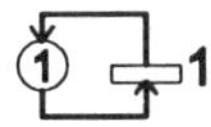

Figure 7: A "block" control

8. An example

In this section the control problem relating to a super-elevated conveyor with multiple feed-lines is taken into account. We will focus to the PN design of the logic controller. In the PN designs proposed here, input and output signals will be omitted for the sake of clarity.

The process is sketched in fig. 6. The conveyor is basically composed by power-driven rollers, nearly 5 meter high off the ground. The materials are carried to the conveyor from four separate production lines by four elevators (E_i, i=1,..,4). The actual material transfer from elevators to the conveyor rollers is performed by suitable devices (R_j, j=1,..,3).

Because of the presence of such transfer devices R_j, the roller conveyor is divided into four lumps (L_i, i=1,..,4). The transfer devices R_j "decide" whether to transfer objects from the lump L_j or from the elevator E_{j+1}, according to a scheduling policy (inessential in the present context), taking into account the full state of all the lumps. Each elevator has got one slide to carry one object at a time. To reduce the rise and fall time, the elevator motor can work at two speeds (say, *slow* and *fast*), so that the rise and fall motion is performed according to the following criterion: slow → fast → slow. Finally, observe that at the top of the first elevator (E_1) the transfer device is much simpler than the others: this is why it was omitted in fig. 6.

Because of the concurrent nature of the process, the controller design may be divided into four parallel parts, each one devoted to the control of a single block. The first block is composed of E_1 only, while the second, third and fourth are constituted of L_j, E_{j+1}, R_j, with j=1,..,3. Thus, the generic block control can be designed, at the most abstract level, like in fig. 7, where the place represents

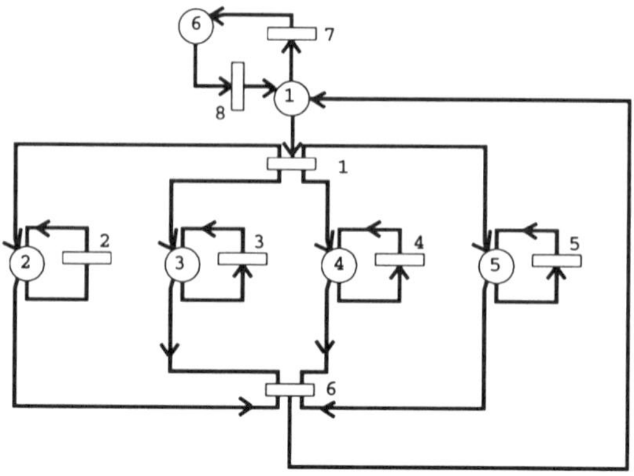

Figure 8: Global conveyor controller

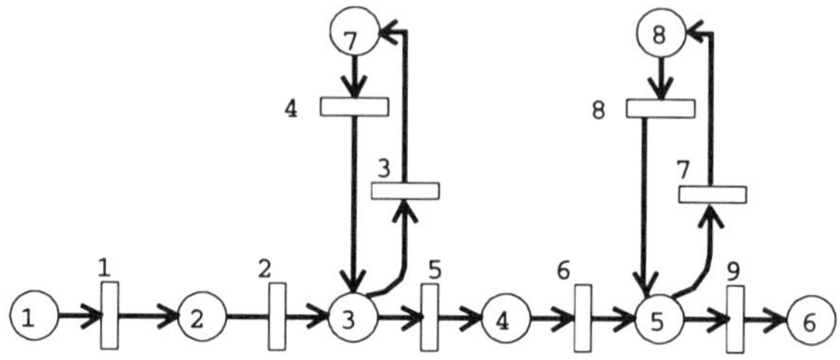

Figure 9: Refinement of p_2 of fig. 8 (p_1=start; p_2=load; p_3=up-slow; p_4=unload; p_5=down-slow; p_6=end; p_7=up-fast; p_8=down-fast)

the control sequence. In the following figures, a place p_i is represented by a circle with the number i inside, while transition t_j by a bar with the number j next to it.

In order to synchronize the start and the stop of the control program, the global conveyor design is performed as in fig. 8, where the place p_1 is the start place, while p_2, p_3, p_4 and p_5 are the four block controls and p_6 is the fault management routine.

The place p_2 represents the block controlling E_1 only. It can be detailed as in fig. 9. This is the first level refinement of the place p_2. On the contrary, places p_3, p_4 and p_5 of fig. 8 represent both the control sequence of the elevators E_{j+1} and the transfer devices R_j, j=1,..,3. The refinement of such places are quite similar to one another, so in fig. 10 only the refinement of them is reported. The refinement

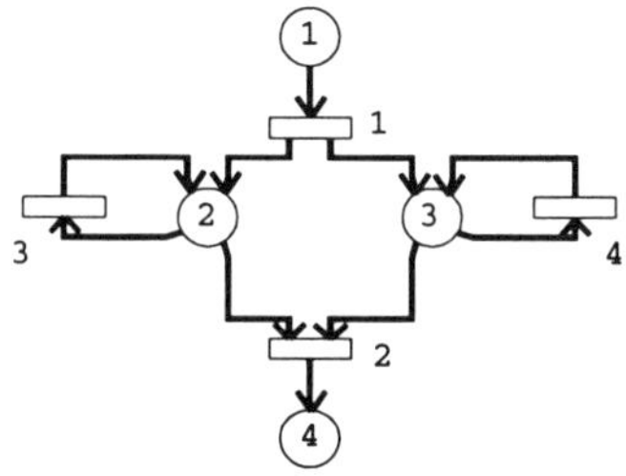

Figure 10: Refinement of p_3 of fig. 8

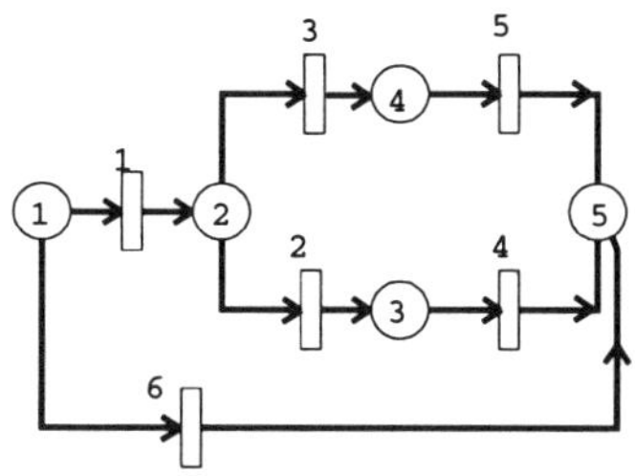

Figure 11: Refinement of p_3 of fig. 10 (p_1=start; p_2=piece loaded ready to move; p_3=transf. from lump; p_4=trans. from elevator; p_5= end)

of place p_2 in that figure is, in turn, quite similar to that of p_2 of fig. 8, so the subnet representing its refinement is not reported here. Observe that this omitted subnet belongs to the second level of abstraction while that of fig. 9 attains to the first one.

The place p_3 (the transfer device control sequence) of fig. 10 is detailed in fig. 11. Observe that places p_3 and p_4 in fig. 11 and places p_2 and p_5 in fig. 9 could be further detailed iterating the refinement procedure, to represent the activation of the relating motors with more details. Finally, observe that at any time the designer deals with simple subnets only; the management of the various subnets could be inefficient, or even impossible, if not supported by an integrated CAD tool.

9. Concluding remarks

In the present chapter, it is shown how theoretical and methodological tools can be exploited in the design of logic control systems. A design methodology and an analytical framework based on Petri nets is also illustrated. In particular, it is shown how the application of Computer Aided Control System Design principles is mandatory for the solution of complex control system designs problem, like that of logic control systems, which are characterized by an intrinsic interdisciplinary feature that requires particular knowledge in a variety of sectors besides that of the process under control (for example, those concerning communication protocols, operating systems, microprocessors and their programming) and where the type of problems to be solved may be very different (from alarm and protection management, local control of workstations, supervision and coordination, monitoring and diagnostic, quality control).

Acknowledgments

The author is particularly grateful to Prof. C. Maffezzoni for his helpful suggestions and stimulating discussions on the subject.

References

[1] Åstrom K.J. and Kreutzer, W., 1988. System Representation, *25th IEEE CDC*, Athens, 1986.

[2] Barker. H.A. et al.. 1993. Open Architecture for Computer-Aided Control Engineering, *IEEE Control System Magazine,* Special Issue on Computer Aided Design in Control Systems, 1993, Vol. 13, n°2, pp. 17-27.

[3] Crockett, D., Desrochers A. et al.,1987, Implementing a Petri Net Controller for a Machining Workstation. *Proc. of IEEE Conf. on Robotics and Automation*, Raleigh (North Carolina), 1987, pp. 1861-1867.

[4] Ferrarini L. . 1992. An Incremental Approach to Logic Controller Design with Petri Nets, *IEEE Trans. on System, Man and Cybernetics*, 1992, Vol.22-I n°3, pp. 461-473.

[5] Ferrarini L., 1994. On the Semantics of Petri Nets Models for Manufacturing Systems. *Simulations in Manufacturing and Communications*, Tokyo (Japan), 1994, pp.148-153.

[6] Ferrarini L. and Ciancimino R., 1994, Modular Simulation for Logic, Distributed, Real-Time Control Systems. *IFAC/IEEE Computer Aided Design in Control Systems*, Tucson (AZ), 1994, vol. 1, pp. 225-230.

[7] Ferrarini, L. and Maffezzoni, C. . 1993. Designing Logic Controllers with Petri Nets. *IEE Intelligent Systems Engineering*, vol.2 N.4, Winter 1993, pp. 246-256.

[8] Ferrarini, L., Narduzzi, M. and Tassan-Solet , M. . 1994. A New Approach to Modular Liveness Analysis Conceived for Large Logic Controller's Design, *IEEE Transactions on Robotics and Automation*, 1994, Special Issue on Computer Integrated Manufacturing. vol. 10, n0 2, 1994, pp. 169-184.

[9] Ferrarini, L. and Tomasini, R. . 1994, An Extensible Modular Tool for Distributed Control Systems. *IFAC/IEEE Computer Aided Design in Control Systems*, Tucson (AZ), 1994, vol. 1, pp. .219-224.

[10] Ferrarini, L. and Trioni, M. . 1994, Modeling Shared Resources with Generalized Synchronization within a Petri Net Bottom-Up Approach. *IEEE IECON'94*, Bologna (Italy), 1994, vol. 3, pp. 1105-1110.

[11] Ferrarini, L. and Trioni, M. . 1994. Optimizing structural analysis of extended Petri nets models. *11th Int. Conf. on Analysis and Optimization of Systems - Discrete Event Systems* (G. Cohen and J.P. Quadrat Eds.), Lecture Notes in Control and Information Science n. 199, Springer-Verlag, London 1994, pp. 183-190.

[12] Gilles, M.. 1990. *Programmable Logic Controllers,* John Wiley and Sons, 1990.

[13] Grevet JM L., Jandura L. Brode J. and Levis A.H., 1988, Execution Strategies for Petri Net Simulations. *12th World Congress on Scientific Computation* IMACS Paris (France), 1988, pp. 511-516.

[14] Jeng, M.D. and DiCesare, F., 1992, A synthesis method for Petri net modeling of automated manufacturing systems with shared resources. *IEEE 31st Conf. on Decision and Control* Tucson (AZ), 1992, pp.1184-1189.

[15] Maffezzoni, C. and Ferrarini, L., 1992, A Multi-layer CAD System to Design Logic Controller. *Intekama Congress*, Düsseldorf (Germany), 1992, vol.1, 604-612.

[16] Maffezzoni, C., Ferrarini L. and Villa, 1988, Computer Aided Design of Sequence Control by Petri Nets. *Proc. of IFAC Symposium on Low Cost Automation*, Milan (Italy), 1989, vol. 2, pp. F205-F210.

[17] Marce, L. and Le Parc, P.. 1993. Defining the Semantics of Languages for Programmable Controllers with Synchronous Processes, *IFAC J. Control Eng. Practice,* Special Issue on Real-Time Systems. Feb 1993. Vol. 1, n°1, pp. 79-84.

[18] Mattsson, S.V., Andersson, M. and Åstrom. 1993. Object-Oriented Modelling and Simulation, in *CAD for Control System*, D.A. Linkens, Ed., Marcel Dekker, Inc., New York, 1993.

[19] Papelis, Y.E. and Casavant, T.L., 1992, Specification and Analysis of Parallel/Distributed Software and Systems by Petri Nets with Transition Enablig Functions. *IEEE Trans. on Software Engineering*, vol.18, n.3, pp.252-261.

[20] Petri, C.A.. 1987. Forgotten Topics in Net Theory. *Petri Nets: Central Models and and Their Properties* (Brauer, Reisig and Rozenberg, Springer-Verlag, Berlin 1987.

[21] Reisig, W. 1982. Petri Nets: an Introduction. Springer-Verlag.

[22] Silva M., Velilla S., Programmable Logic Controllers and Petri nets: a comparative study, *IFAC Conference on Software for Computer Control*, E.A. Puente and Ferrate (Eds.), pp. 83–88, Pergamon Press, 1983.

[23] Tamaki, H., Mukumoto S., Nishikawa Y. and Araki M., 1994, Simulation-Based Scheduling Package: models and solutions. *IEEE IECON'94*, Bologna (Italy), 1994, vol. 3, pp. 1521-1526.

[24] Valavanis, K.. 1990. On the Hierarchical Modeling, Analysis and Simulation of FMS with Extended Petri Nets. *IEEE Trans. on Sys., Man and Cyb.*, vol. 20 n.1, 1990, pp. 94-110.

[25] Valette, R.. 1979. Analysis of Petri Nets by Stepwise Refinements. *Journ. of computer and system sciences.* 1979. pp. 35-46.

[26] Valette, R. "Nets in Production Systems" in *Petri nets: Central models and their properties,* W. Brauer, W. Reisig and G. Rozenberg, Ed., n°1, Springer-Verlag, pp. 191-217.

[27] Zhou, M.C., DiCesare, F. and Desrochers, 1989, A Top-Down Approach to Systematic Synthesis of Petri Nets Models for Manufacturing Systems. *Proc IEEE Robotics and Automation Conf.*, Scottsdale (AZ), 1989, pp. 534-539.

[28] Zhou, M.C., DiCesare, F. and Desrochers. 1992. Hybrid Methodology for Synthesis of Petri Nets Models for Manufacturing Systems, *IEEE Trans. on Robotics and Automation*, 1992, vol. 8, no.3, pp. 350-361.

[29] Zhou, M.C., F. DiCesare and Rudolph. 1992. Design and Implementation of a Petri Nets Based Supervisor for a Flexible Manufacturing Systems, *IFAC J. Automatica*, 1992, vol. 28, no.6, pp. 1199 - 1208.

4

Automatic Generation of Sequence Control Programs via Petri Nets and Logic Tables for Industrial Applications

Takashi Sato[1] and Kazuo Nose[2]
[1] Industry Machinery Group, Kobe Steel, Ltd.
3-1, 2-chome, Shinhama, Arai-cho, Takasago 676, Japan.
Tel :0794-45-7704 Fax:0794-45-7715 Email:GFD03414@niftyserve.or.jp
[2] Electronics Research Laboratory, Kobe Steel, Ltd.
5-5, 1-chome, Takatsukadai, Nishi-ku, Kobe 651-22, Japan.

ABSTRACT

This chapter describes a new method for automatic generation of a ladder list program for a programmable logic controller used for sequence control of industrial machinery. To enhance the readability of specification of the sequence control, which is expressed by a sequence of motions and logic, the sequence of motions combined with the on/off readings of sensors is specified by Petri nets and the on/off logic of switches and actuators is specified by logic tables. The control specifications which are independently expressed by Petri Nets and the logic table are converted to the ladder list program by rules. The generated program can be linked to the simulator of the machine motions to simplify verification and debugging of the described control specifications. The proposed method has proved to be effective through the application to sequence control of an press machine and it has reduced the period for making a ladder list program by about 75% as compared with the conventional method which directly prepares a ladder list program.

1. INTRODUCTION

Sequence control by a programmable controller (PLC) is important to promote automation in the factory. In particular, it is essential to control complicated functions of the industrial machinery. The sequence control specifications are given by the on/off sequence of switches, sensors and actuators, as well as the related logic. Only a few control specifications of machine function are deduced theoretically, and the rest is given based on the designer's intentions. Consequently, control specifications are frequently expressed as the sequence control algorithm on the basis of past experience and know-how. Thus the conventional method to express the sequence control algorithm directly using the ladder list program yield the following problems.

1) Since the ladder list program has extremely high freedom in description, the program lacks readability and it is difficult for another programmer to understand the program.
2) It is difficult to completely debug at the time of programming, inevitably requiring modification of the program during the commissioning of the machine. The readability is further degraded in this program modification process, and the commissioning period tends to be longer.
3) It is difficult to apply the same ladder list program to the similar type of machines, making it difficult to improve the program quality.

In recent years, description and verification methods of control specifications using network expressions such as SFC (Sequential Function Chart) [1], MFG (Mark Flow Graph) [2], or Petri Nets [3], [4] have been proposed, and are expected to solve the above problems. In order to improve the readability of control specifications, these methods contain not only expressions of control specifications but also expressions of the sequence of motions of the machine to be controlled. These methods are suitable for describing the sequence of machine motions. However, the expression of the on/off logic of switches, sensors and actuators in detail, seems to be too complicated. As a result, it is difficult to visually understand how sequential motions of the machine actually take place.
When sequential control specifications are described, it is important to determine which part of the on/off sequence of switches, sensors and actuators, the logic, and the sequence of machine motions should be visually expressed and which portion should be expressed in a non-graphical method. This chapter investigates the description method of control specifications and develops an interface between control specifications and conventional ladder list program, i.e., an automatic program generation system.
It is important to visually understand the progress of sequence of machine motions. Therefore, only the sequence of machine motions is described using Petri Net based expression called "sequence flow," combined with on/off state of the sensors. The on/off logic of switches and actuators is separated from the sequence flow and is described by the given logic table called "sequence table." In order to automatically generate the ladder list program, the rules are given which convert the sequence flow and table to the ladder list programs and link both ladder list programs generating the final executable ladder list program. In addition, to simplify verification and debugging of described control specifications, i.e., to examine the program without real operation of the machine, the generated ladder list program is linked to the simulator of the machine motions.

2. FEATURES OF SEQUENCE CONTROL OF MACHINE

When the sequence control algorithm is constructed, the following seven items must be taken into account. [5] :

1) Sequence of motions : Sequence of machine motions and operations.
2) Selection of motions : To select specific motions from multiplicity of selectable motions in accordance with conditions.
3) Parallel motions : Motions which are carried out independently and in parallel.
4) Timing of motion : Timing of machine motions and sensors which detect them.
5) Logic of control : Logics related to PLC I/O and information processing.
6) Connection between components : To express motion as connective relation between components.
7)Restrictive relationship : Restrictive relationship between motions of components.

The ladder list program can express these items. Items 1),3),4),6) express the flow of the machine motions, while items 2),5),7) express timing and conditions of machine motion. Designers can freely express the motion sequence and on/off logic expression of switches, sensors and actuators, i.e., dynamic and static contents as a ladder list program. The resulting program often lacks of readability. Therefore, it is important that the specifications of Items 1),3),4),6) are expressed using a visual expression, while specifications of items 2),5),7) are separated from visual expression and expressed using the logical expression. In this chapter, the former is called the upper level expression and the latter the lower level one.
The following sections describe the developed system in terms of the upper and lower level.

3. UPPER LEVEL EXPRESSION

At the upper level, the sequence of machine motions is expressed together with the on/off status of motions and sensors. Consequently, describing the specifications using visual expression improves readability of the design.
Petri nets are a visual presentation to describe concurrent systems, and their effectiveness is widely recognized. Therefore, they are adopted as the upper level expression.

3.1 Petri net and Sequence of Motions

In the sequence flow, places, transitions, tokens and arcs which comprise a Petri Net correspond to the following specifications of motions, respectively.

Place : Motion of each unit component in the machine or device as well as motion information from limit switch, pressure switch, etc.

Transition: The end or start of motion of a unit component.

Token : State of motion of a unit component.
A token in a place standing for a motion means the motion of the unit component is running.

Arc : Connection of motion sequence and information of a unit component.

As an example, the sequence flow shown in Fig. 1 has the following meaning. Places m_1 and m_2 in the figure stand for machine motions, places p_1 and p_2 stand for motion information, and transitions t_1 and t_2 stand for the end of motions. When sensor p_1 turns on as a result of motion m_1, motion m_1 completes and motion-end t_1 fires. When motion m_1 completes, motion m_2 commences. When sensor p_2 turns on as a result of motion m_2, motion m_2 completes and motion-end t_2 fires.

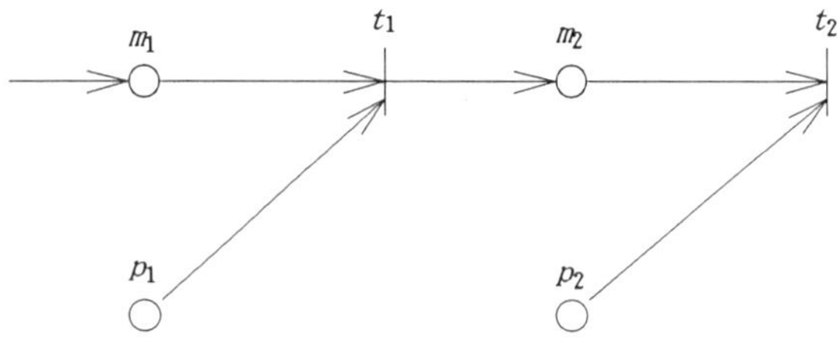

Fig. 1 Petri Net Representation

Place p_1 and p_2 may have a token initially or receive it as a result of motion of unit component. In same way they lose the token during following motion.

3.2 Transformation of Place

A place with n input arcs and m output arcs can be converted to the ladder list program as shown in Fig. 2. If any of input transitions (t_{i1},t_{i2}, •••, t_{im}) fires the place p receives a token and holds it until any of output transitions (t_{o1},t_{o2},••• ,t_{on}) fires. This Petri Net is analogous to the logical expression as follows.

p = (t_{i1} OR t_{i2} OR ••• OR t_{im} OR p) AND NOT t_{o1} AND NOT t_{o2} ••• AND NOT t_{on}

p keeps ON state even after input transitions' fire because logical expression of p includes p itself, until any of output transitions (t_{o1}, t_{o2}, •••, t_{on}) fires. This logical expression can be converted to the ladder expression.

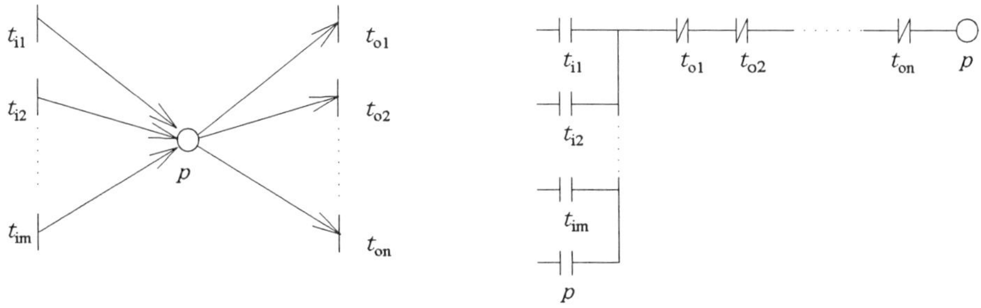

Fig. 2 Transformation of Place

3.3 Transformation of Transition

A transition which has *n* inputs can be converted to the ladder list program as shown in Fig. 3. The input places of the transition are all combined with AND. And a pulse is the output. The pulse output remains energized for just one scanning time of PLC.

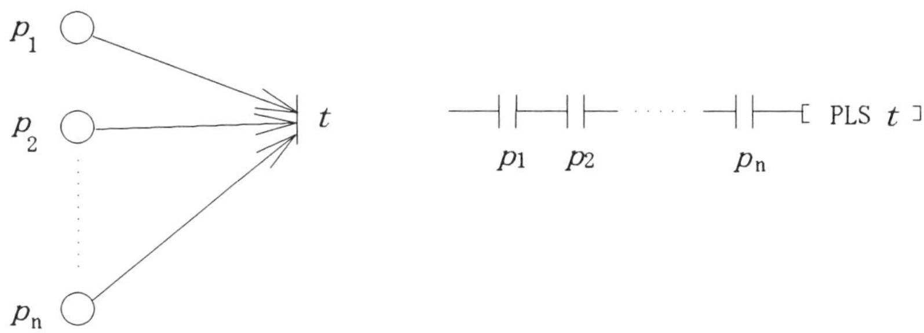

Fig. 3 Transformation of Transition

When it is necessary to delay transition to the next motion, “timer” is used to express delayed time for the firing of a transition in the sequence flow.

3.4 Difference from Petri Nets

The sequence flow can be converted to the ladder list program by these conversion rules,. However, the sequence flow differs from the ordinary Petri Net in the interpretation. The following describes the difference.

1) Conflict: In a Petri Net, when a place has multiple output arcs, a token moves one of the places through one arc. However, as shown in Fig.4, all output transitions can receive a token and send them to their output places. If the conditions must be branched, dummy places as shown in Fig. 4 must be added.

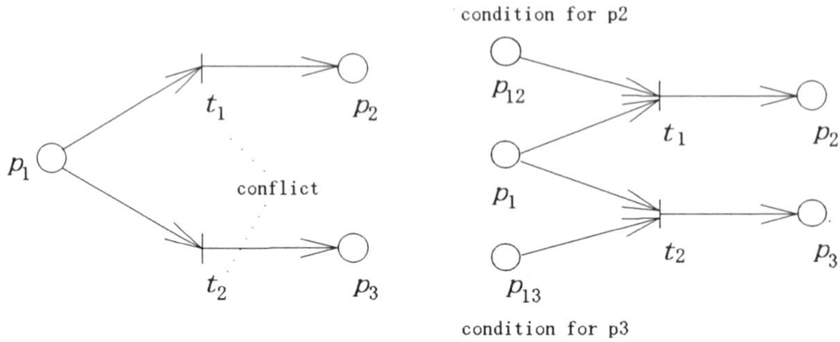

Fig. 4 Places in Conflict

2) Selection switch: When input information from a selection switch for setting the operation mode is expressed as a place directly by the Petri Net, the following inconveniences will occur. In the transfer to the ladder list program from the sequence flow, a transition is converted to a pulse output. In such an event, the pulse continues energized only once when conditions are satisfied. However, the selection switch remains energized and does not dissipate energy until it is turned off. On the other hand, to execute the sequence program from the second scanning time, no pulse is output. This is apparently not the action the designer intends. Therefore, we adopt a concept of a stable place. That is, the transition which has only input places not representing motion are defined as a stable transition and the place which has only stable input transitions is defined as a stable place. The transition having only stable input places is also stable. The stable transition should not be converted to AND NOT combination of its output. On the other hand the stable place should not have it self as OR combination. It is apparent from the illustration of Fig. 5 that the above problems are solved by this implementation.

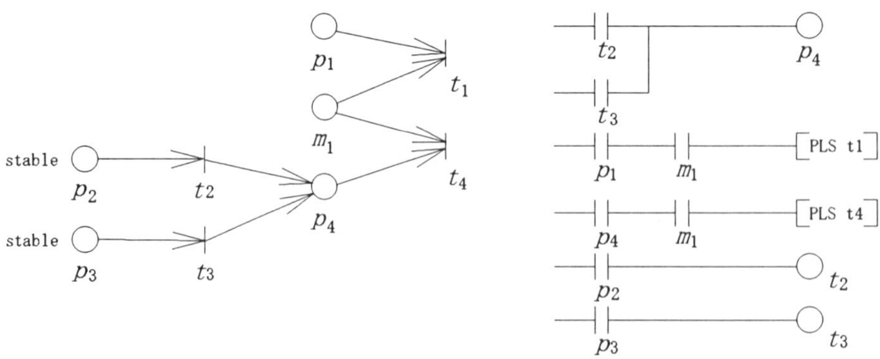

Fig. 5 Transformation of Selection Switch

4. LOWER LEVEL EXPRESSION

Information of the actuators, which are either *on* or *off* to carry out motions of each unit component in the machine and device as well as conditions and output required to carry out motions are expressed at the lower level . These items include 1) interlock conditions restricting motions of machines and devices, 2) start signals to commence motions of unit components, 3) on/off signals to actuators, 4) on/off signals to indicators.

We adopt a method to generate automatically a ladder list program by preparing the given sequence table for the places expressing motions. Fig. 6 shows one example of the sequence table. Table 1 shows the basic logic symbols used in the sequence table.

Symbol	Logic	Ladder Expression
*A	Inverese	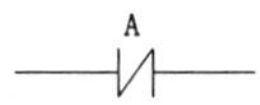
A, B	AND	
A+B	OR	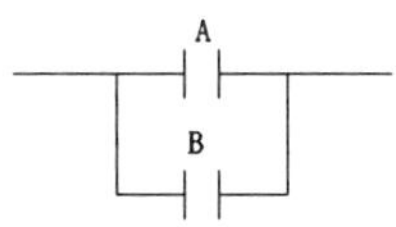

Table 1 Basic Logic Symbols

Tagged Name

1) ITL (Interlock Condition)

A, (B + *C)

2) STC (Start Condition)

Manual	D
End	E
Inching	F
Auto	G

3) MTN (Motion)

Process	(A)	(B)	(C)	(D)	(E)	....
A->B	NEXT1					
B->C		NEXT2				
C->D			END			
.....						

4) Output

OUT1	ON	ON	ON
OUT2	ON	ON	
OUT3	ON		

Fig. 6 Sequence Table

1) Interlock condition (ITL)

Interlock conditions describe conditions to carry out motions to insure the safety of the operator and to prevent damage to the machines, and consist of logical expressions of bit information on switches and sensors. The logical expressions expressed by AND, OR, NOT can be converted to the ladder list program. Fig. 7. 1) shows the example with A•(B + *C) in Fig. 6.

2) Start signals (STC)

Start signals to commence motions have three modes, Manual, Inching and Automatic modes. Manual is the mode in which the start signal is maintained once the conditions are satisfied until end or stop conditions are satisfied. Inching is the mode in which the start signal is maintained only when the conditions are satisfied. Automatic is the mode in which the start signal is generated in accordance with the sequence flow. Each signal for each mode can be expressed by logical expression, which can be converted to a ladder list program. As a result a start signal can be expressed, by combination of OR blocks which is made of start signals for each mode. A stop signal is inserted to logical expression of the manual mode. When this logical expression becomes false the manual start signal becomes de-energized. The start signal shown in Fig. 6 can be converted to the ladder list program of Fig. 7 2).

3) Motion signal (MTN)

In general, one motion can be divided into several phases. For example, the cylinder moving speed is set to be fast at the beginning and slow at the end of stroke. This indicates two phases. In this chapter, the division of this kind of motions is called "process." The special process executed at the end of motions is called "ending process." Transition condition (logical expression) is described for each process, which can be converted to a ladder list again.

Fig. 7 (3) shows a result of conversion of processes of Fig. 6. The process "B" starts with transition condition of process "A" , in this case "NEXT1". Process "C" occurs after process "B" sequentially by the condition of "NEXT2" in same way. Process "C" is the last process of this motion, which is terminated by the condition of "END." Logical expression can be adopted for these process transitions.

4) Output Condition

A motion of unit component is realized by some actuators, for example, solenoid valves, motors, etc. Output condition is described by logic table as shown Fig. 6 "OUT1" is energized at all processes (A, B and C), on the other hand "OUT3" is energized at the first process (A) only. The ladder list program for this logic table can be generated by conjunction of the process conditions and output conditions.

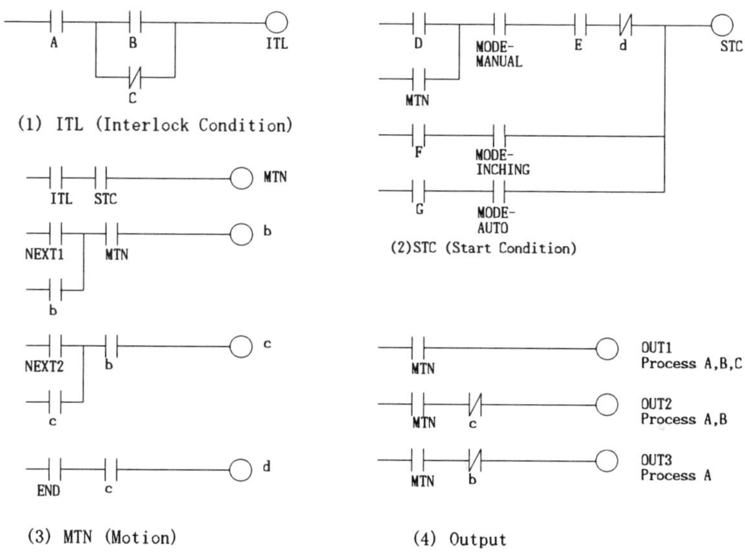

Fig. 7 Transformation of Sequence Table

5. LINKAGE OF UPPER AND LOWER LEVELS

The upper level defines the sequence of machine motion, while the lower level defines specifications of motion timing and conditions. Therefore, the overall motion specifications will be completed by linking these two level expressions.
The automatic generation system links the flow charts and logic tables using "tagged name." "Tagged name" is assigned for each logic table, which correspond to "tagged name" on the flow chart. If a place standing for a motion of unit component has a token, the start signal of automatic mode becomes true. Every logic table of automatic motions has automatic start condition. When the place loses a token, start condition of corresponding motion becomes false, as a result the motion of unit component is terminated.
In the flow chart every motion place has the special name, which can be allocated to the logic table. Thus the ladder programs of two levels can be linked.
Other modes of start condition, manual and inching mode, are not related to automatic mode directly. These modes are used to operate each unit component independently.

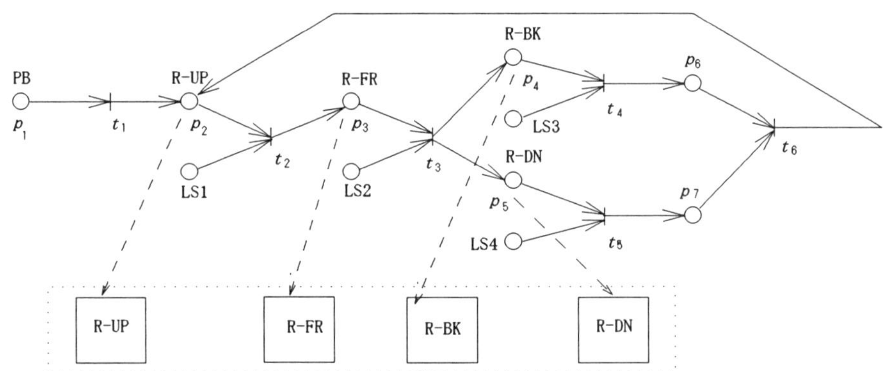

Fig. 8 Example of Linkage of Sequence Flow and Sequence Table

6. AUTOMATIC GENERATION SYSTEM

6.1 System Configuration

Fig. 9 shows the configuration of the developed automatic generation system of the sequence program. The software consists of the sequence flow and sequence table editors written in the C language, sequence flow and sequence table files, conversion program, linkage program, transfer program, document making program, and motion simulator. The hardware is composed of the MS-DOS personal computer and PLC, and both are linked using the serial interface.

6.2 Editor

The editors are used to enter and correct the sequence flow and sequence table that define specifications of the sequence control. The sequence flow editor draws the sequence flows on the CRT using icons and a mouse. On the other hand, the sequence table editor defines the logic.

6.3 Simulator

A simulator is made to verify that the automatically generated sequence program satisfies the required specifications and the program evades problems such as interaction between machines moved in parallel. That is, the simulator reads the motion commands from PLC and writes in PLC the conditions of the sensors which vary depending on motions, and carries out motion simulation step by step. It monitors the interaction between parallel motions in accordance with the preset interaction conditions.

6.4 Usage of System

The specifications of machine motions are entered using the sequence flow and table editors. The entered data are saved as the sequence flow and table files. The files are converted into the ladder list programs by the conversion program. Then, using the tagged name as a keyword, the final execution ladder list program is generated by the linkage program. Finally, it is downloaded to the PLC by the transfer program. The document is automatically made by the document making program. In addition, the PLC is also linked with the motion simulator so that motion control by the generated program can be verified.

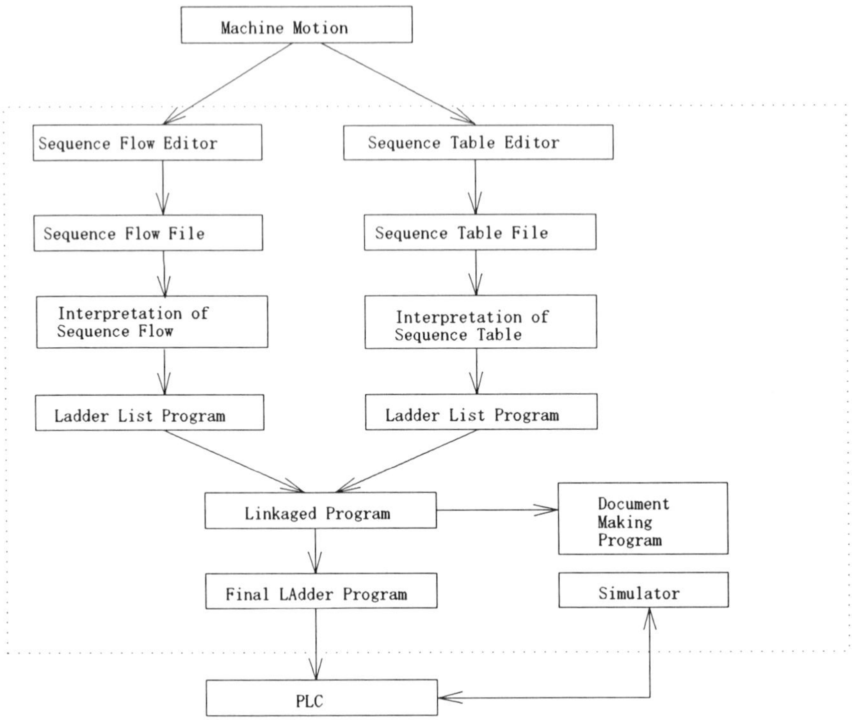

Fig. 9 Configuration of Automatic Generation System

7. APPLICATION RESULTS

The system described above was applied to designing and programming of the sequence control of a press, a typical industrial machine, the results of which are discussed as follows.

The machine selected to evaluate the effect was a 2500 ton press shown in Fig. 10, which is used for making construction materials. This machine has about a hundred motion components, almost of which are operated by hydraulic system. It has about two thousands discrete I/Os (sensors, solenoids and push buttons), and the sequence flow is expressed with more than a hundred places and transitions and more than a thousand Arcs. The number of sequence tables reached more than a hundred.

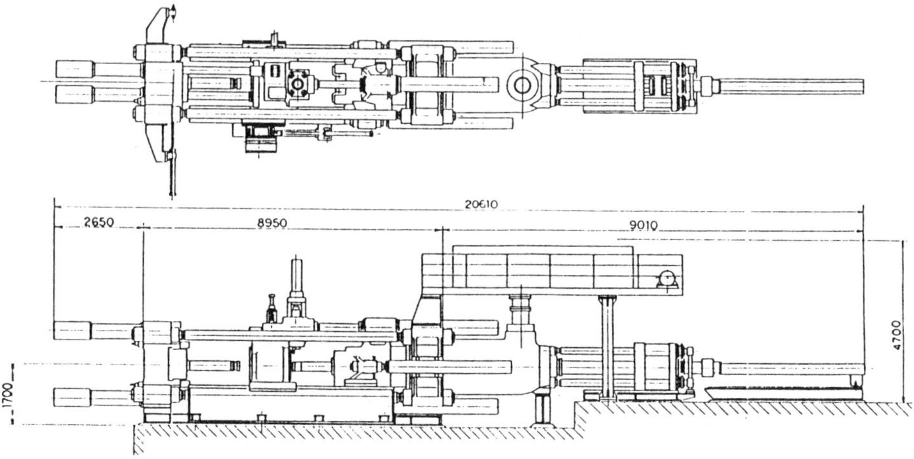

Fig. 10 2500 ton Press Machine

Table 2 shows the results of applying this system to a series of control system making processes from general design to documentation. The application of this system has made it possible to generate the programs in 1.7 months, which previously took 6 months, reducing the program generation period by 75%.

	Conventional Method	Automatic Method
General Design	1.5 month	0.1 month
Program Coding	2.0 month	0.5 month
Program Check	0.5 month	0.0 month
Program Debug	1.5 month	1.0 month
Documentation	0.5 month	0.1 month
TOTAL	6.0 month	1.7 month

Table 2 Effects of Automatic Generation System

Presently, this system is applied to a large number of industrial machines and is effectively used for designing and making the control systems, automating the making of the documents, and standardizing and reusing the programs.

8. CONCLUSION

An automatic generation system of ladder list programs of sequence control using PC was developed. The features are summarized as follows:.

1) It is important to understand the progress of sequential motions of the machine through visualization, using Petri Nets, the specification of the sequence of machine motions is described by including the on/off status of switches and sensors.
2) The specification of on/off logic of switches and actuators is separated from the visual expression to maintain the readability of the Petri Net expression and is expressed by a logic table.
3) To generate automatically a ladder list program, the rules which convert specifications via both Petri net and logic table to two ladder list programs are given and both programs are linked to generate automatically the final executable ladder list program.
4)To facilitate verification and debugging of control specifications, the automatically generated program is linked to the motion simulator of the machine to be controlled.

As a result of applying the sequence control of the actual industrial machines, the following effects have been achieved:

1) Compared with the conventional method to prepare the ladder list program, the developed method has reduced the program making period by about 75 %.
2) Documents has also been automatically made.
3) Readability of software is improved and description of control specifications can be brought into a module for each sequence table, enabling the repetitive use of the software.

REFERENCES

[1] H. Sasajima, "PC Programming Language and SFC" , *Automation*, Vol. 32, No. 7, pp. 15-23, 1987 .
[2] K. Hasegawa, K. Takahashi, R. Masuda and H. Ohno, "Proposal of Mark Flow Graph for Discrete System Control", *trans. of SICE*, vol. 22, No. 2, pp. 122-129, 1984.
[3] Y. Nakamura, S. Abe and S. Takeda, "A Functional Verification for Design Modification of Industrial Sequence Control Systems", *trans. of SICE*, vol. 21, No. 10, pp.1036-1042, 1985.
[4] K. Tsuji, S. Kumagai, S. Kodama, T. Yamada and S. Huang, "Modelling and Verification of Sequential Control Systems by Petri Nets", *IEICE*, vol. J69-A, No. 7, pp. 840-849, 1986.
[5] K. Hasegawa, "Sequential Control Overview", *J. of SICE*, Vol. 27, No. 8, pp. 665-672, 1988.
[6] J. L. Peterson, "Petri Net Theory and the Modelling of Systems", *Prentice-Hall*, 1981.

[7] T. Sekiguchi, H. Takahashi, M. Aoki, K. Shimokawa and N. Komoda, "Sequence Control Engineering - New Theory and Design Method ", *The institute of Electrical Engineers of Japan*, 1988.

5

Planning and Scheduling based on Petri Nets

Jean-Marie PROTH[1-2] and Ioannis MINIS[2]

[1] INRIA-Lorraine, Technopôle Metz 2000, 4 rue Marconi, F-57070 Metz, FRANCE
[2] Institute for Systems Research, University of Maryland, College Park, MD 20742, USA

Abstract

The objective of this chapter is to show that Petri nets facilitate a comprehensive approach to production management and reduce the complexity of the problems involved at the expense of some constraints imposed on the decision making system. The introduction provides a content outline of the chapter. In the second section, we introduce different aspects related to Petri nets which will be used in the approaches proposed for the short-term planning and scheduling of manufacturing systems. This section includes, but is not limited to, event graphs, decomposable Petri nets and controlable Petri nets. The third section focuses on cyclic manufacturing systems. For this type of systems, it is always possible to propose an event graph model which represents both the physical and the decision making systems. We use such a model to propose a near-optimal scheduling algorithm that maximizes productivity while minimizing the work-in-process (WIP) in the deterministic case. The approach used for non-cyclic manufacturing systems in the fourth section is different in the sense that only the manufacturing processes (i.e. the physical part of the system) and the related constraints are modeled using Petri nets. We use such a Petri net model to propose a short-term planning process which results in a trade-off between the computation burden and the level of resource utilization. The short-term planning model is then enhanced to obtain the scheduling model. The latter is used to develop an efficient scheduling algorithm that is able to satisfy the requirements imposed by short-term planning. Section five introduces a modular approach for modeling and managing manufacturing systems.

1. Introduction

During the 70s and early 80s, Petri nets were considered mainly as modeling and simulation tools. This is the reason why research studies were first oriented toward small-sized models based on high-level nets such as the predicate/transition nets [6], coloured Petri nets [9], and Petri nets with individual tokens [17]. More recently, researchers became interested in the analytical properties of elementary nets, also called black-and-white nets. Very important results were proposed by Commoner et

al. [3], Murata [13], Chrétienne [2], Zhou [19], and many others. As far as we know, the application of those results to cyclic manufacturing systems, and their enhancement in order to model, evaluate and manage manufacturing systems were initiated by Hillion et al. [8]. Several studies are currently in progress and deal with both cyclic and non-cyclic manufacturing systems. The purpose of this chapter is to present the most important results in this field which can reduce the complexity of the management problems in cyclic and non-cyclic manufacturing systems.
The second Section of this chapter is devoted to the definitions, concepts and properties to be used in the remainder of the paper. We introduce the definition of elementary nets, the state equation, the definitions of p-invariants and t-invariants, and the qualitative properties which are desirable in manufacturing. We also introduce the event graphs (also called marked graphs), a special type of Petri nets, which are used to model cyclic manufacturing systems. The second Section concludes by presenting the decomposable nets, which are the basic tools in the planning and scheduling of non-cyclic manufacturing systems.
In Section three, we show that it is possible to model both the physical and the decision making systems (DMSs) of cyclic manufacturing systems with an event graph. Using the properties of event graphs and a given cyclic control, we show how to maximize the productivity of a manufacturing system (i.e. to minimize the cycle time) while minimizing the work-in-process. The properties of event graphs are also used to develop a near-optimal scheduling algorithm.
Section four focuses on non-cyclic manufacturing systems. We present a short-term planning problem and develop a solution approach which takes advantage of the fact that the system can be modeled as a decomposable Petri net. We then propose a scheduling model and develop a scheduling algorithm which allows us to satisfy the requirements imposed by short-term planning.
In Section five, we introduce a modular approach for modeling and evaluating real-life systems from the point of view of production management.
Section six presents the concluding remarks.

2. Petri nets (PNs): definitions, concepts and properties

2.1. Qualitative properties of elementary Petri nets

We restrict ourselves to the properties which are the most important when manufacturing systems are concerned.

a. Structural liveness

Definition 1: A PN Q = (P, T, I, O) is structurally live if there exists an initial marking m_0 such that Z = (P, T, I, O, m_0) verifies the following property: for any $t \in T$ and $m \in R(Z, m_0)$ there exists a firing sequence which leads from m to a marking which enables t.

In manufacturing systems, structural liveness implies that it is always possible to perform any operation for which the system was designed, assuming that the initial state of the system is properly chosen. In other words, any operation which can be performed by the system will remain possible in the future, irrespective of the past sequence of decisions.

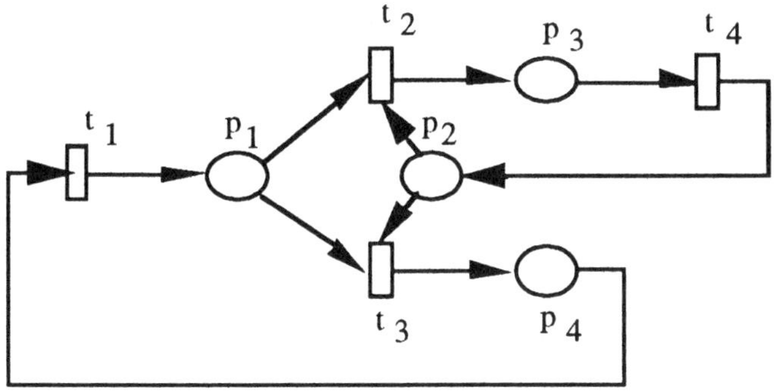

Fig. 1: Petri net which is not structurally live

For instance, let us consider Figure 1. Whatever the marking of this Petri net, and assuming that this marking is finite, it is always possible to empty places p_3 and p_2 by firing t_4 and t_3 an adequate number of times. When p_2 and p_3 are empty, it is no more possible to fire neither t_2 nor t_4: thus, this Petri net is not structurally live.

b. Reversibility and home state

Definition 2: A PN $Z = (P, T, I, O, m_0)$ is reversible if, for any $m \in R(Z, m_0)$, there exists a firing sequence e such that $m [e > m_0$.

This property is also very important for manufacturing systems. It guarantees that it is always possible to return to the initial state, no matter what the current state is. This is often necessary for maintenance, tool adjustments or changes in production. The next definition generalizes Definition 2.

Definition 3: A marking m' of a Petri net $Z = (P, T, I, O, m_0)$ is a home state for m_0 if it can be reached from any marking reachable from m_0, i.e. $m' \in R(Z, m)$ for any $m \in R(Z, m_0)$.

The use of a home state is the same as the one of m_0 in Definition 2.
According to these definitions, any marking reachable from the initial marking in a reversible PN is a home state, but a PN with a home state may be not reversible.
As an example, the Petri net represented in Figure 2 is not reversible. Let us consider marking $m_0 = [1,0,0,0,1]$. If we fire t_2 and t_3, two tokens appear in p_4, and it is no more possible to reduce the total number of tokens in the set of places $\{p_4, p_5\}$. The system generates one more token in $\{p_4, p_5\}$ each time t_3 is fired. This is true for any initial marking m_0 containing at least one token in $\{p_1, p_2, p_3\}$ Thus the Petri net of Figure 2 is not reversible. There is also no home state for any marking m_0. Figure 3 presents a Petri net which is not reversible, but has two home states: $m' = [0,1,0]$ and $m'' = [1,0,0]$ as soon as t_1 is fired.

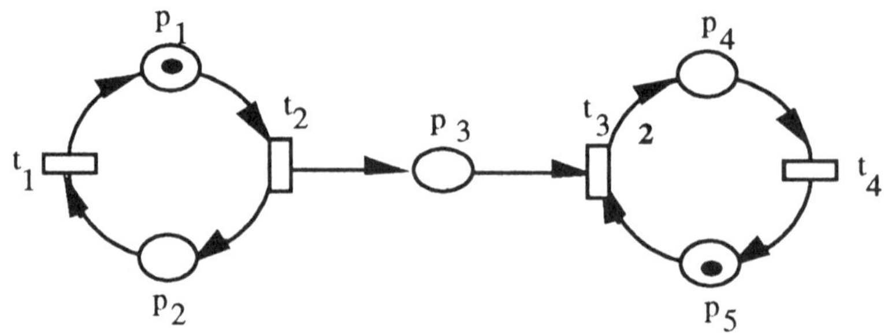

Fig. 2: Non reversible Petri net

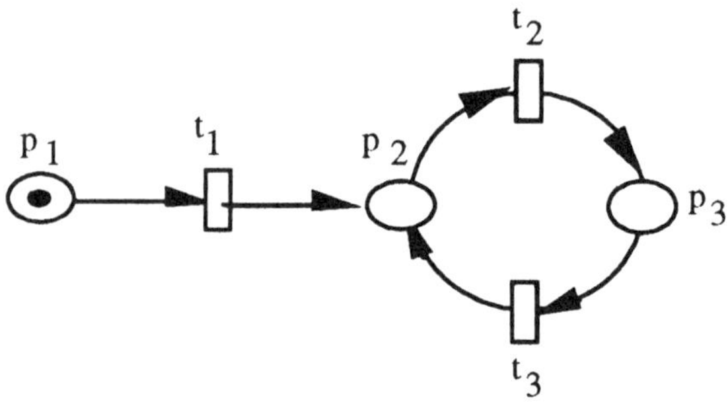

Fig. 3: Non reversible Petri net with home states

c. Boundedness

Definition 4

(i) A marked PN Z = (P, T, I, O, m_0) is said to be k-bounded if $m(p) \le k$ for any $p \in P$ and $m \in R(Z, m_0)$.

(ii) A marked PN Z = (P, T, I, O, m_0) is said to be bounded if it is k-bounded for some integer $k > 0$.

(iii) A PN Q = (P, T, I, O) is structurally bounded if the marked PN Z = (P, T, I, O, m_0) is bounded for any initial marking m_0.

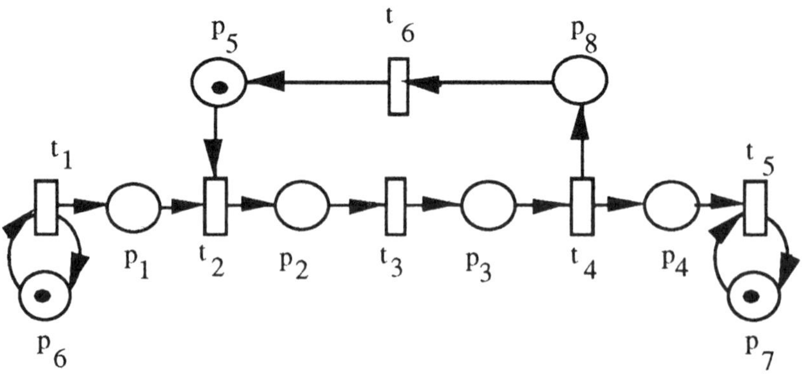

Fig. 4: A flow-shop composed of two machines

Boundedness is not necessary in manufacturing, but may be desirable when fully automated systems are concerned. Nevertheless, it is always necessary to be able to keep the PN model of a manufacturing system bounded; for example, a manufacturing system whose model cannot be kept bounded may result in WIP that increases to infinity. Let us consider the Petri net presented in Figure 4. This net models a sequence of two machines, the transportation between the machines being performed by a cart. This system manufactures only one type of products.

We assume that a transition is fired as soon as it is enabled. Transition t_1 represents the operation performed on the first machine. The firing time of t_1 is the time needed to perform the operation on one part. When t_1 fires, the token located in p_6 disappears, which means that the first machine processes at most one part at a time. Place p_1 is the buffer located at the exit of the first machine. Transition t_2 models the loading of one part on the cart, and the firing time of t_2 is the time required for the loading operation. Note that, when t_2 fires, the token in place p_5 disappears, which means that the cart carries one part at a time. As soon as the firing of t_2 is completed, t_3 starts firing: it represents the transportation of the part, i.e. the movement of the cart from the first to the second machine. Transition t_4 starts firing as soon as the firing of transition t_3 is completed. The firing of t_4 represents the unloading of the cart, and the firing time of t_4 is defined accordingly. When the firing of t_4 ends, one token appears in p_4 and p_8. The token in p_4 represents a part waiting for an operation on the second machine while the token in p_8 represents the fact that the cart is idle. The token in p_8 is used to fire t_6. This models the fact that the idle cart returns to the first machine to restart the previous cycle. Firing t_5 represents an operation performed on the second machine. The token in p_7 is introduced to ensure that at most one token is used at a time to fire t_5, guaranteeing that at most one part is processed by the second machine at a time.

Note that there is only one token in the elementary circuit $<t_2, p_2, t_3, p_3, t_4, p_8, t_6, p_5, t_2>$ at any time, and this token models the cart.

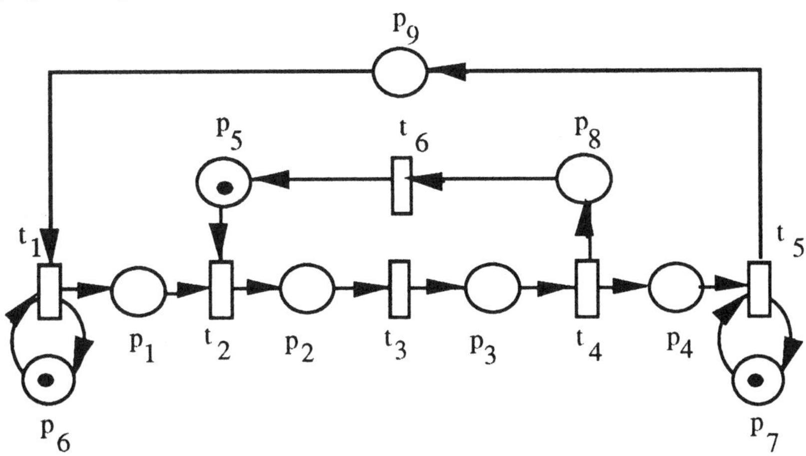

Fig. 5: Bounded flow-shop

Let θ_i be the firing time of t_i (i = 1,2,3,4,5,6). Under the assumption that a transition fires as soon as it is enabled, it is easy to understand that the model shown in Figure 4 is bounded if and only if:

$$\theta_1 \geq \theta_2 + \theta_3 + \theta_4 + \theta_6 \geq \theta_5$$

To ensure that the model remains bounded for any values of the firing times, it is possible to complete the model as shown in Figure 5. The number of tokens initially introduced in p_9 represents the maximal number of parts in the system at any time. If this number is k, then the Petri net model is k-bounded.

2.2. Event graphs (or marked graphs)

a. General properties

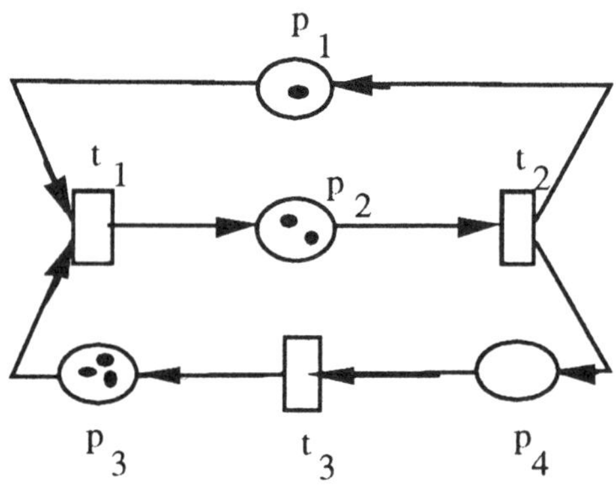

Fig. 6: An event graph

An event graph is a Petri net such that each place has exactly one input and one output transition. Furthermore, the weight associated to each transition is 1. Such a Petri net is represented in Figure 6. Note that event graphs may contain elementary circuits. For example, the event graph of Figure 6 contains two elementary circuits: namely $\gamma_1 = \langle p_1, t_1, p_2, t_2, p_1 \rangle$ and $\gamma_2 = \langle p_2, t_2, p_4, t_3, p_3, t_1, p_2 \rangle$.

It is easy to prove that:

(i) If we assign 1 to each place belonging to one of the elementary circuits and 0 to all other places, then the vector with elements of these values is a p-invariant —more precisely a minimal p-invariant. Another way to express the foregoing is to say that the number of tokens in any elementary circuit is invariant by any transition firing. We owe this result to Commoner et al. [3].

(ii) The n-vector whose components are equal to 1 is a t-invariant. It means that the marking returns to the initial marking after firing each transition exactly once.

The following result is due to Commoner et al. [3].

Result 1: A strongly connected event graph is guaranteed to be deadlock-free if and only if every elementary circuit contains at least one token.

b. Deterministic event graphs

In this section, we consider timed event graphs where the times associated to the transitions are deterministic. For any elementary circuit γ belonging to the event graph $Z = (P, T, I, O, m_0)$, we define the **cycle time** as:

$$C(\gamma) = \mu(\gamma) / m(\gamma) \quad (1)$$

where:

$\mu(\gamma)$ is the sum of the firing times of the transitions belonging to γ,

$m(\gamma)$ is the number of tokens circulating in γ ($m \in R(Z, m_0)$).

If m_0 is the initial marking, we know that $m(\gamma) = m_0(\gamma)$, $\forall\, m \in R(Z, m_0)$. Thus $C(\gamma)$ is an invariant. Let C^* be the maximal cycle time among all elementary circuits, i.e.:

$$C^* = \max_{\gamma \in \Gamma} C(\gamma) \quad (2)$$

where Γ is the set of elementary circuits of a strongly connected event graph. Any $\gamma \in \Gamma$ such that $C(\gamma) = C^*$ is called a **critical circuit**.

In order to guarantee that at most one firing of a certain transition may occur at any time, we introduce a self-loop to each transition with initially one token in each self-loop place.

From now on, we consider only the firing policy called "Earliest Operating Mode" (EOM). It is the policy according to which transitions fire as soon as they are enabled. This corresponds to the common policy applied to fully automated systems as we will see in Section 3. Chrétienne [2] showed that, under an EOM, the operation of the system becomes periodic after a finite time. Periodicity means that there exist two integers n_0 and K such that:

$$S_t(n+K) = S_t(n) + KC^*, \quad \forall n \geq n_0, \quad \forall t \in T$$

where $S_t(k)$ is the starting time of the k-th firing of transition t, and K is the period. The following result holds.

Result 2: Provided that the event graph is strongly connected and that the EOM policy applies, the cycle time of the model is C^* in the steady state. In other words, the firing rate of all transitions in steady state is $\lambda = 1 / C^*$.

This result provides the productivity of a manufacturing system which is modeled by a strongly connected event graph, assuming that the WIP is known.

Several algorithms have been proposed to reach a given cycle time while minimizing a linear combination of the markings, the coefficients of which are the components of a p-invariant. They can be found in Laftit et al. [11]. In particular, an adjustment heuristic algorithm which can be used in practice for models of any size, is proposed in this chapter.

We summarize this algorithm hereafter. The following assumptions are used:

(i) A transition fires as soon as it is enabled (EOM),

(ii) A transition cannot start firing while it is in the process of firing. In other words, a self-loop which contains exactly one token initially is associated to each transition.

The consequence of (ii) is that the desired cycle time α must be greater than or equal to $\max_{t \in T} \theta_t$, where θ_t is the firing time of transition t.

The problem $\mathcal{P}$ to be solved can be written as follows:

$$\mathcal{Minimize}\ x.m_0^t$$

s.t.

$$\mu(\gamma)/m_0(\gamma) \leq \alpha, \forall \gamma \in \Gamma, \text{ set of elementary circuits}$$

where x is a p-invariant.

Constraint $\mu(\gamma) / m_0(\gamma) \leq \alpha$ can be rewritten as:

$$m_0(\gamma) - \mu(\gamma)/\alpha \geq 0$$

or:

$$\omega(m_0, p, \alpha) \geq 0, \forall p \in P \tag{3}$$

where $\omega(m_0, p, \alpha) = \mathcal{Min}_{\gamma \in \Gamma_p}\{m_0(\gamma) - \mu(\gamma)/\alpha\}$, Γ_p being the set of elementary circuits containing p.

The adjustment algorithm is based on Result 3.

Result 3 : Let $\mathcal{M}(\alpha)$ be the set of markings m_0 such that $\omega(m,p,\alpha) \in [0,1]$, $\forall\ p \in P$. According to inequalities (3), any $m_0 \in \mathcal{M}(\alpha)$ is a feasible solution of the problem.

The following properties hold:

(i) Let $m_0 \in \mathcal{M}(\alpha)$ and m' such that $m'(p) \leq m(p)$, $\forall\ p \in P$ and there exists at least one $p^* \in P$ such that $m'(p^*) < m(p)$. Then m' is not feasible and thus cannot be an optimal solution to problem $\mathcal{P}$.

(ii) $\mathcal{M}(\alpha)$ contains at least one optimal solution to $\mathcal{P}$.

According to Result 1, the algorithm starts with a feasible marking (for instance one token in each place) and iteratively removes tokens from places until a solution belonging to $\mathcal{M}(\alpha)$ is obtained.

The basic idea of the adjustment heuristic algorithm is to remove tokens which reduce the value of the criterion $x.m_0^t$ as much as possible while decreasing the degree of freedom of the whole system as little as possible, taking into account the elements of the p-invariant x.

At a given iteration, the place $p^* \in P$ from which we remove one token transforming the current marking m^c into the next marking m^n is a place such that:

$$\left[\sum_{q \in P} x_q\left(\omega(m^c, p^*, \alpha) - \omega(m^n, p^*, \alpha)\right)\right] \Big/ x_{p^*}$$

$$= \mathcal{Min}_{p \in E[m^c]} \left\{\left[\sum_{q \in P} x_q\left(\omega(m^c, p, \alpha) - \omega(m^n, p, \alpha)\right)\right] \Big/ x_p\right\} \tag{4}$$

where $E[m^c] = \{p / p \in P \text{ and } \omega(m^c,p,\alpha) \geq 1\}$, (5)

x_p is the component of x corresponding to $p \in P$.

Notes

(i) The determination of $\omega(m,p,\alpha)$ apparently requires the computation of all the elementary circuits containing p for any $p \in P$. However, it is possible to transform this problem into the computation of a shortest path in a directed graph $G = (T,V,d)$ where the set of nodes T is the set of transitions of the strongly connected event graph, $V \subset T \times T$ is the set of arcs, and $d: V \rightarrow R$ is the length of the arcs defined from:

$$d(t_i, t_j) = \begin{cases} m(p) - \theta_{\circ p} / \alpha \text{ if } t_i \in {\circ p} \text{ and } t_j \in p{\circ} \\ +\infty \text{ otherwise} \end{cases}$$

G does not contain circuits with negative length for any feasible marking m. Therefore, the shortest distance between any pair of transitions exists, and:

$$\omega(m_0, p, \alpha) = d(\circ p, p\circ) + v(p\circ, \circ p)$$

where $v(p^\circ, {}^\circ p)$ is the length of the shortest path connecting p° and ${}^\circ p$.

(ii) Let $p^* \in P$ be a place given by relation (4). If $m^c(p^*) = 0$, we know that there exists $m'^c \in R\ (Z, m^c)$ such that $m'^c(p^*) \geq 0$. We thus compute m'^c and remove one token from p^*.

Thus, the adjustment algorithm can be summarised as follows.

Adjustment algorithm

1. Let m_0 be the initial feasible marking.
   For instance, we can choose m_0 such that $m_0(p) = 1$, $\forall p \in P$.
2. If $E[m_0] = \varnothing$, stop the computation. m_0 is kept as the solution of the problem.
   See relation (5) for the definition of $E[m_0]$.
3. If $E[m_0] \neq \varnothing$, select p^* using relation (4).
4. If $m_0(p^*) \geq 1$, do:
   $m_0(p) = m_0(p)$, $\forall\ p \in P$, $p \neq p^*$
   $m_0(p^*) = m_0(p^*) - 1$
   Otherwise, compute $m' \in R\ (Z, m_0)$ such that $m'(p^*) \geq 1$ and do:
   $m_0(p) = m'(p)$, $\forall\ p \in P$, $p \neq p^*$
   $m_0(p^*) = m'(p^*) - 1$
5. Go to 2.

c. Stochastic event graphs

c_1. Basic results and problem formulation

In a stochastic event graph, the firing times are generated at random. We denote by $X_t(k) \in R^+$ the random variable generating the time required for the k-th firing of transition t. We assume that the sequences of transition firing times $\{X_t(k)\}$, $k = 1,2,\ldots$ for $t \in T$ are mutually independent sequence of i.i.d. (independent identically distributed) integrable random variables. Under these assumptions, it was proven in Baccelli [1] that there exists a positive constant $\pi(m_0)$ depending on the initial marking m_0 such that:

$$\pi(m_0) = \lim_{k \to +\infty} S_t(k) / k = \lim_{k \to +\infty} E[S_t(k)] / k \quad \text{almost surely } \forall t \in T \tag{6}$$

From a practical point of view, "almost surely" can be interpreted as "always".
Result 4 is of significant importance for the algorithm proposed hereafter.

Result 4 : Let Z be a strongly connected stochastic event graph and Z* the deterministic event graph derived from Z by associating the mean value of its stochastic firing time to any $t \in T$. Then the cycle time of Z* is a lower bound of the mean cycle time of Z, assuming that Z and Z* have the same initial marking and that the EOM applies to both.

Result 5 claims that it is possible to reach a cycle time as close as we want to the largest mean value of the firing times of the transitions.

Result 5 : Let $u^* = \underset{t \in T}{\mathbf{Max}}\, u_t$, where u_t is the mean value of the firing times of t. Then, for any $\varepsilon > 0$, it is possible to find an initial marking m_0 which leads to a mean cycle time equal to or less than $u^*+\varepsilon$.

The fundamental problem to be solved in order to reach a given cycle time by minimizing a linear combination of the markings can be written as follows:

$$\mathbf{Minimize}\ x.m_0^t$$

s.t. the mean cycle time of the strongly connected event graph be less than or equal to $u^*+\varepsilon$. x is a p-invariant.

We propose two heuristic algorithm to solve this problem.

c_2. An algorithm based on the elementary circuits

The first phase of this algorithm consists of computing the optimal solution of the deterministic problem obtained by assigning to each transition the mean firing time. In other words, we solve the optimal problem for Petri net Z* introduced in Result 4. According to Result 4, the criterion related to the optimal solution is a lower bound of the optimal value of the criterion for the problem at hand.

The second phase of the algorithm is a stepwise procedure. At each step of the procedure, we select a place belonging to as many elementary circuits having a large mean cycle time as possible among the places corresponding to the smallest components of the p-invariant x. We add one token in the selected place, and we restart the process as long as the mean cycle time of the system remains greater than $u^*+\varepsilon$.

We can summarize this algorithm as follows:

Algorithm 1

First phase: Solving the deterministic problem

1. Compute marking m_0, which leads to a cycle time less than or equal to $u^*+\varepsilon$ when the firing times are the mean values of the random variables assigned to the transitions.

   m_0 can be computed using the adjustment algorithm.

2. Simulate the system with m_0 to evaluate the mean cycle time C_0 of the stochastic problem.

   C_0 is greater than the cycle time of the deterministic problem according to result 4.

Second phase: Adding tokens

3. Let $\gamma_0 \in \Gamma$ be the elementary circuit having the largest mean cycle time and $\mathcal{P}$ the set of places belonging to γ_0 and corresponding to the components of x which have the smallest value. We then select $\gamma_1 \in \Gamma\text{-}\{\gamma_0\}$ which has the largest mean cycle time and at least one place in $\mathcal{P}$. We further select $\gamma_2 \in \Gamma$ - $\{\gamma_0 \cup \gamma_1\}$ which has the largest mean cycle time and at lest one place in $\mathcal{P} \cap \gamma_1$, and so on until we reach the first elementary circuit γ_q such that:

$$\mathcal{P} \cap \left(\bigcap_{k=1}^{q} \gamma_q \right) = \varnothing$$

4. Add one token in one of the places belonging to $\mathcal{P} \cap \left(\bigcap_{k=1}^{q-1} \gamma_q \right)$. We still denote the new marking by m_0.
5. Evaluate, using simulation, the mean cycle time C_0 when the new marking applies.
6. If $C_0 > u^*+\varepsilon$, go to 3, otherwise, stop the computation.
   m_0 is the solution to the problem and C_0 the related mean cycle time.

The main drawback of this algorithm is that it requires computation of all the elementary circuits. Although the computation of these circuits is straightforward, their number can be very large, even for small nets. For this reason we developed the next algorithm, which is based on perturbation analysis.

c_3. An algorithm based on perturbation analysis

The only difference between this algorithm and the previous one is the way the place in which a token is added is chosen. Before introducing a new procedure to make this choice, let us consider an efficient way to simulate the stochastic event graph. As shown in Chrétienne [2], the starting time of a transition firing can be defined using the following evolution equations:

$$S_t(k) = \underset{\tau \in in(t)}{\mathcal{M}ax} \left\{ S_\tau[k - m_0(\tau, t)] + X_\tau[k - m_0(\tau, t)] \right\} \qquad (7)$$

where:

$S_t(k)$ is the time the k-th firing of transition t begins,

in(t) is the set of transitions which immediately precede transition t,

i.e. $in(t) = \{s \in T \,/\, \exists\, p \in P,\ {}^\circ p = s \text{ and } p^\circ = t\}$

(t,s) is the place p such that $s = p^\circ$ and $t = {}^\circ p$.

The simulation of the stochastic event graph is based on the evolution equation (7): we compute the values $S_t(k)$ and use the ergodicity relation (6). This approach has been shown to be more efficient than the classical "simulation by events".

Simulation algorithm

1. Choose the number N of cycles.
   N should be large enough so that the ergodicity relationship is valid.

2. Starting from the initial marking m_0, define a sequence $e = <t_{r(1)}, \ldots, t_{r(|T|)}>$ firable from m_0 and in which each transition of T appears exactly once.
   We know that such a sequence exists if m_0 is such that there is at least one token in each elementary circuit.
3. Set $S_t(k) = 0$ for $k \leq 0$ and $t \in T$.
   Here, the process is initialized.
4. <u>For k = 1 to N</u>
   - 4.1. Generate at random the k-th firing time of transition t for all $t \in T$.
     These times are generated using the random variables associated to the transitions.
   - 4.2. For $t = t_{r(1)}$ to $t_{r(|T|)}$ apply the evolution equation (7).
5. Compute $\pi(m_0) = S_{t_{r(1)}}(N) / N$.

This simulation algorithm is the basis for the computation of the place in which a token will be added at a given iteration. Let us denote this place by p*. Place p* maximises the quantity:

$$\Delta\pi(m_0, p) / x_p \tag{8}$$

where:

$\Delta\pi(m_0, p) = \pi(m_0) - \pi^+(m_0, p)$ ($\pi^+(m_0, p)$ is the cycle time of the system when, starting from m_0, we add one token in p),

x_p is the component of the p-invariant x corresponding to p.

To compute the quantity (8) for all the $p \in P$, we should make $|P|+1$ simulations, that is one simulation for $\pi(m_0)$ and one simulation for $\pi^+(m_0, p)$ for each $p \in P$.
We propose to evaluate all the values (8) using a single simulation. Note that our goal is simply to **evaluate** and not compute the exact values of the $\Delta\pi(m_0, p)$. This evaluation is carried out by using additional information obtained along the simulation.
We consider a place p of the event graph. This place is the output place of a transition τ and the input place of a transition t, as shown in Figure 7.

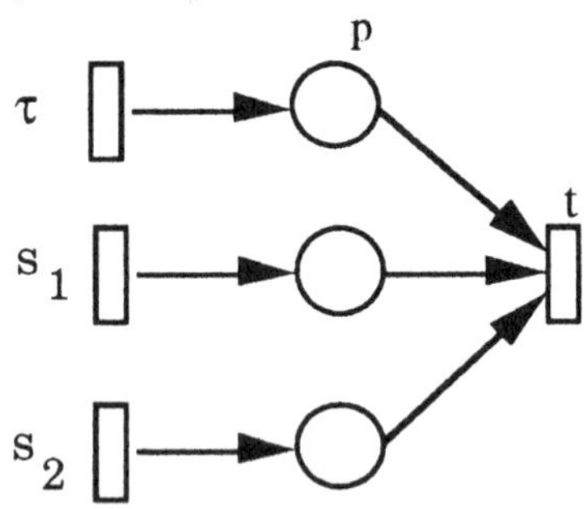

Fig. 7 Evaluation of $\Delta\pi(m_0, p)$

We denote by X the duration of the simulation. X is assumed to be large enough such that the average values obtained from the simulation are good estimations of the mean values in steady state.

A good evaluation of $\Delta\pi(m_0, p)$ is given by:

$$\Delta\pi(m_0, p) = \psi(m_0, p) \,.\, h(m_0, p)$$

Definition of $\psi(m_0, p)$

$\psi(m_0, p)$ is the mean time $t = p°$ is waiting for a token to arrive in place p in order to start firing. More precisely:

(i) if there is at least one token in p before t stops firing, then the contribution to $\psi(m_0, p)$ is zero,

(ii) otherwise,

- if there is at least one token in each $s \in °t$, $s \neq p$, before t stops firing, then the contribution to $\psi(m_0, p)$ is the difference between the time the first token arrives in p and the time t stops firing, divided by X. Thus, if the k-th firing of t is concerned by this case:

  $$\psi(m_0,p) \leftarrow \psi(m_0,p) + \left[a_p(k) - S_t(k) - \theta_p(k)\right] / X$$

  where:

  $a_p(k)$ is the time the first token arrives in p after the end of the k-th firing of t,

  $\theta_t(k)$ is the k-th firing time of t.

- if p and some other places of $°t = °(p°)$ are empty after the k-th firing of t and if $a_p(k) > a_s(k)$ for $s \in E_k$, set of places of $°t$ which are empty after the k-th firing of t, then $\psi(m_0, p)$ increases as follows:

  $$\psi(m_0,p) \leftarrow \psi(m_0,p) + \left[a_p(k) - \mathop{\mathbf{Max}}_{s \in E_k} a_s(k)\right] \Big/ X$$

  Of course, $\psi(m_0, p)$ is initialized at 0 at the beginning of the simulation.

Definition of $h(m_0, p)$

$h(m_0, p)$ is the average number of times $t = p°$ is waiting for a token delivered by $\tau = °p$ to start firing. In other words:

$$h(m_0,p) \leftarrow h(m_0,p) + 1/X$$

in the case (ii) above.

d. Qualitative properties of event graphs

It is easy to verify that a strongly connected event graph Z is:

* Structurally live for any marking m_0 such that each elementary circuit contains at least one token.
* Reversible for any marking m_0 such that each elementary circuit contains at least one token. Furthermore, under the same condition, any $m \in R(Z, m_0)$ is a home state.
* Structurally bounded, assuming that the initial marking m_0 is finite.

2.3. Decomposable nets

a. Definitions

Let $y = [y_1, \ldots, y_n]$ be a t-invariant of a Petri net $Q = (P, T, I, O)$.

Definition 5: The set of transitions $t_i \in T$ such that $y_i > 0$ is called the support of the t-invariant y and is denoted by $\|y\|$.

In a non-cyclic manufacturing system, parts enter and exit the system; thus, its PN model contains source transitions (also called input transitions of the model, or input transitions for short), which are used to model the entrance of parts into the system, and sink transitions (or output transitions of the model or output transitions for short), which are used to model the exit from the system.

Definition 6: Let y be a t-invariant of the PN model $Z = (P, T, I, O, m_0)$ of a manufacturing system. Let $Z_y = (P_y, T_y, I_y, O_y, m_{0_y})$ be a PN defined as follows:

$$P_y = \left\{ p \,/\, p \in P \text{ and } \exists\, t_1, t_2 \in \|y\| \text{ s.t. } p \in {}^{\circ}t_1 \text{ and } p \in t_2^{\circ} \text{ in } Z \right\},$$

$T_y = \|y\|$,

m_{0_y} is the restriction of m_0 to P_y,

I_y (resp. O_y) is derived from I (resp. O) by removing the rows corresponding to the places of P which do not belong to P_y and by removing the columns corresponding to the transitions of T which do not belong to T_y. Thus, $Z_y \subset Z$. The PN Z_y is refered to as **y-related subnet of Z**.

If, in addition:

(i) the cardinality of p° is 1 for every $p \in P_y$,

(ii) there exists at least one $t_1 \in \|y\|$ and one $t_2 \in \|y\|$ such that ${}^{\circ}t_1 = \varnothing$ and $t_2^{\circ} = \varnothing$,

(iii) Z_y is acyclic,

then Z_y is refered to as **y-CFIO of Z**, where CFIO stands for Conflict Free net with Input and Output transitions.

b. Decomposability of a net

Decomposability is the key concept in the planning and scheduling approaches for non-cyclic manufacturing systems we are proposing in this chapter.

Definition 7: Let Z be a PN model of a manufacturing system with input and output transitions. Let $\{y^1, \ldots, y^r\}$ be a set of t-invariants of Z such that:

$Z = \bigcup_{i=1}^{r} Z_{y^i}$, where Z_{y^i} is a y^i-CFIO of Z (we say that the set of y^i-CFIOs Z_{y^i} covers Z).

Then Z is said to be decomposable.

Example

Let us consider the PN Z given in figure 8a and the t-invariants:

$y^1 = [1,1,1,0,0,0,0,0]$, $y^2 = [1,0,0,1,1,1,0,0]$, $y^3 = [0,0,0,0,0,1,1,1]$

Z is decomposable since $Z = Z_{y^1} \cup Z_{y^2} \cup Z_{y^3}$, where Z_{y^1}, Z_{y^2} and Z_{y^3} are CFIOs.

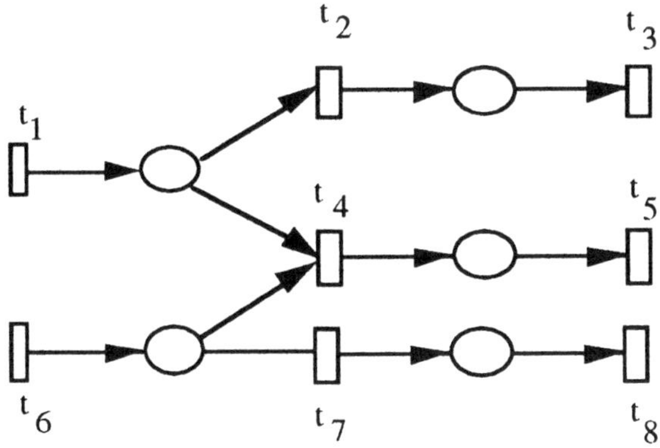

Fig. 8a: Petri net Z

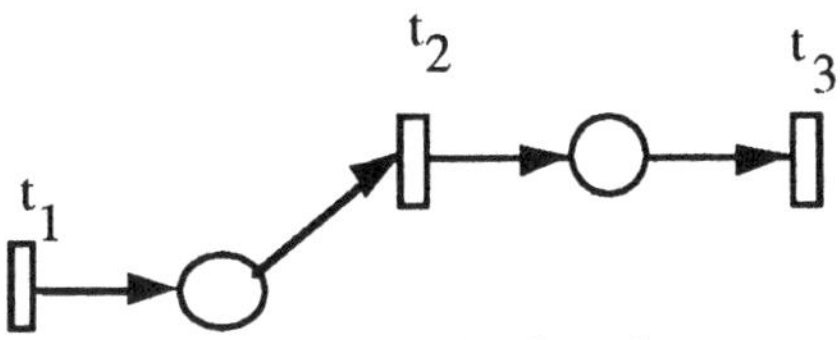

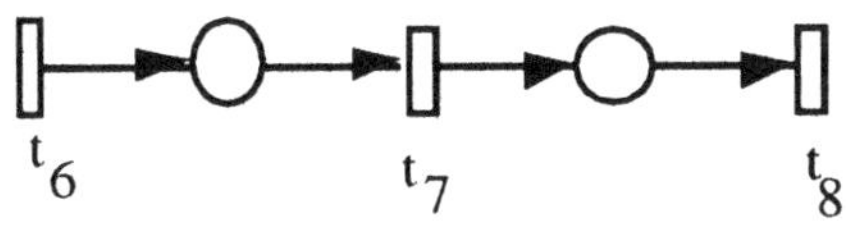

Fig. $8b_1$: CFIO Z_{y^1}

Fig. $8b_2$: CFIO Z_{y^3}

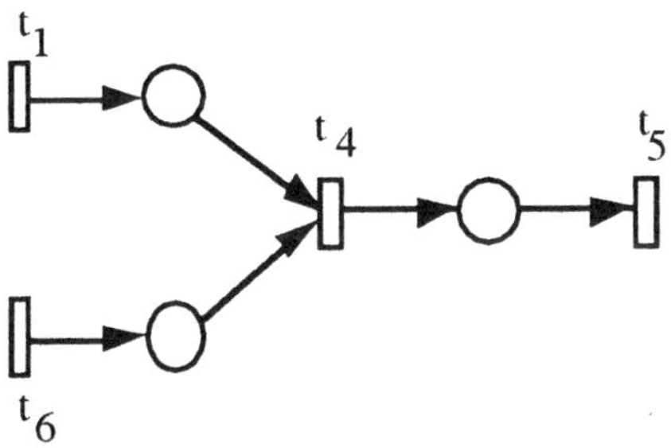

Fig. $8b_3$: CFIO Z_{y^2}

Fig. 8: A decomposable net

c. Qualitative properties of a decomposable net

Result 6: A decomposable net Z is structurally live for any initial marking m_0.

Proof:

α. Let $m \in R$ (Z, m_0) and e a firing sequence which leads to m starting from m_0. Since Z is a decomposable PN, there exists a set $\{y^1, \ldots, y^r\}$ of t-invariants of Z such that the corresponding set of y-CFIOs Z_{y^i} covers Z. Thus, it is possible to assign each element of e to one Z_{y^i}, designing firable sequences which maintain the order of the elements in e. Let us call e_{y^i} (i = 1,2,...,r) the sequence related to Z_{y^i}. Note that some of these sequences may be empty.

β. Let us now consider $t \in T$ and let Z_{y^k} be a y-CFIO containing t. Due to the definition of y-CFIOs, there exists $\tilde{e}_{y^k}$ such that firing $e_{y^k} \circ \tilde{e}_{y^k}$ from m_0^k (restriction of m_0 to Z_{y^k}) leads to m_0^k again, and this sequence contains each transition of Z_{y^k} at least once ($\circ$ represents the concatenation). As a consequence, $e_k^* = \tilde{e}_{y^k} \circ e_{y^k} \circ \tilde{e}_{y^k}$ is a firing sequence which applies to m and contains t.

This proof holds for any m_0.

Q.E.D.

Result 7: A decomposable PN is reversible for any initial marking m_0.

Proof:

α. This first part of the proof is identical to the first part of the proof of Result 6.

β. Due to the definition of y-CFIOs, there exists $\tilde{e}_{y^i}$ such that $e_{y^i} \circ \tilde{e}_{y^i}$ leads to m_0^i from m_0^i for $i = 1,2,\ldots,r$. Thus if $\tilde{e} = \underset{i=1,2,\ldots,r}{\bigcirc} \tilde{e}_{y^i}$ then $e \circ \tilde{e}$ leads to m_0 from m_0.

Q.E.D.

A decomposable net is not structurally bounded; for instance, the number of tokens in such a system increases indefinitely if we keep firing only the input transitions. Nevertheless, the following result holds.

Result 8: A decomposable net $Z = (P, T, I, O, m_0)$ can be kept bounded by appropriate firings of its input transitions, no matter how many times the output transitions are fired, provided that the number of the output firings is finite.

Proof:

Let $k_1, \ldots, k_s$ be the minimal numbers of times the output transitions $t_{0,1}, \ldots, t_{0,s}$, must be fired. Let $y^1, \ldots, y^r$ be a set of t-invariants such that the corresponding y-CFIOs Z_{y^i} cover Z. Let n_j^i be the component of the t-invariant y^i which corresponds to $t_{0,j}$. At least one of the n_j^i, $i = 1,2,\ldots,r$, is strictly positive for any $j \in \{1,\ldots,s\}$, and at least one of the n_j^i, $j = 1,2,\ldots,s$, is strictly positive for any $i \in \{1,\ldots,r\}$. Thus the integer linear programming problem

$$\mathcal{Min} \sum_{i=1}^{r} s_i$$

s.t.

$$\sum_{i=1}^{r} s_i . n_j^i \geq k_j, \quad j = 1,2,\ldots,s$$

$$s_i \in \{0,1,\ldots,\}, \quad i = 1,2,\ldots,r$$

has a finite solution. The integer s_i is the number of times the set $\|y^i\|$ of transitions must be fired according to y^i in order to fire the output transitions for the required number of times. Thus, the marking m(p) of $p \in t_v^\circ$ is bounded from above by:

$$m_0(p) + \sum_{i=1}^{r} s_i x_v^i, \quad \text{where } x_v^i \text{ is the v – th componant of } y^i$$

This relation holds for any $p \in P$.

Q.E.D.

2.4. Controlable nets

a. Definitions

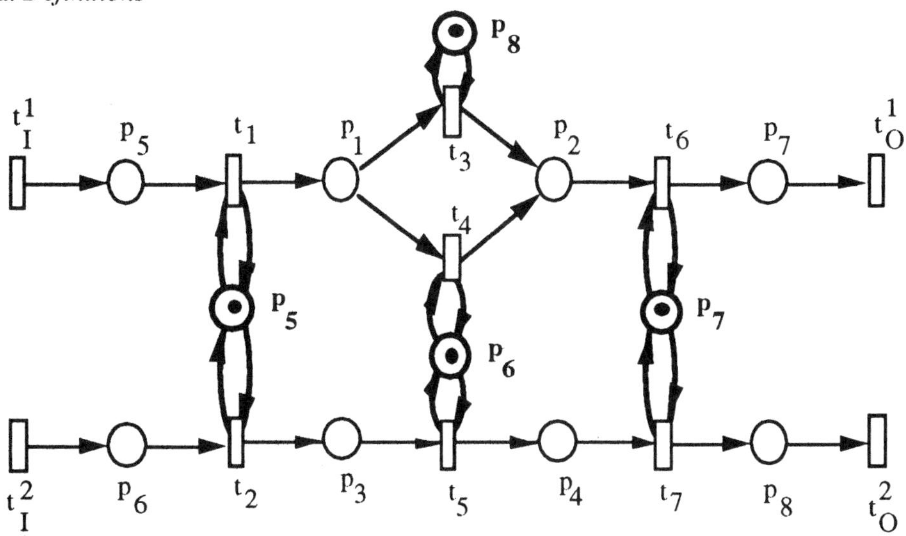

Fig. 9: A controlable net Z

A Petri net Z = (P∪R, T, I, O, m_0) where:

P (resp. R) is a set of **process** (resp. **resource**) places and $P \cap R = \emptyset$,

each arc is weighted at 1,

is called a **Petri net with controlable outputs**, or a **controlable Petri net** for short, if:

(H1) $(t,p) \in Z \Leftrightarrow (p,t) \in Z, \forall\, t \in T$ and $p \in R$,

(H2) $m_0(p) \geq 1, \forall\, p \in R$,

(H3) The sub-net $Z' = (P, T, I', O', m')$ where:

I' and O' are derived from I and O respectively by removing the rows corresponding to the places belonging to R,

m' is the restriction of m_0 to P,

is a decomposable net,

(H4) Z' has source and sink transitions, but no source or sink places,

(H5) There exists a set $\chi = \{y^1, ..., y^k\}$ of t-invariants of Z' such that the Z'_{y^i} nets are y^i-CFIOs which cover Z' and have only **one sink transition each**.

The Petri net Z given in Figure 9 is a controlable net where $P = (p_1, p_2, p_3, p_4)$ and $R = (p_5, p_6, p_7, p_8)$.

The subnet Z' of Z is shown in Figure 10.

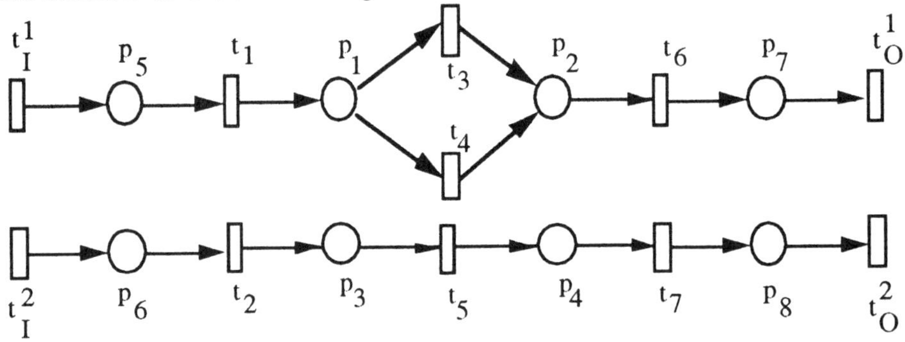

Fig. 10: The subnet Z' of Z: a decomposable net

A controlable net differs from a decomposable net in two main aspects. Unlike a decomposable net, a controlable net:

(i) has resource places,

(ii) can be reduced to a decomposable net which, in turn, can be decomposed into y-CFIOs which have only one sink transition.

These two aspects will allow us to model resources —and thus to model scheduling problems— and to model the fact that products can be manufactured independently from each other.

b. Properties

The properties of a controlable net are summarized in Result 9.

Result 9 : A controlable net Z is:

(i) consistent if $R = \varnothing$,

(ii) live for any initial marking m_0 if $R = \varnothing$,

(iii) not structurally bounded if $R = \varnothing$; nevertheless, it can be kept bounded,

(iv) reversible for any initial marking m_0 if $R = \varnothing$;
furthermore, any $m \in R(Z, m_0)$ is a home state.

The proof of these results is obvious if we observe that, for $R = \varnothing$, a controlable net is a decomposable net, and these results hold for a decomposable net, as shown in

Section 2.3.c. But, it is easy to show that Result 9 also holds for $R \neq \emptyset$, i.e. for any controlable net.

3. Cyclic manufacturing systems

3.1. Problem formulation

In this section, we emphasize the usefulness of event graphs to model **cyclic manufacturing systems**, also called **off-line** or **ratio-driven** manufacturing systems. Such a system produces a given set of part types at given ratios. The objective of the control problem is to maximize the system's productivity while minimizing the WIP.

In a cyclic system, there exists an optimal control which is periodic. From the management point of view, a periodic control can be expressed as a sequence of part types associated to each machine, cell or transportation system. Such a sequence is refered to as the input sequence of the resource. It is emphasized that the same part type can appear several times in the same input sequence in order to satisfy the production ratios.

Let us consider, for instance, a set of three machines denoted by M_1, M_2 and M_3, which manufacture three types of parts, say R_1, R_2 and R_3. Assume that the manufacturing routings of these part types are as follows (the number in parenthesis provides the manufacturing time of each operation):

R_1: $M_1(2)$, $M_2(1)$, $M_3(4)$

R_2: $M_3(2)$, $M_1(3)$

R_3: $M_2(4)$, $M_1(1)$

Assume also that the production ratios are 0.25, 0.25 and 0.5 for R_1, R_2 and R_3, respectively. In this case, a set of input sequences associated to M_1, M_2 and M_3 could be:

$$\sigma(M_1) = \langle R_1, R_2, R_3, R_3 \rangle,\ \sigma(M_2) = \langle R_1, R_3, R_3 \rangle,\ \sigma(M_3) = \langle R_1, R_2 \rangle.$$

We do not claim that this periodic control is optimal, but that there exists an optimal control with respect to the objective defined above, say $(\sigma^*(M_1), \sigma^*(M_2), \sigma^*(M_3))$ where $\sigma^*(M_i)$ is obtained by applying a permutation to $\sigma(M_i)$, $i = 1,2,3$.

The next subsection is devoted to the modeling of cyclic manufacturing systems. In Subsection 3.3, we show that it is possible to maximize the productivity of the system for any cyclic control. Finally, Subsection 3.4 proposes a heuristic algorithm to reach a near-optimal cyclic control.

3.2. Modeling of cyclic manufacturing systems

The event graph model of a cyclic manufacturing system is developed as follows:

(i) Model the manufacturing process of each part type. In such a model, a transition represents an operation and a place a buffer. Each transition firing corresponds to the execution of an operation and the firing time of the transition is the time required to perform this operation. Each manufacturing process is reproduced as many times as necessary to yield the given production ratios. For instance, in the example introduced in the previous subsection, the manufacturing processes of R_1 and R_2 are represented once, and the manufacturing process of R_3 is represented twice.

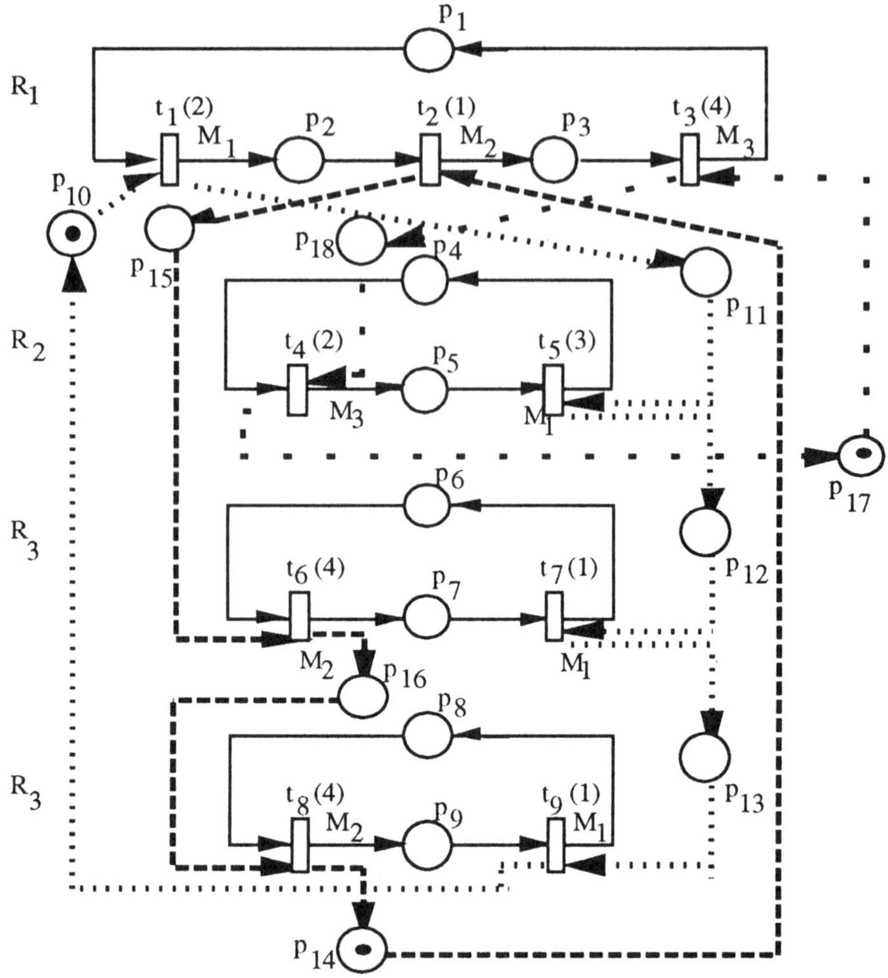

Fig. 11: A job-shop model

(ii) Model the cyclic operation mode of the system by assuming that a new part is launched in the system as soon as a part of the same type is completed. This is modeled by adding an output place to the last transition of each manufacturing process model, and by connecting this place to the first transition of the same model. Thus, we obtain elementary circuits called **process circuits**.

(iii) The final step is to model the sequencing of the part types for each machine. This is done by connecting in a unique circuit all the transitions corresponding to operations performed by the same machine. The order of transitions in these new elementary circuits (called **command circuits**) is determined by the sequencing of the jobs on the corresponding machines.

Tokens circulating in the process circuits represent parts, whereas tokens in the command circuits represent machine status information. Note that there is exactly one token in each command circuit, since a machine is assumed to manufacture at most one part at a time, and since an elementary circuit without a token blocks after a finite number of transition firings.

a. Model of a cyclic job-shop

The above steps were employed to model the job-shop introduced in Subsection 3.1 with the control ($\sigma(M_1)$, $\sigma(M_2)$, $\sigma(M_3)$). The resulting model is shown in Figure 11.

b. Model of a cyclic assembly system

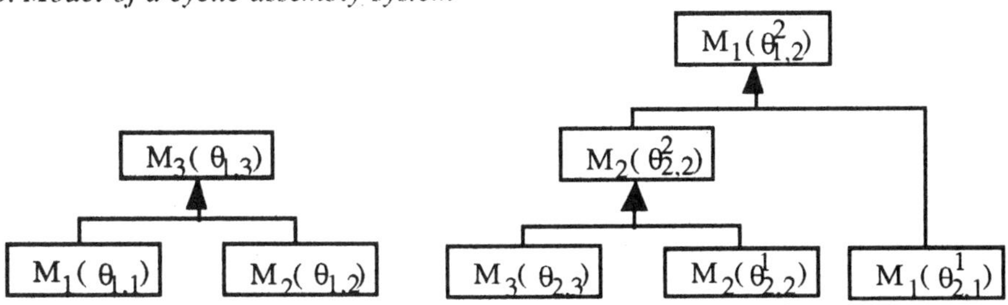

a. Part type R_1 b. Part type R_2

Fig. 12: Two assembly manufacturing processes

As a second example, let us consider two product types, the bills of materials and manufacturing processes of which are represented in Figure 12. Each box in the figure corresponds to a make item (part). In addition, each box represents an operation and contains the machine which performs this operation. The parameters in parentheses denote operation times.

We assume that the production ratios are 0.5 and 0.5. Following the general approach presented in Subsection 3.1, we obtain the model shown in Figure 13.

In this model, we only show the command circuit related to machine M_2 for simplicity. This circuit is $<w_2, t_2, w_1, t_6, w_3, t_8, w_2>$.

Places p_i ($i = 1,2,\ldots,10$) represent buffers. Places q_i ($i = 1,2,\ldots,10$) are used to control the number of components of each type in the system. For instance, q_1 contains initially n_1 tokens, which guarantees that at most n_1 units of the component of R_1 that is manufactured by M_1 can be found simultaneously in the system.

As shown in Figure 13, some transitions which do not represent operations have been introduced in the model. These transitions are t_3, t_7, t_{10} and t_{11}. The time associated to these transitions is 0. They have been introduced to allow for WIP control.

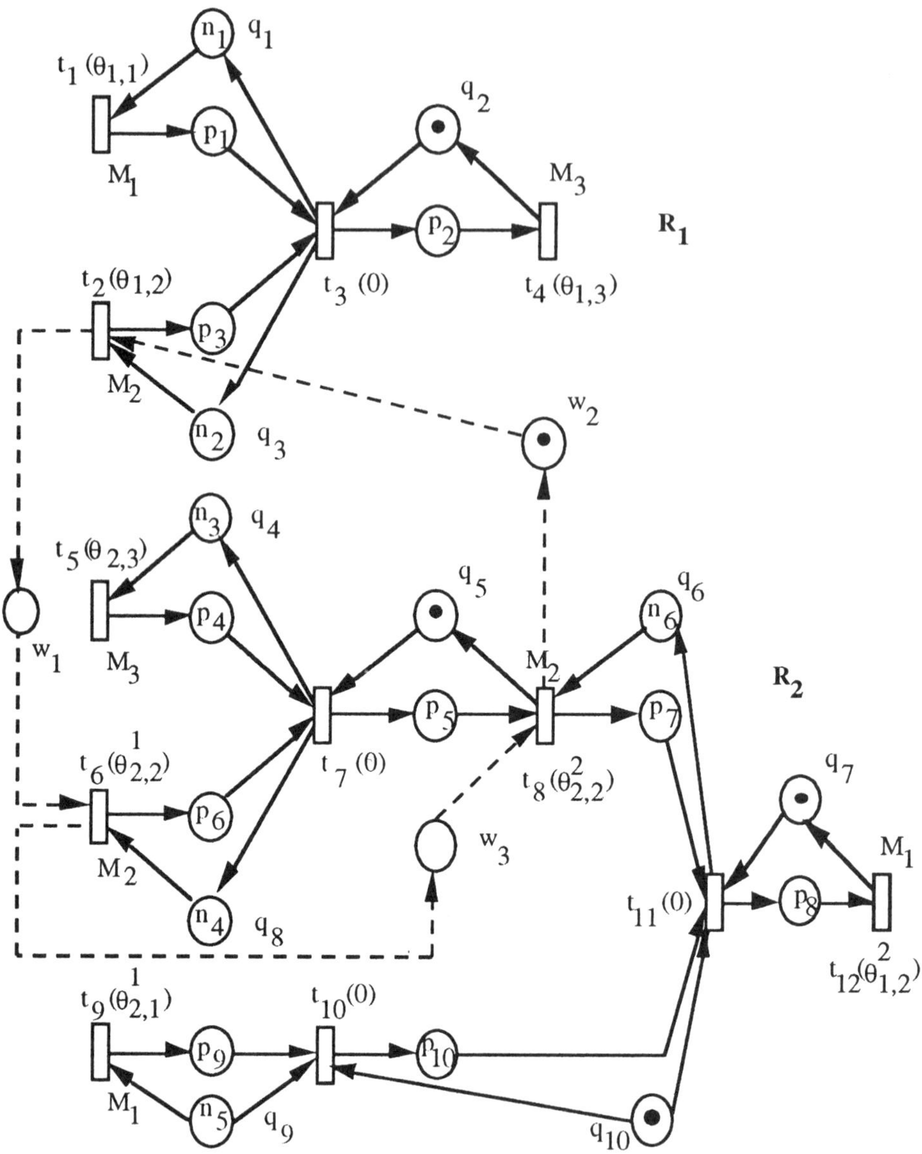

Fig. 13: The assembly system model

The following result holds for cyclic manufacturing systems:

Result 10

Assuming that:

(i) a new part is launched in the manufacturing system as soon as a part of the same type is completed,

(ii) the control applied to the system is cyclic and can be expressed in terms of input sequences which satisfy the production ratios,
(iii) one manufacturing process is associated to each part type,

we claim that the corresponding Petri net model encapsulates both the physical and the decision making systems, and is an event graph.

Remarks

1 - If more than one manufacturing process alternatives are available to manufacture some part types, then it is possible to split up each part type into as many sub-part types as the number of manufacturing processes available, and to distribute the production ratio of the part type among the sub-part types in order to optimize a criterion of interest, such as work load balancing among the resources. This leads to a problem that satisfies condition (iii) of Result 10.

2 - If the model is not strongly connected, then it is always possible to introduce transitions with zero duration which represent the entrance of the parts in the system. These transitions are then connected by means of a command circuit, which transforms the model into a strongly connected event graph.

3.3. Evaluation of a cyclic manufacturing system

Since a Petri net model of a cyclic manufacturing system is, or can always be transformed into, a strongly connected event graph. The results presented in Section 2.2 apply when the firing times are deterministic. In such a model, three types of elementary circuits exist, namely:

(i) **process circuits**, which can contain as many tokens as desirable,
(ii) **command circuits**, which contain one token each, the initial position of which is defined by the input sequence of the related resource,
(iii) **hybrid circuits**, which are composed of portions of command and process circuits; for instance, in Figure 11, the circuit $\gamma = <t_1, p_{11}, t_5, p_4, t_4, p_{17}, t_3, p_1, t_1>$ is a hybrid circuit.

Since a command circuit contains exactly one token and since it is possible to put as many tokens as desirable in the places which belong to process circuits, it is possible to find a finite initial markings such that the critical circuit is a command circuit, and more precisely the command circuit γ^* such that:

$$C(\gamma^*) = \mathop{\mathbf{Max}}_{\gamma \in \Gamma_C} C(\gamma) \tag{9}$$

where Γ_C is the set of command circuits and $C(\gamma)$ is the cycle time of γ.

In other words, **it is always possible to fully utilize the bottleneck machine** if the initial WIP is large enough, and thus to reach the maximal productivity of the system. However, the objective is to reach the maximal productivity while minimizing the WIP. This problem can be written as follows for a given cyclic control:

$$\mathbf{Minimize} \sum_{i / p_i \in P_{\overline{C}}} n_i \tag{10}$$

s.t.

$$\sum_{i / p_i \in \gamma \cap P_{\overline{C}}} n_i \geq \mu(\gamma) \Big/ C_0 - \sum_{i / p_i \in \gamma \cap P_C} n_i, \quad \forall \gamma \in \Gamma_{\overline{C}},$$

$n_i \in \{0,1,\ldots\}$, $\forall i$ such that $p_i \in P_{\overline{C}}$.

where $n_i = m_0(p_i)$, P_C is the set of places belonging to the command circuits, $P_{\overline{C}}$ is the set of places which do not belong to the command circuits, $\Gamma_{\overline{C}}$ is the set of elementary circuits which are not command circuits, and $C_0 = C(\gamma^*)$ is defined by relationship (9). Having determined all elementary circuits, problem (10) is solved by linear programming.
Note that $n_i = m_0(p_i)$ is known for $p_i \in P_C$; this is derived from the input sequences of the resources. The difficulty in solving problem (10) is to compute all elementary circuits of the model, the number of which increases often exponentially with the size of the model. For this reason, a heuristic algorithm, called adjustment algorithm, was given in Section 2.2.b. In practice, this heuristic algorithm is not limited by the size of the model.
So far, we have shown that it is possible to reach the maximal productivity of the system while minimizing the WIP when the periodic control is known. However the problem of finding the periodic control which results in the maximal productivity with a WIP which is the minimum among all WIP values corresponding to all periodic controls remains open. This problem is considered in the next sub-section.

3.4. Optimization of the cyclic control

The problem to be considered is a scheduling problem which is NP-hard. Thus, only heuristic algorithms can be used to solve large-sized problems. Two algorithms have been shown to provide near-optimal solutions within reasonable times.

a. The simulated annealing algorithm

This algorithm has been presented in numerous papers, and in particular in [10] and [12]. The part of the algorithm which is specific to the problem at hand is the generation of alternatives by perturbation of a given schedule, i.e. a given set of input sequences. We generate a new schedule in the neighborhood of a previous one by permutating two elements in each input sequence of the resources.
The results obtained by applying simulated annealing to this problem have been found to be better, on the average, than those obtained from Tabou and genetic search approaches.

b. Construction approach

The basic idea of this algorithm is to (i) determine, for each process circuit, the minimal number of tokens (≥ 1) required such that a command circuit becomes the critical circuit; (ii) assign to each process circuit as many periods of duration $C_0 = C(\gamma^*)$(see (9)) as the number found in (i); and (iii) fit the firing periods of the transitions within these C_0 periods, taking into account the scheduling constraints, i.e. the fact that two firing periods of transitions corresponding to the same machine do not overlap, and the fact that the partial order imposed by the manufacturing processes is verified.
A sufficient (but not necessary) optimality condition is to find a schedule with the appropriate number of C_0 periods initially assigned to the manufacturing process circuits.

Algorithm

1. Initialization
   1.1. For each process circuit i (i = 1,...,n)
      1.1.1. Set $k_i = 0$
      k_i is a counter equal to the number of firing periods already placed in the C_0-period related to manufacturing process i.
      1.1.2. Set $N_i(0)= 1$
      $N_i(k_i)$ is the number of C_0-periods available for the next firing period.
      1.1.3. Set $\varphi_i(0) = 0$
      $\varphi_i(k_i)$ is the ending time of the first firing of the k_i-th transition of i.
      1.1.4. Compute $d_i(0) = 0$
      $d_i(k_i)$ is the degree of freedom corresponding to i given the schedule of the k_i first firing periods.

$$d_i(k) = 1 - \sum_{t \in T_i - T_i(k)} \mu(t) \Big/ \left(C_0 - \varphi_i(k) + \left[N_i^*(k) - N_i(k)\right] C_0\right)$$

      where: *T_i is the set of transitions of i,*
      $T_i(k)$ is the set of the first k transitions of i,
      $N_i^*(k) = \mathrm{Max}[N_i(k), \alpha_i]$, where $\alpha_i = \lceil \mu(i) / C_0 \rceil$,
      $\mu(i)$ is the sum of the firing times of the transitions of i,
      $\mu(t)$ is the firing time of transition t.
   1.2. Set m = 1
   m is the iteration counter.
   1.3. Set $E_m = \{1,2,\ldots,n\}$
   E_m is the set of process circuits for which some firing periods have not been placed.
2. Scheduling
   2.1. Compute j such that $d_j(k_j) = \underset{i \in E_m}{\mathrm{Min}}\, d_i(k_i)$
   j is the elementary circuit to be considered next.
   2.2. If $d_j(k_j) < 0$
   Add $g_j(k_j)$ C_0-periods to the process circuit j: $N_j(k_j) = N_j(k_j) + g_j(k_j)$ where $g_j(k_j)$ is the smallest integer such that

$$\left(\sum_{t \in T_j - T_j(k_j)} \mu(t)\right) \Big/ \left(C_0 - \varphi_j(k_j) + \left[N_j^*(k_j) - N_j(k_j) + g_j(k_j)\right] C_0\right) \le 1$$

   2.3. Choose the transition to be fired
   If $k_j > 0$, t_j is the next transition in the order given by the manufacturing process. Otherwise, we choose the transition which yields the minimal value of $d_j(1)$.
   2.4. Place the firing period of t_j at the earliest time, taking into account the manufacturing constraints.
   2.5. Set $T_j(k_j +1) = T_j(k_j) \cup \{t_j\}$.

2.6. Compute $\varphi_j(k_j+1)$ and $d_j(k_j+1)$.
2.7. Set $k_j = k + 1$ and $m = m + 1$.
2.8. Update E_m.
2.9. If $E_m \neq \varnothing$ go to 2.1, otherwise stop.

To illustrate this algorithm, let us consider the model represented in Figure 11. The bottleneck machine is M_2 and $C_0 = 9$. To ensure that the critical circuit of the model is the command circuit corresponding to machine M_2, we need a minimum of:

* one token in the process circuit corresponding to R_1,
* one token in the process circuit corresponding to R_2,
* one token in each of the process circuits corresponding to R_3.

Applying the previous algorithm leads to the schedule represented in Figure 14.

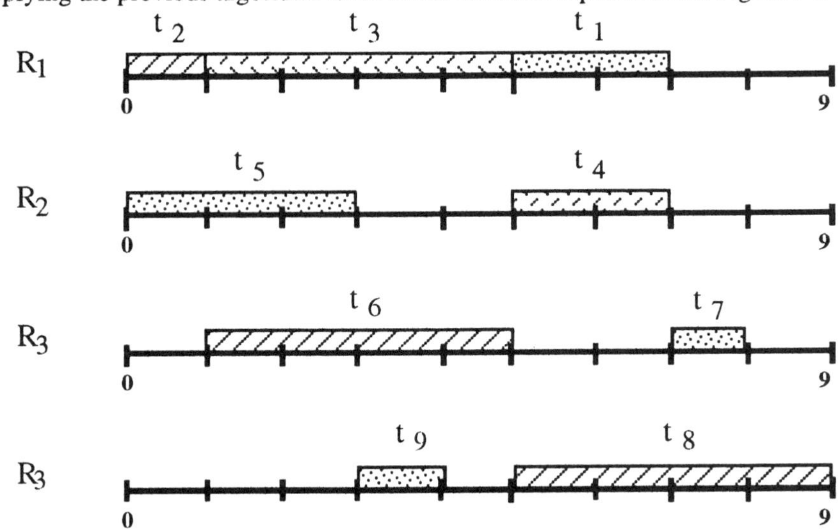

Fig. 14: An optimal schedule

Since no C_0-period was added to the initial ones, the schedule provided by the algorithm is optimal. This indicates that the optimal input sequences are:

$\sigma(M_1) = \langle R_2, R_3, R_1, R_3\rangle$, $\sigma(M_2) = \langle R_1, R_3, R_3\rangle$, $\sigma(M_3) = \langle R_1, R_2\rangle$.

Furthermore, the initial marking is:

$m_0(p_2) = m_0(p_5) = m_0(p_6) = m_0(p_9) = 1$

$m_0(p_1) = m_0(p_3) = m_0(p_4) = m_0(p_7) = m_0(p_8) = 0$

Finally, the command circuit and the initial locations of its tokens are fixed by the optimal input sequences.

Note

When the manufacturing times are stochastic, the algorithm presented in Section 2.2.c applies when the schedule is known. In this case, we dispose of a powerful heuristic algorithm which allows to reach a given productivity while minimizing the WIP. However, to-date, no satisfactory algorithm exists to find the optimal schedule. An alternative, non-optimal, approach is to replace the stochastic event graph into a deterministic one by replacing the stochastic firing times by their mean values, and to apply the algorithm proposed in Section 3.4.

4. Non-cyclic manufacturing systems

4.1. Problem formulation

We are interested in a job-shop comprising n machines $M_1, M_2, \ldots, M_n$ that manufacture q types of parts denoted by $R_1, R_2, \ldots, R_q$. The demand for each part type is known at the end of each of R consecutive elementary periods. For instance, an elementary period could be a day and R = 5; in this case, we are interested in managing the system over a working week on a day-to-day basis. In the remainder of the paper, the union of the R elementary periods is refered to as the sub-period. Let us denote by d_i^j, i = 1,2,.,q, j = 1,2,...,R, the demand for part type R_i at the end of the j-th elementary period. s_i^0 is the inventory level of part type i at the beginning of the first elementary period. We also define $\mathcal{M}(R_i)$ as the manufacturing process (routing) of part type R_i. $\mathcal{M}(R_i)$ provides:

(i) the sequence of operations to be performed on a part of type R_i;
(ii) the type of each operation, which can be either "assembly" or "regular"; disassembly operations are not considered;
(iii) the list of machines on which each operation can be performed;
(iv) the time required to perform each operation on each alternative machine.

The first problem to be solved is the short-term planning (STP) problem on the sub-period. Knowing the capacity of the system (i.e. the time available within each elementary period), we seek to determine the number of parts of each type to be manufactured during each elementary period in order to optimize a given criterion. Commonly used criteria include the number of delayed products, the maximal delay, the weighted sum of delays, or the sum of the inventory and backlogging costs. In the remainder of this paper, we seek to minimize the sum of the inventory and backlogging costs to illustrate the proposed approach.

Note that the time assigned to manufacturing tasks within each elementary period is bounded above by the duration of the period. The difference between the duration of an elementary period and the time assigned to tasks within this period is the maximal idle time of the machines during the period. It represents the flexibility of the system: the smaller the time assigned to tasks within an elementary period, the more likely it is that a feasible schedule exists. However, in this case, the productivity of the system becomes lower.

Starting from the number of parts of each type to be manufactured during the first elementary period (provided by the solution of the STP problem), the scheduling process (S) consists of assigning operations to their alternative machines and computing the starting time of each operation in order to satisfy the usual manufacturing constraints, namely:

(i) operations should be performed according to the partial order specified by the manufacturing processes (routings): two operations belonging to the same manufacturing process should be performed according to the required order;

(ii) a given machine performs at most one operation at a time.
We do not try to optimize some criterion, but simply to find a feasible schedule, i.e. a schedule which meets the requirements of the STP for the first elementary period. If such a feasible schedule does not exist, the only solution is to reduce the time assigned to tasks within the elementary periods, and to re-compute the STP: since such a change reduces the amount of parts to be manufactured during each elementary period, it is more likely that a feasible schedule exists.
In Figure 15, we summarize the procedure to obtain a feasible schedule. The STP problem can usually be solved using classical optimization software tools whose choice depends upon the type of criterion to be optimized. However, the SP is NP-hard and, therefore, only heuristic procedures are practical.
In the next section, we introduce basic concepts used to re-visit this problem in the light of PNs.

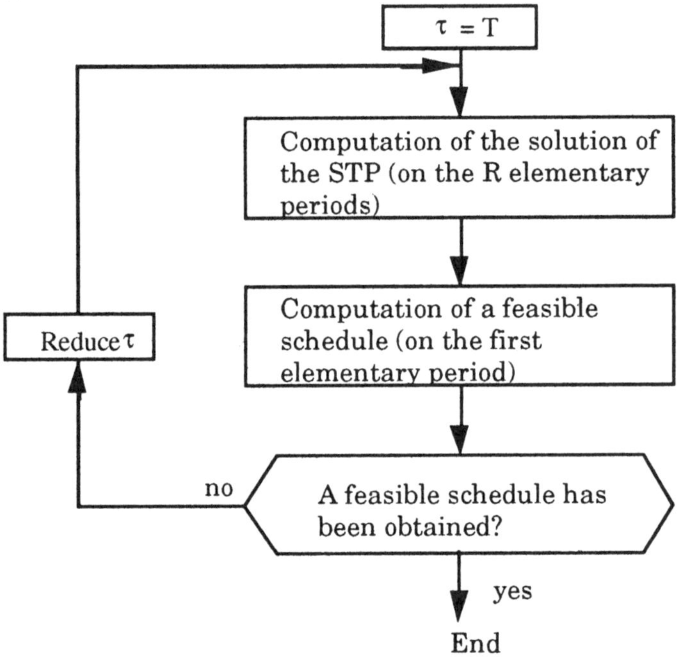

Fig. 15: General flow-chart for production management

4.2. Short-term planning

a. Modeling examples
Figure 16 shows the manufacturing processes (routings) for three part types R_1, R_2 and R_3. Alternative machines for each operation are separated by commas in the corresponding boxes. The corresponding PN models are shown in Figure 17. The numbers in squares are the manufacturing times of the operations (i.e. the firing times of the corresponding transitions). The duration of all input transitions of the models in Figure 17 is zero. These transitions represent the launching of components to production. Each of the remaining transitions represents an operation

on a machine. The machines corresponding to these transitions are included in parentheses. At most one firing is in progress at each transition at a time, which implies that a self-loop, with one token in the corresponding place, is associated to each transition. These self-loops are not represented in Figure 17 for simplicity.

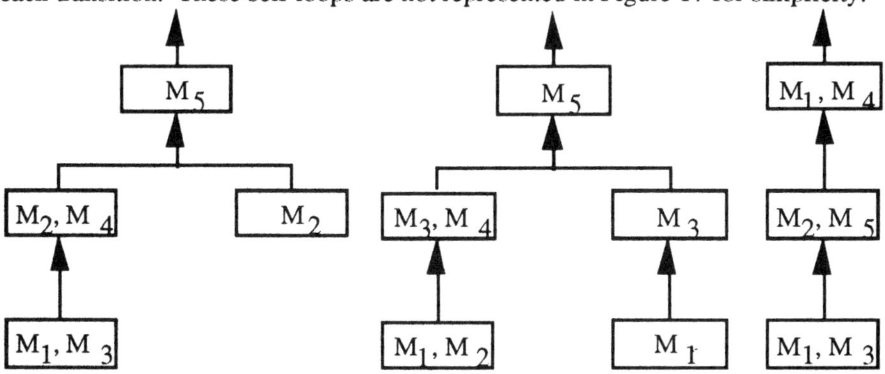

a. Part-type R_1 b. Part-type R_2 c. Part-type R_3

Fig. 16: Sample manufacturing processes

The PN model Z presented in Figure 17 is obviously a decomposable net. It is possible to find several sets of t-invariants for which the corresponding CFIOs cover Z. We can, for instance, choose the minimal t-invariants of the PN, which are obtained by combining the minimal t-invariants of each of the manufacturing process models. Since the models corresponding to R_1, R_2 and R_3 have respectively 4, 4 and 8 minimal t-invariants, we would obtain $4 \times 4 \times 8 = 128$ minimal t-invariants for the complete model. We can also choose some linear combinations of the minimal t-invariants provided that the derived CFIOs cover N.

b. Planning process

Let $\{y^1, \ldots, y^r\}$ be a set of t-invariants such that:

$$Z = \bigcup_{i=1}^{r} Z_{y^i}$$

Let O_k be the set of output transitions of the manufacturing process model of part type R_k, $k = 1,\ldots,q$. The demands for parts of type R_k are known at the end of each of R consecutive elementary periods. We denote, as in Section 4.1, by d_k^j the number of parts of type R_k required at the end of the j-th elementary period; d_k^j is also the total number of times the transitions of O_k must be fired to satisfy the demand by the end of the j-th elementary period. T is the duration of an elementary period and τ is the time assigned to an elementary period (see Section 4.1).

If v_i^j is the number of times the transitions of $\|X_i\|$ fire according to the components of y^i during the j-th elementary period, the following relations hold.

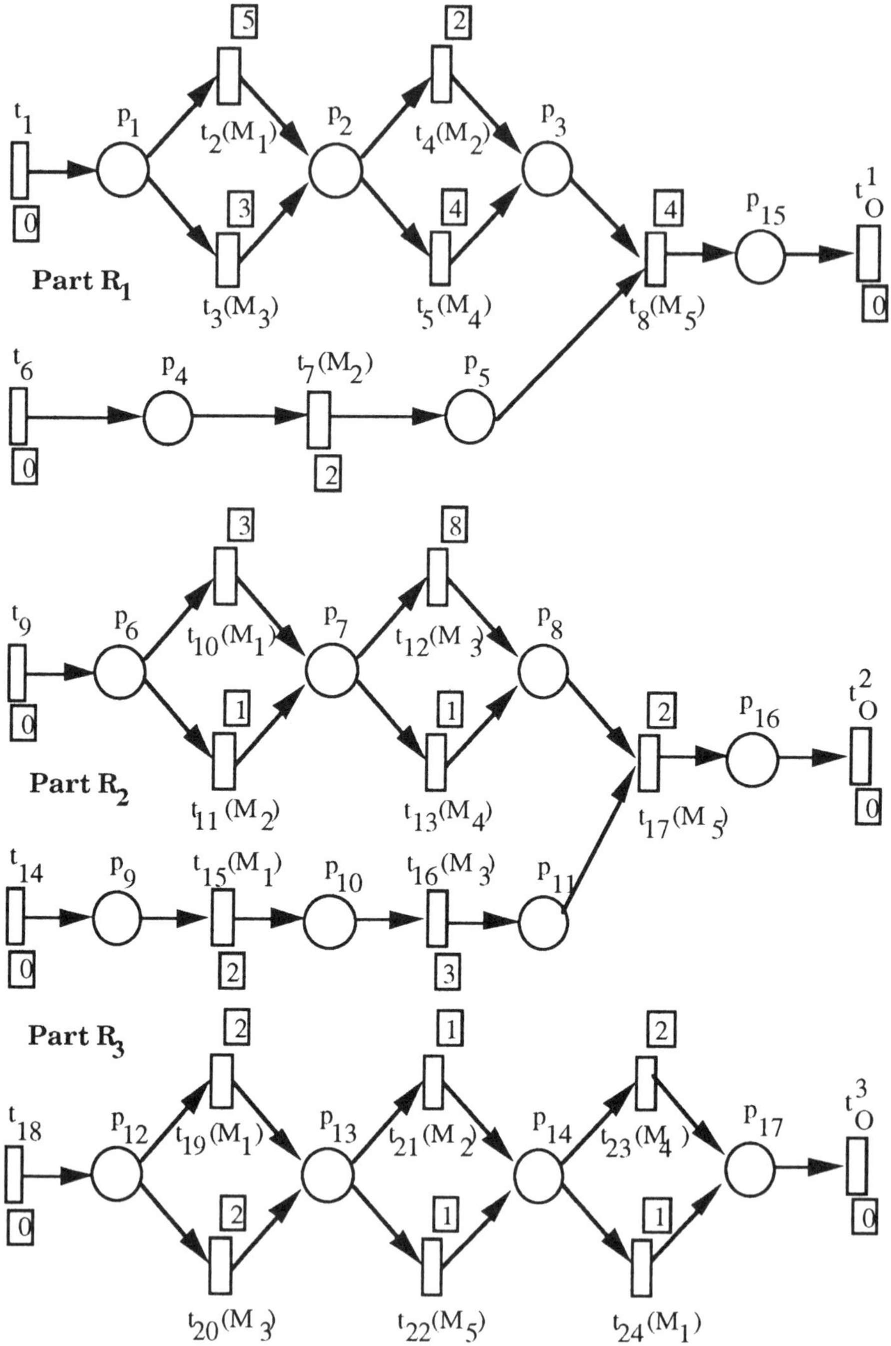

Fig. 17: The decomposable PN model of the system of Figure 16 from the short-term planning point of view

$$s_k^{j+1} = s_k^j - d_k^j + \sum_{i=1}^{r} \left[v_i^j \sum_{t \in O_k} y_t^i \right] \tag{11}$$

where y_t^i denotes the component of y^i corresponding to t.

This state equation holds for k = 1,...,q and j = 0,...,R-1. s_k^0 is the initial inventory of parts of type R_k.
Furthermore, if Z(t) is the firing time of t:

$$\sum_{i=1}^{r} v_i^j y_t^i Z(t) \leq \tau, \quad j = 1,\ldots,R, \ t \in T \tag{12}$$

are the capacity constraints.
The criterion to be minimized is the sum of the backlogging costs and inventory costs, and can be expressed as:

$$\mathcal{M}in \sum_{k=1}^{q} \sum_{j=1}^{R} \left[b_k (-s_k^j)^+ + i_k (s_k^j)^+ \right] \tag{13}$$

where b_k (resp. i_k) is the backlogging cost (resp. the inventory cost) of one unit of part type R_k during one elementary period. Note that the problem which consists of minimizing (13) under constraints (11) and (12) can be re-written as a linear programming problem.

c. Remarks
The key to this approach is the choice of the set of t-invariants $\{y^1, \ldots, y^r\}$. Depending on this choice, Z_{y^i} may be the model of one of the manufacturing processes or the model of a set of manufacturing processes corresponding to different part types. In the first case, the short-term planning procedure provides a better result, but requires extensive computation. In the second case, the number of v_i^j variables may be very small, and thus the computation required may be very little; however, the result will be certainly inferior to that of the previous case. In general, selecting a smaller number of t-invariants (assuming that the CFIOs derived from them cover the PN model) offers the potential for reducing the computational burden at the expense of the quality of the results (i.e. the productivity). It is noted that the short-term planning process introduced in this section guarantees that the qualitative properties presented in the previous section hold.

4.3. Scheduling

a. Problem formulation and definitions
The goal of the scheduling process is to assign operations to resources (in the case that several resources are available) and to define the starting time of each operation for the first elementary period in order to meet the firing requirements of short-term planning.

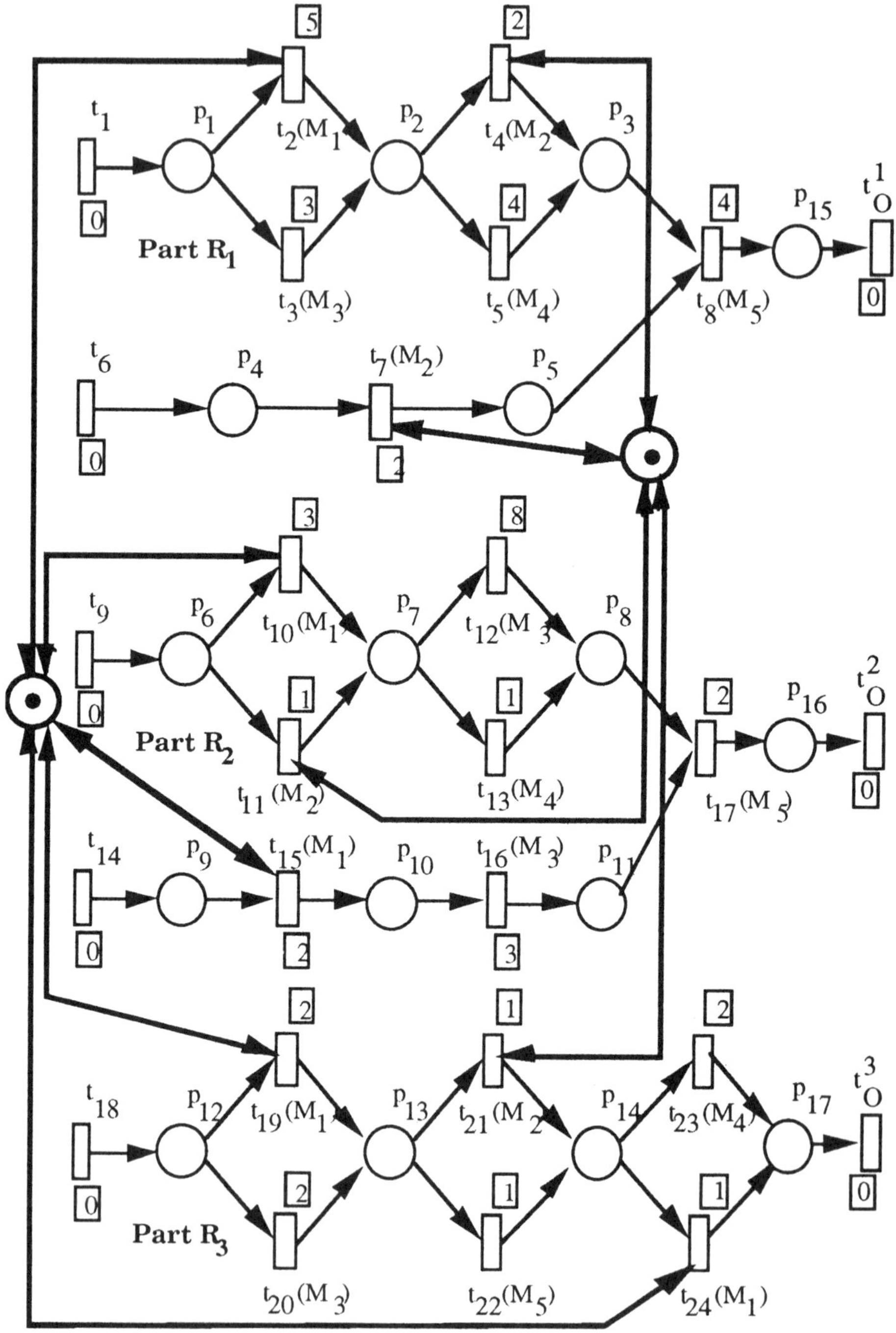

Fig. 18: Completion of the model of Figure 17 with the resource places corresponding to machine M_1 and machine M_2

The model used in the scheduling process is obtained by adding resource sharing places to the model used for short-term planning. Such a place:
(i) initially contains one token,
(ii) forms a self-loop with each transition corresponding to the same machine.

These places guarantee non-reentrance; i.e. the fact that machines can perform at most one operation at any given time. For instance, we should have five resource sharing places in the model given in Figure 17. In Figure 18, the resource place corresponding to machine M_1 and M_2 are represented. The same kind of resource places should be introduced for machines M_3, M_4 and M_5.

Thus, there are two types of decisions that have to be made, namely:

(i) The decisions related to the selection of a resource, when several resources are available to perform the same task. We define such a decision as an RU decision, where RU stands for Resource Use. An RU decision must be made in the model of Figure 19, in which the next operation on the part represented by the token can be performed by M_1 or M_2 or M_3.
In Figure 17, this situation can be observed in p_1, p_2, p_6, p_7, p_{12}, p_{13} and p_{14}.
(ii) The decisions related to the sequencing of part types on resources, called PS decisions, where PS stands for Product Sequencing. Figure 20 represents such a situation in this example: we can fire either t_1, which represents the manufacturing of a part of type R_1 on M_1, or t_2, which represents the manufacturing of a part of type R_2 on M_1.

Hereafter, places related to the RU (resp. PS) type of decisions will be called RU (resp. PS) places.

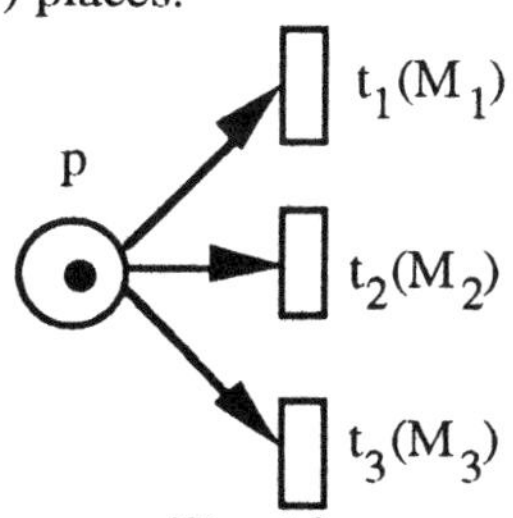

Fig. 19:
Modeling of a RU type of decision

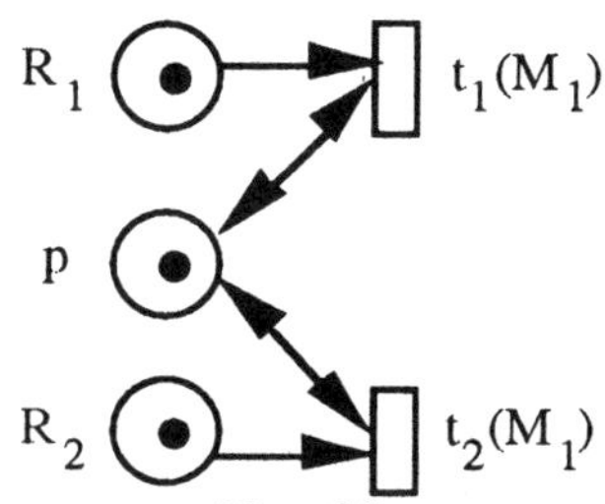

Fig. 20:
Modeling of a PS type of decision

We assume that transitions are fired as soon as they are enabled. Consequently, a schedule is defined as soon as a sequence of transitions is assigned to each RU and each PS place. The transitions belonging to such a sequence are the output transitions of the place, and they appear in the sequence as many times as specified by the short-term planning solution for the first elementary period. Let us for instance consider the RU place p in Figure 19. If, according to the short-term planning solution, t_1, t_2 and t_3 have to be fired twice, once and three times, respectively, a valid sequence could be $\sigma_p = < t_1, t_3, t_3, t_2, t_1, t_3 >$.

As we can see, the PN modeling of the scheduling problem comprises the RU and PS decisions, thus illustrating the decisions to be made explicitly. Nevertheless, it should be noted that these decisions are not independent and that improper sequences may lead to blocking. Such a case is shown in Figure 21 which contains two PS places, q_1 and q_2: if t_1, t_2, t_3 and t_4 are fired once, then $\sigma_{q2} = < t_2, t_3 >$ leads to a blocking situation.

Having established the appropriate system model, the objective of the scheduling problem can be expressed as follows: determine the sequences to be assigned to each RU and PS places that result in a makespan which is less than the duration of an elementary period, knowing that:

(i) the transitions belonging to a sequence are the output transitions of the related RU or PS place,

(ii) a transition appears in a sequence as many times as specified by the short-term planning solution for the first elementary period.

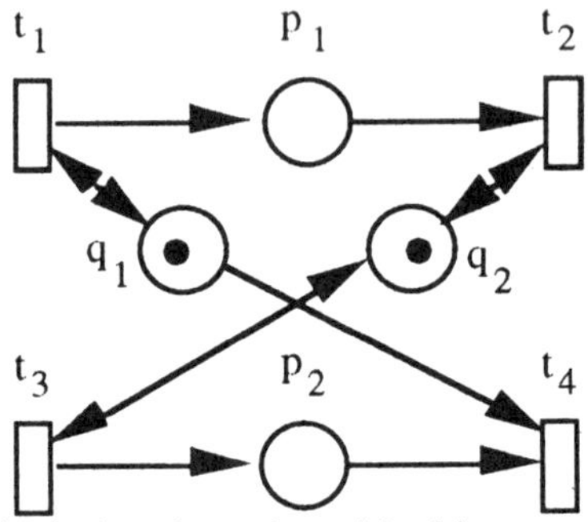

Fig. 21: A situation where blocking may occur

b. Scheduling algorithm

It is well-known that the scheduling problem of the general job-shop is NP-hard. As a consequence, only heuristic algorithms can be considered for solving large-sized problems. We have developed two heuristics: a simulated annealing approach and an approach based on the improvement of a critical circuit. For the sake of brevity, only the second algorithm is presented in this section.

The algorithm starts with a feasible set of sequences, i.e. with a set of valid sequences which do not lead to blocking. An easy way to build such a feasible set of sequences consists of assigning any valid sequence to each PS place, simulate the system according to these sequences by firing the transitions as soon as they are enabled, and assign to the RU places the sequences resulting from the simulation.

Let T be the duration of the elementary period, n_t the number of times t should be fired during the first elementary period, $S_t(k)$ the instant when the k-th firing of t starts and $F_t(k) = S_t(k) + Z(t)$. This notation refers to the initial feasible set of sequences, assuming that the starting time of the first transition firing is 0 and that a transition fires as soon as it is enabled. A critical path is a sequence of pairs:

$$\mathcal{C} = <(t_{i_1}, k_{\alpha_1}), \ldots, (t_{i_r}, k_{\alpha_r})>$$

such that:

(i) $S_{t_{i1}}(k_{\alpha 1}) = 0$;

(ii) $F_{t_{i_r}}(k_{\alpha_r}) = \underset{(t,k)}{\mathrm{Max}}\{F_t(k)\}$ (this value is the makespan);

(iii) $F_{t_{ij}}(k_{\alpha j}) = S_{t_{ij+1}}(k_{\alpha j+1})$, for j = 1,...,r-1.

A necessary (but not sufficient) condition to reduce the makespan is to reduce $F_{t_{i_r}}(k_{\alpha_r})$, and thus to bring forward the finishing time of one of the transition firings that belongs to the critical path. To do this, we will have to delay some transition firings which do not belong to the critical path. Note that in doing so it is not allowed to violate the machine precedence constraints (i.e. the constraints related to the manufacturing process, taking into account the initial marking). Furthermore, in order to reduce the computation burden, we introduce the **float time**, which is the maximal time a transition can be delayed without increasing the makespan. The calculation of the float time takes into account the schedule and machine precedence dependencies of the operations. The objective is to move earlier a transition firing which belongs to the critical path only if this delays another transition firing for less than its float time. The two transition firings which are involved in the scheduling perturbation are not taken into account in the computation of the float times. The following algorithm is derived from the above remarks.

Algorithm

1. Compute a feasible set of sequences and the initiation times of the transition firings.
2. Compute the critical path and the makespan.
3. Select (t_u, k_u) and (t_c, k_c) such that:
   * $t_u \neq t_c$,
   * (t_u, k_u) belongs to the critical path,
   * the k_c-th transition t_c just preceding the k_u-th transition t_u in one of the sequences belonging to the feasible set of sequences,
   * swapping the k_u-th transition t_u and the k_c-th transition t_c in the sequence does not violate the machine precedence constraints,
   * the float time associated to (t_c, k_c) is greater than or equal to the delay resulting from the swapping.

   If several pairs (t_u, k_u) and (t_c, k_c) are candidates for swapping, select the one for which the delay is the closest to the float time.
4. If a pair has been selected, go to 2, else stop.

Note that modeling a manufacturing process by means of Petri nets leads to a comprehensive approach of the scheduling problem by clearly distinguishing the two types of decisions to be made and their relationships. We believe that this approach may boost the research for efficient scheduling algorithms.

5. A modular modeling approach

In this section, we address exclusively non-cyclic manufacturing systems. We assume that these systems can be decomposed into modules, each of which can be modeled using controlable nets. As we showed in the previous section, this type of model is common in the management of manufacturing systems.

5.1. Integration of controlable nets

Let us consider a set $\mathbf{Z} = \{Z^1, \ldots, Z^n\}$ of controlable nets. Let $Z^i = (P^i \cup R^i, T^i, I_i, O^i, m_0^i)$. Let T_I^i (resp. T_O^i) be the source (resp. sink) transitions of Z^i. The Petri nets Z^i are integrated by means of a set Q of **interface places** and a set Γ of **inter-module** arcs. The resulting Petri net Z is defined as follows:

$$Z = \left[\bigcup_{i=1}^{n} (P_i \cup R_i) \cup Q, \bigcup_{i=1}^{n} T_i, I, O, \left(\bigcirc_{i=1}^{n} m_0^i \right) \circ m_0^Q \right]$$

where:

∘ represents the concatenation,

m_0^Q is the initial marking of the places of Q,

I (resp. O) is the union of the I^i (resp. O^i) matrices completed by the rows corresponding to the places of Q and the arcs of Γ.

Every $t \in T_O^i$ has at most one output place belonging to Q (i.e. $|t°| = 1$). We also assume that $q° \neq \emptyset$ and $°q \neq \emptyset$ for any $q \in Q$. Figure 22 shows an integrated system. The module models are represented by rectangles for simplicity.

Result 11 shows that the qualitative properties of the modules are preserved in the integrated net.

Result 11 : A Petri net Z obtained by integrating the controlable Petri nets Z^1, ..., Z^n as shown above remains a controlable net. Thus, Result 9 applies to Z.

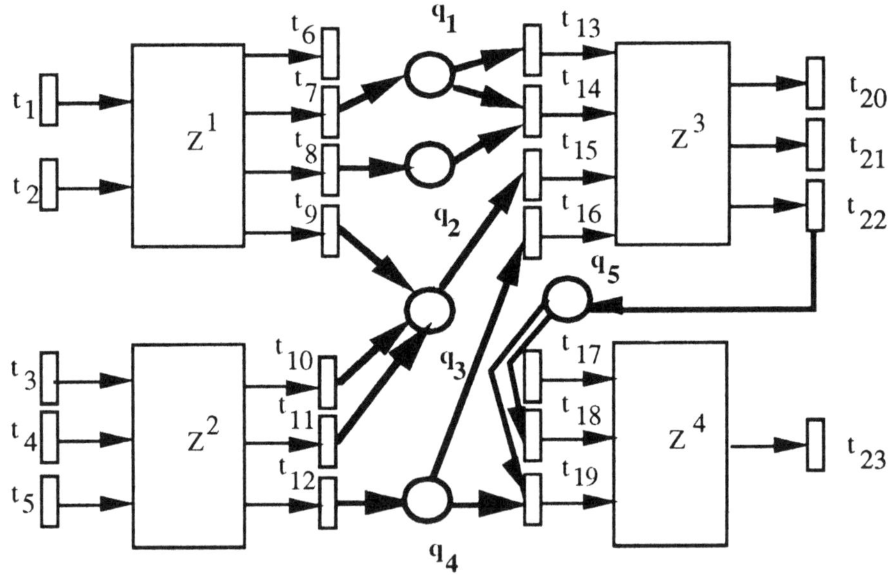

Fig. 22: Integrated Petri net

5.2. Simplification of a controlable net

The goal of the simplification is to replace each module model Z^i by a Petri net Z^{i*} such that:

(i) Z^{i*} is a controlable Petri net with no resource places,

(ii) the source and sink transitions of Z^{i*} are the same as those of Z^i,

(iii) for every t-invariant y^i of Z^i there exists a t-invariant y^{i*} of Z^{i*} such that the components of y^i and y^{i*} corresponding to the source and sink transitions are the same; the reverse also holds.

Such Petri nets Z^{i*} exist since the Z^i nets are controlable. It can be obtained by computing the minimal t-invariants of Z. Alternative iterative processes have already been developed to obtain the simplified net.

Example

We consider the controlable Petri net Z given in Figure 9. The simplified Petri net Z* is presented in Figure 23.

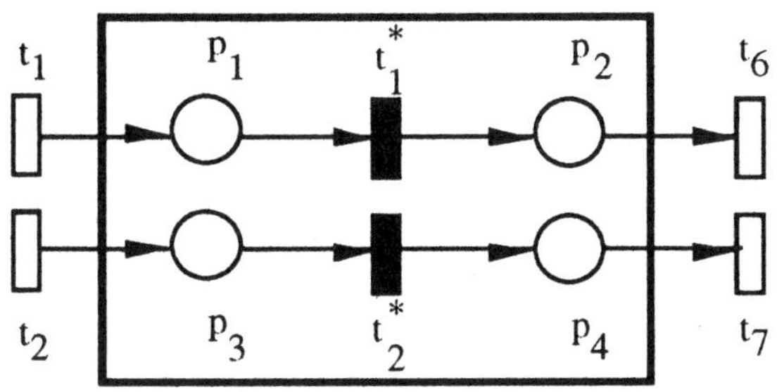

Fig. 23: The simplified net of the Petri net represented in Figure 9

The firing times to be associated to the transitions of the simplified net N* which are neither sink nor source places depend on the firing times of the transitions of Z **and** on the mean firing ratios of the output transitions of Z. The latter holds since we know that it is always possible to fully utilise the bottleneck machine in a cyclic production system. Note that we still remain in the non-cyclic case. Nevertheless, we use the cyclic environment to compute the firing times since this computation is based on the mean values of the parameters provided by the history of the system. We thus start with the mean firing ratios to compute the desired firing times of the transitions of Z* assuming that the manufacturing system operates on a cyclic basis. The next step is then to penalize these times in order to make sure that the system will also be able to absorb the demand even if the mean ratios deviate from the initial one.

Consider the model given in Figure 9, and assume that the number in parentheses represent the firing times of the transitions —i.e. the duration of the operations:

$t_1(3)$ and $t_2(4)$ represent operations performed on machine M_1,

$t_3(5)$ corresponds to machine M_2,

$t_4(4)$ and $t_5(4)$ represent operations performed on machine M_3,

$t_6(5)$ and $t_7(2)$ are related to machine M_4.

Assume that the mean firing ratios are 30% for t_6 and 70% for t_7. We assume that a proportion q of tokens arriving in place p_1 is used to fire t_3, and thus that a proportion (1-q) of tokens arriving in place p_1 is used to fire t_4.

In this case, the average time required to fire one output transition is:

$$\theta = \underset{q\in[0,1]}{\mathbf{Min}} \{\mathbf{Max}[3\times 0.3 + 4\times 0.7,\ 5q\times 0.3,\ 4(1-q)\times 0.3 + 4\times 0.7$$
$$5\times 0.3 + 2\times 0.7]\}$$

$$\theta = \underset{q\in[0,1]}{\mathcal{Min}} \left\{ \mathcal{Max}[3.7, 1.5q, 4-1.2q, 2.9] \right\} = 3.7$$

As a consequence, both t_1^* and t_2^* should be timed at 3.7 α, where α is a factor greater than 1 used to penalize the firing times in order to ensure that the performance of the simplified net can be reached by the initial net.
Another way to define a feasible firing time for the transitions of the simplified net is to consider that the mean firing ratios of t_6 and t_7 may vary between given limits. Assume for instance that the mean firing ratio of t_6 is $\beta \in [0.2, 0.4]$. In this case, the average time θ required to fire one output transition is determined from:

$$\theta_\beta = \underset{q\in[0,1]}{\mathcal{Min}} \left\{ \mathcal{Max}[3\beta + 4(1-\beta), 5\beta q, 4\beta(1-q) + 4(1-\beta), 5\beta + 2(1-\beta)] \right\}$$

$$\theta_\beta = \underset{q\in[0,1]}{\mathcal{Min}} \left\{ \mathcal{Max}[4-\beta, 5\beta q, 4-4\beta q, 2+3\beta] \right\}$$

Finally, θ is given by:

$$\theta = \underset{\beta\in[0.2,0.4]}{\mathcal{Max}} \left\{ \underset{q\in[0,1]}{\mathcal{Min}} \left[\mathcal{Max}(4-\beta, 5\beta q, 4-4\beta q, 2+3\beta) \right] \right\} = 3.8$$

Thus, if the average firing time required to fire one output transition is 3.8, this guarantees that the performance of the simplified net can be reached by the initial net, assuming that the ratio β remains in the interval [0.2, 0.4]. It is possible to generalize this approach to any model. This results in computing θ such that:

$$\theta = \underset{B\in D}{\mathcal{Max}} \left\{ \underset{Q\in[0,1]^R}{\mathcal{Min}} \left[\mathcal{Max}(f_1(B,Q), f_2(B,Q), \ldots, f_m(B,Q)) \right] \right\}$$

where:

B is the vector, the components of which are the ratios of the firings of the output transitions;
D is the domain defined as follows:
- each component of B belongs to [0, 1],
- the sum of the components of B is equal to 1;
Q is the vector the components of which are the proportions of tokens used to fire the various transitions;
R is the number of components of Q;
$f_i(B,Q)$ is a bilinear function of the components of B and Q;
m is the number of machines;
θ can be computed using iterative procedures.

5.3. A two-level DMS

The integrated Petri net in which each model Z^i is replaced by its simplified counterpart Z^{i*} is a controlable net with no resource places ($R = \emptyset$). Thus, it can be used to compute a plan on a horizon equal to a subperiod. This provides the number of firings of the output (i.e. sink) transitions of each module of the model allowing the computation of the short term plan for each module. The latter, in turn, provides the basis for the scheduling problem.

6. Concluding remarks

The properties of Petri nets may be exploited in manufacturing systems to cope with the complexity of the planning and scheduling problems.

In cyclic manufacturing systems, the properties of event graphs may be used to propose fast and powerful algorithms which maximize the productivity while minimizing WIP, when a cyclic control is known. An algorithm which proceeds by construction is also available to provide a near-optimal control (i.e. schedule). This algorithm is closely linked with the Petri net model.

Non-cyclic manufacturing systems are modeled using decomposable nets at the short-term planning level. The computation of a short-term plan is based on a set of t-invariants, the choice of which provides for adjustment of the computational burden. A system model applicable to scheduling is determined by enhancing the planning model to obtain a controlable net with resource places. This model was employed in this work to develop a new scheduling heuristic.

Finally, an effective and natural modular approach was introduced to develop models that facilitate the formulation and solution of management problems in manufacturing systems.

References

[1] BACCELLI F., 1992, Ergodic theory of stochastic Petri networks, *Annals of Probability*, Vol. 20, No. 1, pp. 350-374.

[2] CHRETIENNE P., 1983, *Les Réseaux de Petri Temporisés*, Univ. Paris VI, Paris, France, Thèse d'Etat.

[3] COMMONER F., HOLT A., EVEN S. and PNUELI A., 1971, Marked directed graphs, *J. of Comp. and Syst. Sci.*, Vol. 5, No. 5, pp. 511-523.

[4] DI CESARE F. et al., 1993, *Practice of Petri Nets in Manufacturing*, ISBN 0 412 41230 6, CHAPMAN and HALL.

[5] DI MASCOLO M., FREIN Y., DALLERY Y. and DAVID R., September 1989, A Unified Modeling of Kanban Systems Using Petri Nets, *Technical Report No. 89-06*, LAG, Grenoble, France.

[6] GENRICH H.J. and LAUTENBACH K., 1981, System modeling with high-level Petri nets, *Theoret. Comput. Sci.*, Vol. 13, pp. 109-136.

[7] HARHALAKIS G., LEVENTOPOULOS M., LIN C.P., NAGI R. and PROTH J.M., October 1992, A class of conflict free Petri nets used for controlling manufacturing systems, *Technical Research Report No. TR 92-90*, Institute for Systems Research, The University of Maryland at College Park.

[8] HILLION H. and PROTH J.M., January 1989, Performance Evaluation of Job-Shop Systems Using Timed Event-Graphs, *IEEE Trans. Automat. Contr.*, Vol. 34, No. 1, pp. 3-9.

[9] JENSEN K., Coloured Petri nets and the invariant method, *Theoret. Comput. Sci.*, Vol. 14, pp. 317-336, 1981.

[10] KIRKPATRICK S., GELATT C.D. and VECCHI M.P., May 1983, Optimization by simulated annealing, *Science*, Vol. 220, 13.

[11] LAFTIT S., PROTH J.M. and XIE X.L., 1992, Optimization of invariant criteria for event graphs, *IEEE Trans. on Aut. Control*, Vol. 37, No. 5, pp. 547-555.

[12] LUNDY M. and MEES A., 1986, Convergence of an annealing algorithm, *Mathematical Programming*, Vol. 34, pp. 111-124.

[13] MURATA T., April 1989, Petri Nets: Properties, Analysis and Applications, *Proceedings of the IEEE*, Vol. 77, No. 4, pp. 541-580.
[14] PROTH J.M. and XIE X.L., July 1994, Performance evaluation and optimization of stochastic timed event graphs, to appear in *IEEE Trans. on Aut. Control.*
[15] RAMCHANDANI C., 1974, Analysis of Asynchronous Concurrent Systems by Timed Petri Nets, *Technical Report 120,* Lab. Comput. Sci., Mass. Inst. Technol. Cambridge, MA.
[16] RAMAMOORTHY C.V. and HO G.S., 1980, Performance Evaluation of Asynchronous Concurrent Systems using Petri Nets, *IEEE Trans. Software Eng.*, Vol. SE-6, No. 5, pp. 440-449.
[17] REISIG W., 1983, Petri nets with individual tokens, *Informatik-Fachberichte*, Vol. 66, No. 21, pp. 229-249.
[18] J. SIFAKIS, 1982, A Unified Approach for Studying the properties of Transition Systems, *Theoret. Comput. Sci.*, Vol. 18, pp. 227-258.
[19] ZHOU M. and DI CESARE F., 1993, *Petri Net Synthesis for Discrete Event Control of Manufacturing Systems*, Kluwer Academic Publisher, Boston, MA.

6

Petri Net-Based Heuristic Scheduling for Flexible Manufacturing

Doo Yong Lee
Precision Engineering and Mechatronics
Korea Advanced Institute of Science and Technology
Taejon, 305-701, Korea

Frank DiCesare
Electrical, Computer, and Systems Engineering
Rensselaer Polytechnic Institute
Troy, New York 12180-3590, USA

1. INTRODUCTION

A flexible manufacturing system (FMS) can manufacture multiple types of products with relatively small lot sizes, using various resources such as robots and multi-purpose machines. While the increased flexibility of an FMS provides a greater number of choices of routings and resources, and allows greater productivity, it imposes a challenging problem of allocating the given resources to various processes for each product, and scheduling the activities effectively. This chapter presents a Petri net-based heuristic scheduling method for flexible manufacturing environments.

Scheduling problems are complex even for simple formulations [4-5] [10] [20] [38] and are NP-hard in many cases [12] [25]. Various scheduling methods have been developed to tackle the ever increasing complexity and flexibility of manufacturing [13] [39] [46-48] [55].

Traditionally, the routing of a part to complete the sequence of required processes is considered planning, and the allocation of resources according to the determined routing is considered scheduling. Planning and scheduling are often separately carried out with little interaction. Since a machine in the FMS can be used for multiple jobs and there are choices of resources for a process, the allocation of resources, or scheduling, should be considered together with the routing of parts to take full advantage of the flexibility of the system.

Because of the complexity of generative (normative) scheduling methods, evaluative methods such as simulation combined with heuristic dispatch rules have been often employed. However, a comprehensive simulation model that effectively handles a flexible and automated manufacturing environment is expensive and difficult to develop. Simulation models are also often specific to particular applications so that the models and the simulation results are difficult to generalize.

A complete and general scheduling method should formulate a scheduling problem handling constraints explicitly and concisely. It should also provide an efficient way to solve the formulated problem. Current scheduling approaches in the literature for flexible manufacturing environments have difficulties with the formulation and/or the solution methods. Approaches such as linear (integer) programming [15-17] [37] [43], search [11] [21] [35-36] [54] [58], and network models (CPM/PERT, queuing networks, graphs) [1] [6] [9] [22-23] [49] can provide effective solution methods but have formulation difficulties in handling shared resources, routing flexibility, concurrency, lot sizes, etc. Approaches such as algebraic models [2] [48], and control theoretic approaches [13] [19] [47] have difficulties in providing efficient solution methods.

Petri nets can concisely model the dynamics of flexible manufacturing, explicitly handling activities, resources and constraints of systems with a single coherent formulation. Since Petri nets are a graphical tool, one can better understand and formulate problems. Larson and Odoni [24] discuss the merits of graphical models. Petri nets are also widely used for the performance analysis of given schedules [3] [7] [14] [18] [45] [52-53] [56].

Although Petri nets showed promise as an effective tool to formulate and solve scheduling problems, the actual generation of schedules has not been much reported in the literature. There have been some independent efforts to apply Petri nets to generative (normative) scheduling. Shih and Sekiguchi [51] present a scheduling system for an FMS, which simulates the evolution of the FMS modeled by Petri nets. The scheduling system calls for a beam search routine whenever there is a conflict. The beam search routine then constructs partial schedules within the beam-depth and evaluates them to choose the best one. The cycle is repeated until a complete schedule is achieved. This method based on partial schedules does not guarantee global optimization. Onaga et al. [42] present a linear programming approach for periodic scheduling of systems modeled by Petri nets. Shen et al. [50] present a scheme which starts with an arbitrary schedule and applies branch and bound search to find an optimal schedule. Zhang [57] presents a method which translates rules of a rule-based planning system into a timed Predicate/Transition net model and applies the A^* algorithm to find an optimal schedule.

The scheduling method presented in this chapter formulates a scheduling problem using a Petri net model, and employs a global search over the reachability graph of the Petri net model. The search uses heuristic functions to limit the search space [26-34]. The method attempts to generate an optimal or near optimal feasible schedule in terms of a firing sequence of the transitions of the Petri net model. This method is event-driven as opposed to time-driven, i.e., the schedule is provided as an order of the starts of the activities rather than a time table. Most of the current scheduling approaches can be considered as time-driven, i.e., the schedule is a list of time instants when certain activities should happen [41]. The time-driven approach may not always be best for the scheduling of flexible manufacturing systems that are, by nature, discrete event-driven [46]. Event-driven scheduling focuses on the precedence constraints of the activities and is robust to disturbances.

There are many optimization objectives in manufacturing. For example, the minimization of makespan and/or tardiness is one of the frequently adapted objectives. The maximum utilization of critical machines is also often considered. Generating a schedule with the minimum or near minimum makespan is the focus in this chapter.

In timed Petri nets, time can be associated with either places (timed place Petri nets), transitions (timed transition Petri nets), or both. Generally, a timed transition removes tokens from its input places and takes some time before it introduces tokens to its output places. Therefore, between the initiation and the termination of firing, the marking (the state of the system) is uncertain. Depending on whether timed transitions or timed places are used, activities are associated with transitions or places, respectively. In the case of timed transitions, multiple initiation or firing of transitions must be allowed to represent concurrency of activities. Therefore, the time associated with each initiated transition must be tracked in order to correctly update the marking, or state. Since the initiated transitions may not be tracked in applications of Petri net modeling, an additional tracking method is required.

But when timed places are used, multiple marked places naturally represent the concurrency of activities associated with the places and transitions can be fired one at a time. It is necessary to keep track of time for only the marked places. And at any given instant, there is no ambiguity in the marking. In this chapter, time is associated only with places, and all transitions are immediate. Transitions represent the initiation or the termination of the involved activities. Places, therefore, represent activities or resources [19] [51]. Each token in the places represents either the availability of a resource, the readiness of a part for the next process, or a part being processed.

2. PETRI NET MODELING FOR SCHEDULING

A part or partially finished product which enters and moves inside a manufacturing system is called a *part.* A finished part that leaves the manufacturing system is called a *product.* Each product is the result of a sequence of *processes* according to its technological requirements. Resources, such as machines and robots, requirements are not considered in the processes. A sequence of

Table 1: JOB REQUIREMENTS OF EXAMPLE 1.

Process	J_1	J_2	J_3	J_4
1	M_1R/M_2R	M_1	M_3	M_2R
2	M_3	M_2R/M_3	M_1/M_2	M_3
3	M_1R/M_2	M_3	NA	M_1R/M_2

processes defines a *product type* or a *job type.* The manufacturing system can manufacture multiple products of the same product type, i.e., a *lot* of the same job type. It can also concurrently manufacture products of multiple types.

A process can be carried out more than one way. For instance, consider the following scenario of two machines M_1 and M_2, a robot R, and a part. The part can be treated at either M_1 or M_2 with R, accomplishing the same result. Treating the part at M_1 or at M_2 is an *operation.* Hence resource requirements are considered in an operation. An operation $O_{i,j,k}$ represents the j-th process of the i-th job type being performed at the k-th machine. All the operations are assumed non-preemptive. A job, therefore, can be completed by carrying out different sequences of operations although the generic sequence of processes remains the same. These different sequences of operations are vital to the flexibility of the FMS, i.e., the capability to manufacture products of multiple types, and share resources. The requirements of a job can be given by the description of alternative operations and the necessary resources as shown in the following example.

Example 1: An FMS has three machines M_1, M_2, M_3 and a robot R. Four job types J_1, J_2, J_3, J_4 are to be carried out and Table 1 shows the requirements for each job.

The first process of J_1 can be carried out at either M_1 with R or M_2 with R. The second process of J_1 can be carried out only at M_3 but does not require the use of R. This table also gives the technological precedence constraints among the processes. For instance, the second process of J_1 can be carried out only after the first process of J_1 is complete. No technological precedence constraints are assumed among different job types. Notice that J_3 has only two processes to be completed. Figure 1 shows the Petri net sub-model for J_1 with an initial marking. The complete Petri net model contains interconnected sub-models for J_1, J_2, J_3 and J_4.

The transitions are immediate. $t^b_{i,j,k}$ and $t^e_{i,j,k}$ represent the start and the finish of an operation $O_{i,j,k}$, respectively. $p^i_{j,k}$ is the k-th *initial place* of J_j representing the beginning of the job type J_j, i.e., a part for J_j is ready. The number of tokens in an initial place represents the lot size of the corresponding job. $p^f_{j,k}$ is the k-th *final place* of J_j representing the end of J_j, i.e., a product J_j is completed. $p_{i,j,k}$ is an *operation place* representing $O_{i,j,k}$. A token in an operation place represents a specific operation being performed.

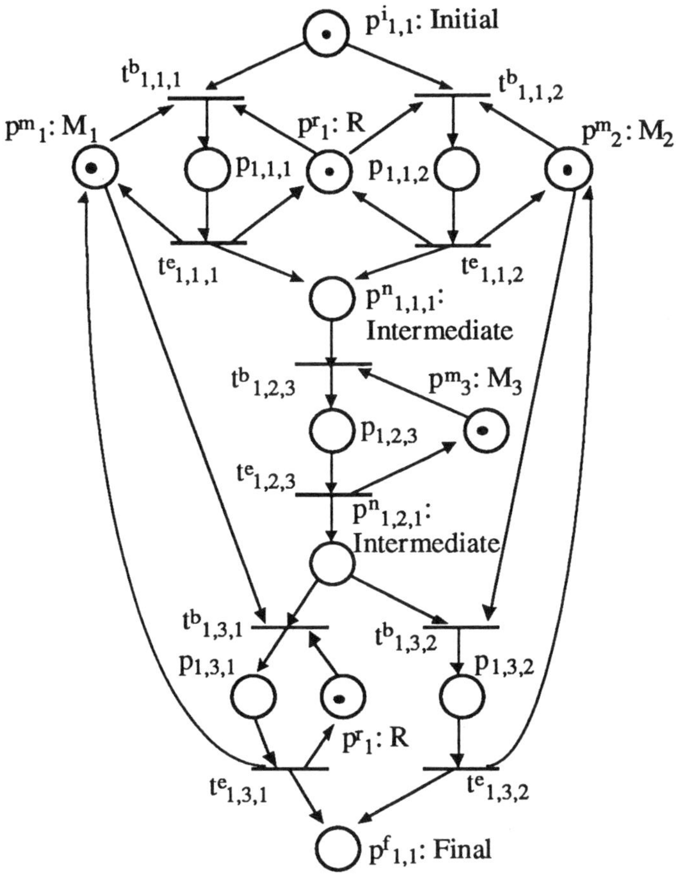

Figure 1: The Petri net sub-model for J_1.

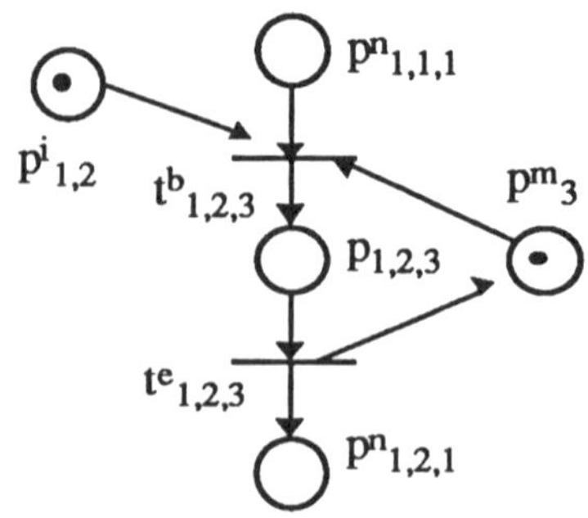

Figure 2: The modified part of the sub-model with an additional initial place.

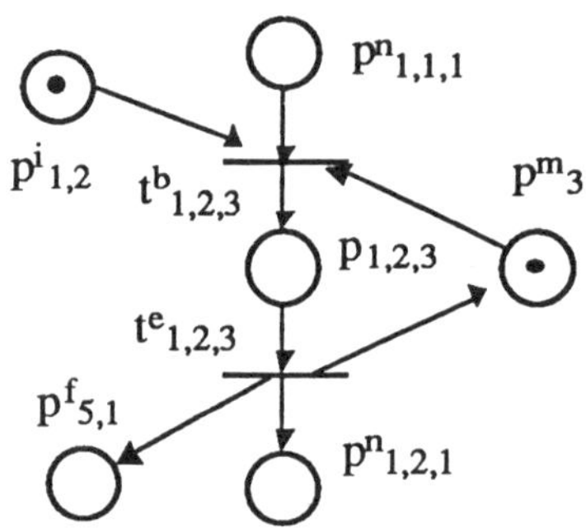

Figure 3: The modified part of the sub-model with an additional final place.

$p^n_{i,j,k}$ is the k-th *intermediate place* of J_i after the j-th process. A token in an intermediate place represents the readiness of the part for the next process, i.e., an intermediate place represents a buffer. p^r_i and p^m_i are *resource places* for the i-th robot and the i-th machine, respectively. A token in a resource place represents the availability of the corresponding resource. In Figure 1, p^r_1 is drawn twice just for clear presentation and is actually a single place.

A job model may have more than one initial and/or final places. For example, assume that the operation represented by $p_{1,2,3}$ requires an additional partially finished part since the operation is an assembly. Then the corresponding Petri net sub-model for J_1 can be modified as shown in Figure 2. $p^i_{1,2}$ is an additional initial place.

In addition, if a new job type J_5 shares the same generic sequence of processes with J_1 up to the operation represented by $p_{1,2,3}$ and does not require further processes, the corresponding part of the Petri net sub-model can be modified as shown in Figure 3. $p^f_{5,1}$ is an additional final place for J_5. This situation can be found in chemical processes where a process generates a byproduct.

A token in an intermediate place or a resource place is readily available. A

Table 2: THE OPERATION TIMES FOR J_1 OF EXAMPLE 1.

Operation	Operation Time
$O_{1,1,1}$	10
$O_{1,1,2}$	12
$O_{1,2,3}$	7
$O_{1,3,1}$	5
$O_{1,3,2}$	8

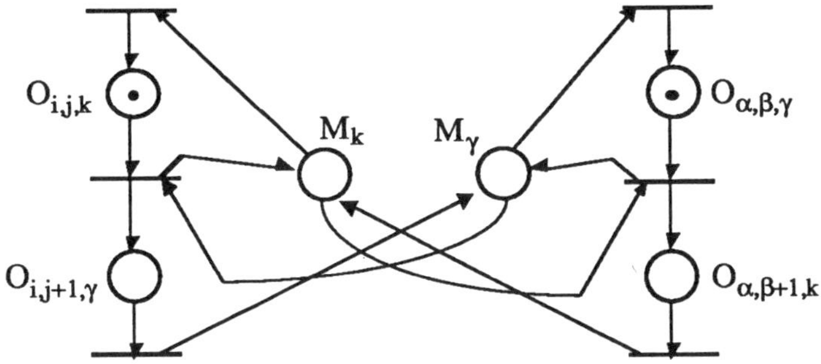

Figure 4: One type of potential deadlock.

token at an operation place becomes available after a delay corresponding to the duration time of the operation. The duration times of operations are given as shown in Table 2.

One type of potential deadlock is implicitly avoided by using the intermediate places. The potential deadlock happens when the operation $O_{i,j,k}$ is completed at the machine M_k and $O_{i,j+1,\gamma}$ requires the machine M_γ, and the operation $O_{\alpha,\beta,\gamma}$ is completed at the machine M_γ and $O_{\alpha,\beta+1,k}$ requires the machine M_k. This is illustrated in Figure 4.

By simply swapping the resources, i.e., M_k and M_γ, the operations $O_{i,j+1,\gamma}$ and $O_{\alpha,\beta+1,k}$ can be initiated. But the model of Figure 4 fails to achieve this. Putting intermediate places can solve this problem as shown in Figure 5. The model of Figure 5 corresponds to having a buffer at the machine. By unloading the part from the machine to the buffer after the operation, the machine becomes available and this particular type of deadlock does not occur. At each machine, the loading (unloading) of the part onto (from) the machine before (after) the operation is regarded as part of the operation.

This systems design and modeling eliminates one particular type of potential deadlock. The scheduling algorithm presented later may still encounter a deadlock during the search. The algorithm keeps on looking for alternative ways to get around the deadlock and chooses those paths. If there is absolutely no way

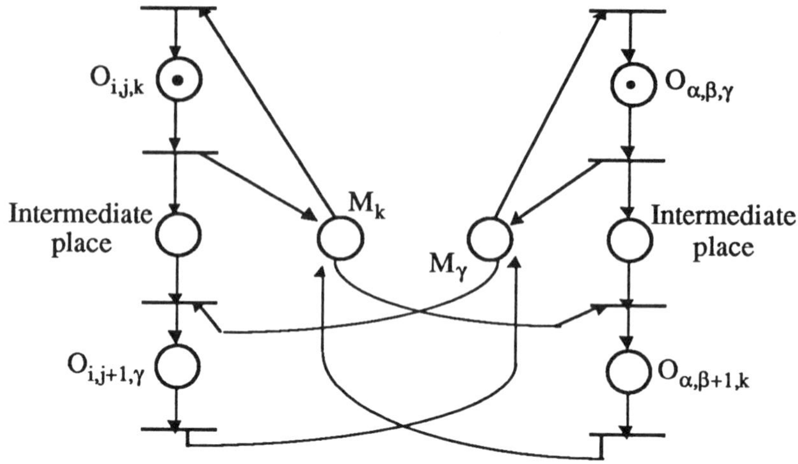

Figure 5: The solution to the potential deadlock of Figure 4.

to get around the deadlock, then, the desired final marking is unreachable from the initial marking. That is, the system can not transit from the given initial state to the given final state. In this case, the algorithm halts and reports it.

In Figure 5, the intermediate places, i.e., buffers, have infinite sizes. When the buffer size is finite, the model can be modified as shown in Figure 6. The buffer size is represented by the number of tokens, e.g., the buffer size of the model shown in Figure 6 is three.

The Petri net models for J_2, J_3 and J_4 of Example 1 can be similarly constructed using the method presented above. Decomposing the entire model by jobs and building the Petri net by sub-nets for jobs facilitate the modeling. Since J_1, J_2, J_3 and J_4 share common resources, i.e., resource places, the sub-nets for J_1, J_2, J_3 and J_4 are connected with each other through the places.

Modeling the job requirements of an FMS using the method presented above gives a Petri net model which incorporates routing flexibility, shared resources, process times, and lot sizes. Since Petri net modeling inherently incorporates concurrency and precedence constraints among the activities, it facilitates maintaining the consistency of the problem formulation.

3. SCHEDULING ALGORITHM

Once the Petri net model of the system is constructed, the evolution of the system can be described by changes in the marking of the Petri net. In other words, all possible behaviors of the system can be completely tracked by the reachability graph of the Petri net. Theoretically, therefore, an optimal schedule can be obtained by generating the reachability graph and finding the

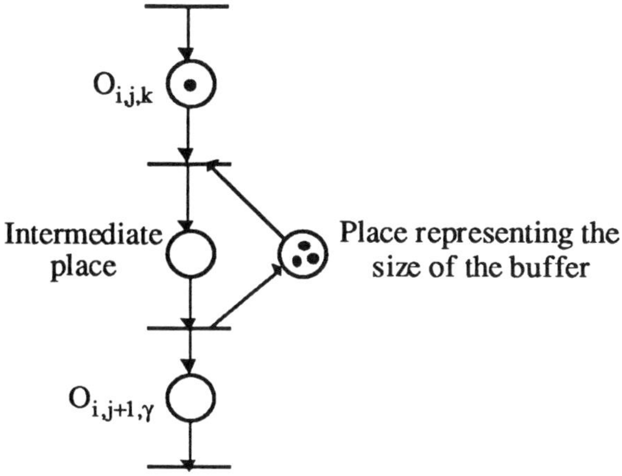

Figure 6: A model for a buffer with a finite size.

optimal path from the initial marking to the final marking. The path is a firing sequence of the transitions of the Petri net model.

But even for a simple Petri net, the reachability graph may be too large to generate in its entirety. Instead of generating the entire reachability graph, a heuristic search algorithm is employed. The algorithm generates only the necessary portion of the reachability graph, and searches for an optimal or near optimal path depending on the heuristic function used. In this way, a good and always feasible, if one exists, schedule can be generated in reasonable amount of time.

The algorithm L1 that seeks an optimal or near optimal schedule is as follows. Given a Petri net model, L1 expands the reachability graph from the initial marking until the generated portion of the reachability graph touches the final marking. Once the final marking is reached, the path is constructed by tracing the pointers that denote the parenthood of the markings, from the final marking to the initial marking. Then the transition sequence of the path provides the order of the starts of the activities, i.e., the schedule.

Algorithm L1:

Step 1: Put the initial marking m_0 on the list OPEN.

Step 2: If OPEN is empty, terminate with failure.

Step 3: Remove the first marking m from OPEN and put m on the list CLOSED.

Step 4: If m is the final marking, construct the path from the initial marking to the final marking and terminate.

Step 5: Find the enabled transitions of the marking m.

Step 6: Generate the next marking, or successor, for each enabled transition, and set pointers from the next markings to m. Compute g(m′) for every successor m′.

Step 7: For every successor m′ of m, do the following.

a: If m′ is already on OPEN, direct its pointer along the path yielding the smallest g(m′).

b: If m′ is already on CLOSED, direct its pointer along the path yielding the smallest g(m′). If m′ requires pointer redirection, move m′ to OPEN.

c: If m′ is not on either OPEN or CLOSED, compute h(m′) and f(m′), and put m′ on OPEN.

Step 8: Reorder OPEN in the increasing magnitude of f.

Step 9: Go to Step 2.

L1 uses a function of the marking m, f(m) = g(m) + h(m). f(m) is an estimate of the cost, i.e., the makespan from the initial marking to the final marking along an optimal path which goes through the marking m. g(m) is the current lowest cost obtained from the initial marking to the current marking m. h(m) is an estimate of the cost from the marking m to the final marking along an optimal path which goes through the marking m.

The scheduling algorithm is admissible [40], i.e., it always finds an optimal path if h(m) satisfies the following condition,

$$h(m) \leq h^*(m) \text{ for all } m,$$

where $h^*(m)$ is the actual cost of the optimal path from m to the final marking. The heuristic functions experimented with in this chapter do not guarantee this lower bound condition. They are designed to get a good solution in reasonable amount of time. Future research includes the development of heuristic functions that guarantee this lower bound condition.

Each arc of the reachability graph corresponds to firing of a transition and has a corresponding cost or time delay. Before a transition can be fired, the tokens of the input places of that transition must be delayed for at least the time delays associated with the input places. Hence, the cost of an arc of the reachability graph, c(m, m′) can be considered to be the maximum time delay of the input places, that is,

$$c(m, m') = \max(d_{1j}, d_{2j}, ..., d_{kj})$$

where d_{ij}, i = 1, 2, ..., k is the time delay of the input place p_{ij} of the transition t_j, and the marking m converts to the marking m′ through the transition t_j. But this argument is not quite correct as explained in the following.

Let d_{max} be max(d_{1j}, d_{2j}, ..., d_{kj}). When the marking m converts to m′, the tokens of the places that are not input places of t_j, are also concurrently delayed by d_{max}. This is true for every marking update. In fact, therefore, the cost of an arc of the reachability graph is not really the maximum time delay of the input places, but rather the maximum of the remaining time delays of the input places, that is,

$$c(m, m') = \max(r_{1j}, r_{2j}, ..., r_{kj})$$

where r_{ij}, i = 1, 2, ..., k is the remaining time delay of the input place p_{ij} of t_j. This observation is accounted for in the implementation of L1 so that the implementation of L1 keeps track of the remaining time delays of the marked places when it generates the next marking.

The list OPEN maintains the markings generated but not yet compared with the final marking. These markings form the frontier of the reachability graph. The reachability graph grows pushing this frontier outward from the initial marking until it touches the final marking. Hence, the program does not generate the entire reachability graph, but rather generates the reachability graph layer by layer until the algorithm finds a path to the final marking.

The list CLOSED maintains the markings generated and compared with the final marking, i.e., the markings that have been checked against the final marking and found not to be the final marking. In Step 7 of L1, if a newly generated marking m′ is not on either OPEN or CLOSED, it is put on OPEN to be explored later. If m′ is on OPEN, it means a new path to m′ from the initial marking has been found. The cost, g(m′) of the new path is compared with the cost of the old path. The path, then, is updated to yield the smallest cost. Since m′ is still on OPEN, it can be explored in the future.

If m′ is on CLOSED, it means m′ is already explored and a new path to m′ has been found. Since m′ is already explored, some descendant markings of m′ are already generated. If the paths leading to the descendant markings are followed all the way down, each path ends with a marking which is on OPEN. If the new path to m′ has a cost lower than the cost of the old path, there are two options to handle this situation. First, the paths to the descendant markings of m′ can be redirected propagating changes all the way down to the markings that are on OPEN. Second, as the algorithm L1 does, just the path to m′ can be redirected and m′ can be put on OPEN for re-exploration.

The first approach has the advantage that the descendant markings of m′ and the paths to them that have already been generated will not be discarded. The disadvantage of the first approach is the extra effort to propagate changes to all the descendant markings of m′ that may be very large in number. The second approach has the advantage of saving the effort of change propagation.

Its disadvantage is that it discards previously generated descendant markings of m′ and repeats the effort of generating again some of the descendant markings of m′. But some of the descendant markings of m′ may not need to be re-generated at all. The advantages and disadvantages of the above two approaches applied to A* is discussed in more detail in [40] [44].

If no enabled transition is found in Step 5, the marking m represents a deadlock. Then, no next marking for the marking m would be generated in Step 6. Subsequently, Step 7 and Step 8 would not result in any effects. The algorithm then returns to Step 2 to explore another marking on the list OPEN, i.e., to explore another alternative path that could eventually lead to the final marking. In this way, a path that avoids the potential deadlocks can be found if one exists. If the markings are exhausted in Step 2, there is no path connecting the given initial and final markings.

Since the algorithm L1 finds a path between any given pair of the initial and the final markings, a schedule from any given state to any desired state can be generated by appropriately setting the initial and the final markings. Therefore, partial scheduling can be handled without any modification of the model.

In this chapter, the cost function g(m) is the accumulation of time because the focus of this chapter is the makespan. But other functions can be used for g(m). For example, if the operating cost of machines is used for g(m), a schedule with the optimal total operating cost can be sought. Therefore, other performance criteria or multiple performance criteria can be adapted by using appropriate functions or combinations of functions for g(m) and h(m) in the scheduling algorithm.

4. THE IMPLEMENTATION OF L1 AND SCHEDULING EXAMPLES

The algorithm L1 is implemented in C on a DECstation 5000/200. Reordering OPEN in Step 8 is actually combined with putting a marking on OPEN. Hence, OPEN is always in the increasing order of f. By picking up the first marking from OPEN in Step 3, L1 always selects the marking that has the smallest estimate of the cost of an optimal path.

In order to find enabled transitions in Step 5, the implementation of L1 identifies marked places. For each marked place, the program identifies its output transitions. For each output transition, the program checks if all its input places are marked. If they are, the transition is enabled and used in Step 6 to generate the next marking. When generating the next marking, the program updates the marking and calculates the cost functions g and f for the next marking. The program also updates the remaining operation times of the places as discussed earlier. Scheduling examples and heuristic functions used are discussed in the following.

Example 2: The problem is to schedule an FMS which has three machines,

Table 3: JOB REQUIREMENTS OF EXAMPLE 2.

Process	J_1	J_2
1	M_1/M_2	M_1/M_3
2	M_2/M_3	$M_1/M_2/M_3$

Table 4: OPERATION TIMES OF EXAMPLE 2.

Op.	Time
$O_{1,1,1}$	3
$O_{1,1,2}$	4
$O_{1,2,2}$	3
$O_{1,2,3}$	2
$O_{2,1,1}$	4
$O_{2,1,3}$	2
$O_{2,2,1}$	3
$O_{2,2,2}$	4
$O_{2,2,3}$	4

M_1, M_2 and M_3. The system will handle two jobs, J_1 and J_2. Table 3 shows the job requirements. The operation times are shown in Table 4. The Petri net model of the system with the initial marking is shown in Figure 7.

L1 is executed using the following four h functions to solve the scheduling problem given in Example 2.

$h_1(m) = 0$ for all marking m. (1)

$h_2(m) = -\ dep(m)$, where dep(m) is the depth of the marking m in the reachability graph. (2)

$h_3(m) = \min(rt_{1m}, rt_{2m}, ..., rt_{km})$, where rt_{im}, i = 1, 2, ..., k is the remaining operation time of the place p_{im} that has a token under the marking m. (3)

$h_4(m) = h_2(m) + h_3(m) = \min(rt_{1m}, rt_{2m}, ..., rt_{km}) - dep(m)$. (4)

$h_1(m)$ makes f(m) equal to g(m) and the program utilizes no heuristic information. It selects the next marking that has the smallest actual cost to reach from the initial marking. This uninformed best-first search corresponds to the uniform-cost search and is guaranteed to find the optimal path [44].

$h_2(m)$ makes the program prefer markings that are deeper in the reachability graph, that is,

$$f(m) = g(m) + h(m) = g(m) + h_2(m) = g(m) - dep(m). \quad (5)$$

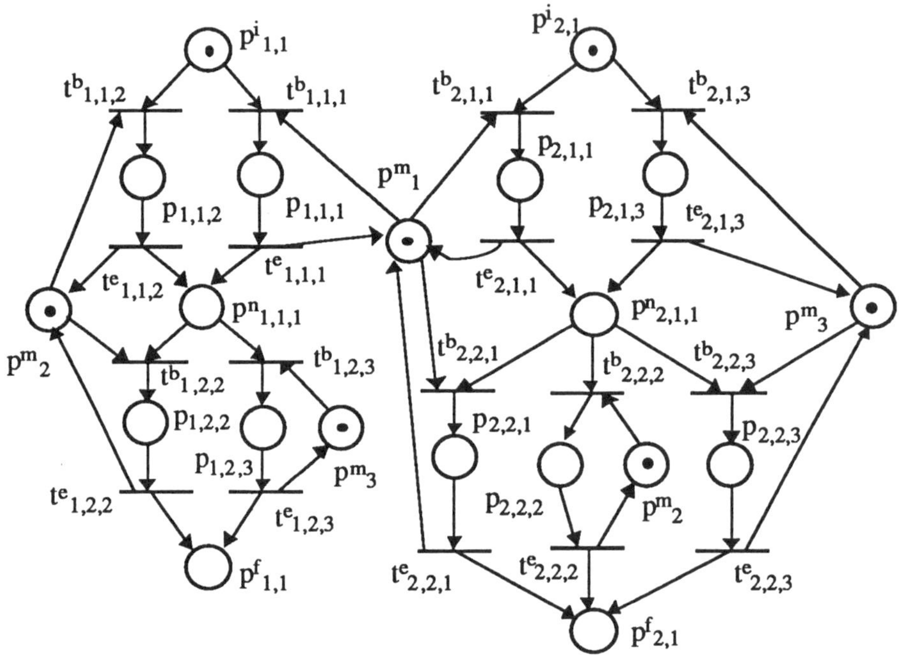

Figure 7: The Petri net model of the system of Example 2.

Hence, even if a marking m has a large cost, the cost is compensated if m is deep in the reachability graph. The compensation takes into account that a marking deeper in the reachability graph may be closer to the final marking.

$h_3(m)$ favors a marking which has an operation close to finish. Completing an operation makes resources available for other operations. $h_4(m)$ combines $h_2(m)$ and $h_3(m)$. Except for $h_1(m)$, the heuristic functions are intended to find a near optimal schedule in a shorter time. Table 5 shows the results of the executions employing the above four h functions.

The number of iterations is the number of times the program repeats its main routine. The depth of the final marking indicates the total number of transition firings to reach the final marking from the initial marking. The cost of the solution path is the makespan of the generated schedule. Since the exploration of a new marking requires an iteration, the number of iterations is always greater than or equal to the depth of the final marking. The transition firing sequence gives the schedule.

The interpretation of the schedule is straightforward from the Petri net model and is given in Tables 6-9. The operation sequence is the order in which each operation should be initiated at the indicated machine.

In Table 5, all the schedules generated have the same total cost, or makespan, of 6. In Tables 6-9, the sum of operation times of each schedule is 10 or 11 which is longer than the solution makespan. This demonstrates that each schedule fosters concurrency of operations.

Notice, in Table 5, $h_2(m)$ reduced the number of iterations quite significantly, i.e., 74% less than the case of $h_1(m)$, that is the uninformed best-first search. This means that the program generated a schedule with the same makespan in a much shorter time using function $h_2(m)$. This suggests the value of heuristic function $h_2(m)$. On the other hand, $h_3(m)$ slightly increased the number of iterations to generate a schedule with the same makespan. This shows that only favoring an operation close to finish, in fact, suggests little heuristic value in a resource-sharing interactive environment. Notice that $h_3(m)$ dominates over $h_2(m)$ as shown in the result of the case of $h_4(m)$.

Example 3: The problem is to schedule an FMS with three multipurpose machines, M_1, M_2, and M_3. There are five jobs, J_1, J_2, J_3, J_4, and J_5 which have four processes each. Each job has a lot size of 10. Table 10 shows the job requirements. Operation times are given in Table 11.

The Petri net model of the system is decomposed by each job and the sub-model for J_1 with the initial marking is shown in Figure 8. The sub-models for J_2, J_3, J_4, and J_5 can be similarly constructed using the method described earlier. The places with the same place notations are actually the same places and are drawn separately for clear presentation. The Petri net model of the system has 69 places and 82 transitions in total. The number inside a place represents the number of tokens.

The problem of Example 3 is solved using the following heuristic function.

Table 5: THE EXECUTION RESULTS.

	Number of Iterations	Depth of Final Marking	Cost of Solution Path	Transition Firing Sequence
$h_1(m)$	50	8	6	$t^b_{2,1,3}$ $t^b_{1,1,1}$ $t^e_{2,1,3}$ $t^e_{1,1,1}$ $t^b_{2,2,1}$ $t^b_{1,2,3}$ $t^e_{1,2,3}$ $t^e_{2,2,1}$
$h_2(m)$	13	8	6	$t^b_{2,1,3}$ $t^b_{1,1,1}$ $t^e_{2,1,3}$ $t^b_{2,2,2}$ $t^e_{1,1,1}$ $t^b_{1,2,3}$ $t^e_{1,2,3}$ $t^e_{2,2,2}$
$h_3(m)$	54	8	6	$t^b_{2,1,3}$ $t^b_{1,1,2}$ $t^e_{2,1,3}$ $t^b_{2,2,1}$ $t^e_{1,1,2}$ $t^b_{1,2,3}$ $t^e_{2,2,1}$ $t^e_{1,2,3}$
$h_4(m)$	53	8	6	$t^b_{2,1,3}$ $t^b_{1,1,1}$ $t^e_{2,1,3}$ $t^e_{1,1,1}$ $t^b_{2,2,1}$ $t^b_{1,2,3}$ $t^e_{1,2,3}$ $t^e_{2,2,1}$

Table 6: THE SCHEDULE FOR EXAMPLE 2 GENERATED USING $h_1(m)$.

Operation Sequence	Operation Time
$O_{2,1,3}$	2
$O_{1,1,1}$	3
$O_{2,2,1}$	3
$O_{1,2,3}$	2
Sum of Operation Times	10
Makespan of Solution Schedule	6

Table 7: THE SCHEDULE FOR EXAMPLE 2 GENERATED USING $h_2(m)$.

Operation Sequence	Operation Time
$O_{2,1,3}$	2
$O_{1,1,1}$	3
$O_{2,2,2}$	4
$O_{1,2,3}$	2
Sum of Operation Times	11
Makespan of Solution Schedule	6

Table 8: THE SCHEDULE FOR EXAMPLE 2 GENERATED USING $h_3(m)$.

Operation Sequence	Operation Time
$O_{2,1,3}$	2
$O_{1,1,2}$	4
$O_{2,2,1}$	3
$O_{1,2,3}$	2
Sum of Operation Times	11
Makespan of Solution Schedule	6

Table 9: THE SCHEDULE FOR EXAMPLE 2 GENERATED USING $h_4(m)$.

Operation Sequence	Operation Time
$O_{2,1,3}$	2
$O_{1,1,1}$	3
$O_{2,2,1}$	3
$O_{1,2,3}$	2
Sum of Operation Times	10
Makespan of Solution Schedule	6

Table 10: THE JOB REQUIREMENTS OF EXAMPLE 3.

	J_1	J_2	J_3	J_4	J_5
1	M_1/M_3	M_1/M_2	$M_1/M_2/M_3$	M_2/M_3	M_1/M_3
2	M_2	M_3	M_2/M_3	M_1/M_3	M_2/M_3
3	M_1/M_3	M_1/M_2	M_1/M_3	M_2/M_3	M_1/M_2
4	M_1/M_2	M_1/M_3	M_1/M_2	$M_1/M_2/M_3$	$M_1/M_2/M_3$

Table 11: THE OPERATION TIMES OF EXAMPLE 3.

Operation	Time	Operation	Time
$O_{1,1,1}$	7	$O_{3,4,1}$	6
$O_{1,1,3}$	4	$O_{3,4,2}$	3
$O_{1,2,2}$	3	$O_{4,1,2}$	9
$O_{1,3,1}$	3	$O_{4,1,3}$	5
$O_{1,3,3}$	6	$O_{4,2,1}$	6
$O_{1,4,1}$	2	$O_{4,2,3}$	2
$O_{1,4,2}$	4	$O_{4,3,2}$	7
$O_{2,1,1}$	8	$O_{4,3,3}$	12
$O_{2,1,2}$	12	$O_{4,4,1}$	9
$O_{2,2,3}$	4	$O_{4,4,2}$	6
$O_{2,3,1}$	7	$O_{4,4,3}$	3
$O_{2,3,2}$	14	$O_{5,1,1}$	10
$O_{2,4,1}$	8	$O_{5,1,3}$	15
$O_{2,4,3}$	4	$O_{5,2,2}$	7
$O_{3,1,1}$	10	$O_{5,2,3}$	14
$O_{3,1,2}$	15	$O_{5,3,1}$	5
$O_{3,1,3}$	8	$O_{5,3,2}$	8
$O_{3,2,2}$	2	$O_{5,4,1}$	4
$O_{3,2,3}$	6	$O_{5,4,2}$	6
$O_{3,3,1}$	2	$O_{5,4,3}$	8
$O_{3,3,3}$	4		

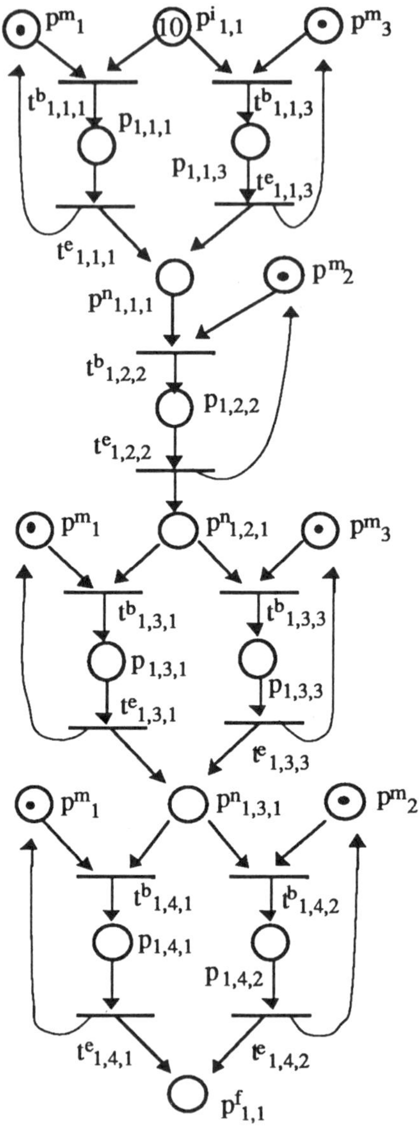

Figure 8: The Petri net sub-model for J_1 of Example 3.

$$h(m) = -w \cdot dep(m), \text{ where w is a weighting factor. } (6)$$

This heuristic function compensates the cost of the marking m by the weighted depth of the marking. A marking with a bigger depth is hypothesized to be nearer to the final marking. The magnitude of the cost of the marking is related to the magnitude of operation times, the lot sizes, and the number of jobs and operations. Therefore, an adequate value of the weighting factor should reflect the mentioned factors.

Table 12 shows the results of executions using different values for w. When the value of w is less than 2, the program does not find a path in reasonable amount of time. This is because the computer used suffers from memory thrashing. The computer used, a DECstation 5000/200, has 16 megabytes of RAM. The execution was killed after running in the background a few days. When the program does find a solution, it takes about 5 to 10 minutes. When the value of w changes from 2 to 5, the number of iterations significantly reduces but the makespan of the solution increases. When the value of w changes from 5 to 10 or greater, the number of iterations reduces a little and the makespan of the solution decreases slightly. The results suggest that the program performs desirably when the value of w is 5 or greater since the number of iterations is close to its lower bound, 400, and the increase in the cost of the solution path is relatively small.

Example 4: The problem is to schedule an FMS which has multi-purpose machines, M_1, M_2, M_3, M_4, M_5, and robots, R_1, R_2, R_3. There are 10 jobs, J_k, k = 1, ..., 10, each with varied number of processes. Tables 13, 14 and 15 show the job requirements. The operation times are shown in Table 16. The Petri net model is constructed using the method discussed earlier. The Petri net model has 154 places and 182 transitions in total.

The heuristic function of (6) is used to solve the problem. Table 17 shows the execution results when the lot size is 5 for each job. Next, an experiment with varied lot sizes is conducted. Table 18 shows the lot size of each job and the execution results are shown in Table 19. In Table 17 and Table 19, the program does not find a path in reasonable amount of time when w is less than 2 and 1.5, respectively.

From the results shown in Table 17 and Table 19, the program performs desirably when the w is greater than or equal to 5. There is little room for improvement in terms of reducing the number of iterations after w reaches 5. A strong correlation between the number of iterations and the w is suggested especially from Table 19.

The following heuristic function is designed through the analysis of various experimental results.

$$h(m) = w \cdot A_1 \cdot (E_1 - depth(m)). \quad (7)$$

depth(m) is the depth of the marking m from the initial marking in the reachability graph. A_1 is the mean of all the possible operation times. E_1 is the sum

Table 12: THE EXECUTION RESULTS OF EXAMPLE 3.

w	Number of Iterations	Depth of Final Marking	Cost of Solution Path
2	758	400	426
5	412	400	447
10	401	400	439
20	401	400	439
50	401	400	439

Table 13: THE REQUIREMENTS OF J_1 TO J_4 OF EXAMPLE 4.

	J_1	J_2	J_3	J_4
1	M_1R_2/M_3	M_2R_2/M_5R_1	$M_2R_1/M_4R_1/M_5$	M_2R_2/M_3
2	M_2R_1/M_5R_1	M_3R_2/M_4	$M_2R_3/M_3R_1/M_4R_1$	M_3R_1/M_5R_2
3	M_2R_3/M_4	M_1R_2/M_5R_3	M_1R_3	M_4R_2
4	NA	$M_1/M_3R_1/M_5$	M_2R_1	NA
5	NA	NA	M_4/M_5R_2	NA

Table 14: THE REQUIREMENTS OF J_5 TO J_8 OF EXAMPLE 4.

	J_5	J_6	J_7	J_8
1	M_2R_1/M_3R_2	M_1R_3/M_3R_2	M_2R_2/M_4R_1	$M_2R_2/M_3/M_4$
2	M_5	M_4R_1/M_3	M_3/M_5	M_4R_3/M_5R_1
3	M_1/M_3R_1	M_3R_3/M_5R_2	NA	$M_2/M_3/M_4$
4	M_2R_1/M_4R_1	$M_1/M_2/M_3$	NA	M_5
5	M_4R_1	M_2R_1/M_4	NA	M_2R_3/M_4R_1
6	NA	$M_2R_1/M_3/M_5$	NA	$M_1/M_2/M_3R_3$
7	NA	NA	NA	$M_1/M_4/M_5R_1$
8	NA	NA	NA	M_2R_1/M_3R_2

Table 15: THE REQUIREMENTS OF J_9 AND J_{10} OF EXAMPLE 4.

	J_9	J_{10}
1	$M_2R_2/M_3/M_5$	M_5
2	M_2	M_1/M_2
3	M_1R_1/M_4	M_5
4	M_4R_1/M_5	M_2R_1/M_3
5	M_1/M_3	NA

of the number of transition firings that each token in the initial places must nominally go through to reach the final places. w is a weighting factor.

h(m) of (7) is a weighted estimate of the cost from the marking m to the final marking. Since E_1 is an estimate of the depth of the final marking, E_1 - depth(m) is an estimate of the remaining number of transition firings. Assume the remaining operations are to be sequentially carried out one at a time. Then, $A_1 \cdot (E_1 - \text{depth}(m))$ is an estimate of the remaining time to reach the final marking since A_1 is the mean of the operation times. Since there is concurrency among operations, the actual remaining time should be less than $A_1 \cdot (E_1 - \text{depth}(m))$. The weighting factor w takes account of this. When full concurrency is assumed, w is the inverse of the number of independent machines. With this heuristic function, L1 is not guaranteed to find an optimal schedule.

A_1 is computed from the table of operation times. E_1 is computed from the Petri net model by counting the number of transitions on a nominal path from the initial places to the final places.

5. A COMPARISON WITH A METHOD OF HEURISTIC DISPATCH RULES

A comprehensive comparison of the scheduling method presented here with

Table 16: THE OPERATION TIMES OF EXAMPLE 4.

Op.	Time	Op.	Time	Op.	Time	Op.	Time
$O_{1,1,1}$	3	$O_{3,5,4}$	7	$O_{6,4,3}$	7	$O_{8,6,3}$	4
$O_{1,1,3}$	5	$O_{3,5,5}$	5	$O_{6,5,2}$	4	$O_{8,7,1}$	9
$O_{1,2,2}$	4	$O_{4,1,2}$	4	$O_{6,5,4}$	5	$O_{8,7,4}$	8
$O_{1,2,5}$	5	$O_{4,1,3}$	6	$O_{6,6,2}$	3	$O_{8,7,5}$	7
$O_{1,3,2}$	2	$O_{4,2,3}$	7	$O_{6,6,3}$	7	$O_{8,8,2}$	5
$O_{1,3,4}$	3	$O_{4,2,5}$	8	$O_{6,6,5}$	8	$O_{8,8,3}$	6
$O_{2,1,2}$	4	$O_{4,3,4}$	5	$O_{7,1,2}$	8	$O_{9,1,2}$	3
$O_{2,1,5}$	4	$O_{5,1,2}$	4	$O_{7,1,4}$	8	$O_{9,1,3}$	4
$O_{2,2,3}$	4	$O_{5,1,3}$	3	$O_{7,2,3}$	9	$O_{9,1,5}$	4
$O_{2,2,4}$	5	$O_{5,2,5}$	6	$O_{7,2,5}$	7	$O_{9,2,2}$	6
$O_{2,3,1}$	8	$O_{5,3,1}$	9	$O_{8,1,2}$	5	$O_{9,3,1}$	5
$O_{2,3,5}$	7	$O_{5,3,3}$	7	$O_{8,1,3}$	7	$O_{9,3,4}$	7
$O_{2,4,1}$	6	$O_{5,4,2}$	10	$O_{8,1,4}$	8	$O_{9,4,4}$	2
$O_{2,4,3}$	3	$O_{5,4,4}$	9	$O_{8,2,4}$	10	$O_{9,4,5}$	4
$O_{2,4,5}$	5	$O_{5,5,4}$	4	$O_{8,2,5}$	10	$O_{9,5,1}$	5
$O_{3,1,2}$	3	$O_{6,1,1}$	6	$O_{8,3,2}$	4	$O_{9,5,3}$	7
$O_{3,1,4}$	4	$O_{6,1,3}$	6	$O_{8,3,3}$	5	$O_{10,1,5}$	4
$O_{3,1,5}$	5	$O_{6,2,3}$	7	$O_{8,3,4}$	6	$O_{10,2,1}$	5
$O_{3,2,2}$	5	$O_{6,2,4}$	5	$O_{8,4,5}$	7	$O_{10,2,2}$	6
$O_{3,2,3}$	5	$O_{6,3,3}$	4	$O_{8,5,2}$	4	$O_{10,3,5}$	7
$O_{3,2,4}$	5	$O_{6,3,5}$	4	$O_{8,5,4}$	3	$O_{10,4,2}$	2
$O_{3,3,1}$	6	$O_{6,4,1}$	7	$O_{8,6,1}$	5	$O_{10,4,3}$	3
$O_{3,4,2}$	3	$O_{6,4,2}$	8	$O_{8,6,2}$	5		

Table 17: THE EXECUTION RESULTS OF EXAMPLE 4 WITH LOT SIZE OF 5 EACH.

w	Number of Iterations	Depth of Final Marking	Cost of Solution Path
2	465	450	304
3	465	450	304
4	453	450	304
5	451	450	298
10	451	450	298
100	451	450	298

Table 18: THE LOT SIZES FOR THE SECOND EXPERIMENT OF EXAMPLE 4.

Job	J_1	J_2	J_3	J_4	J_5	J_6	J_7	J_8	J_9	J_{10}
Lot Size	5	6	4	6	4	5	7	3	5	5

Table 19: THE EXECUTION RESULTS OF EXAMPLE 4 WITH VARIED LOT SIZES.

w	Number of Iterations	Depth of Final Marking	Cost of Solution Path
1.5	1121	420	273
1.51	1012	420	273
1.52	1012	420	273
1.55	885	420	273
1.6	778	420	273
2	499	420	281
5	421	420	293
10	421	420	293

Table 20: JOB REQUIREMENTS OF EXAMPLE 5.

Process	J_1	J_2	J_3
1	M_1	M_3	M_2
2	M_2/M_3	M_1/M_2	M_1/M_3
3	M_4	M_4	M_4

Table 21: OPERATION TIMES OF EXAMPLE 5.

Operation	Time
$O_{1,1,1}$	6
$O_{1,2,2}$	14
$O_{1,2,3}$	10
$O_{1,3,4}$	1
$O_{2,1,3}$	8
$O_{2,2,1}$	15
$O_{2,2,2}$	23
$O_{2,3,4}$	1
$O_{3,1,2}$	4
$O_{3,2,1}$	6
$O_{3,2,3}$	5
$O_{3,3,4}$	1

other methods would be desirable but is not within the scope of this chapter. However, the following example compares the method with results found in the literature.

Previously solved FMS scheduling problems that are well documented with details such as specific job and resource descriptions are not abundant in the literature. Chan and Pak [8] provide some well documented scheduling results simulated using heuristic dispatch rules. On-line heuristic dispatch rules are commonly used in job shop systems. These dispatch rules include "shortest queue length," "shortest process time," etc. The problem presented on pages 111-113 of [8] can be translated into the following example using the notation of this chapter. The problem involves four machines, one automated guided vehicle (AGV), and three job types with three processes each.

Example 5: The manufacturing system has four machines M_1, M_2, M_3 and M_4. It has one AGV and the transportation time between each station is 1 time unit. Buffer sizes are 10 each. There are 3 jobs as shown in Table 20. The operation times are given in Table 21.

In [8], the problem is simulated for 450 time units. Three heuristic dispatch rules, "shortest queue length," "shortest process time," and "shortest waiting time," are used in the simulation. "Shortest queue length" produces 21, 21 and 14 products of job type 1, 2 and 3, respectively. The production rate is 56/450 = 0.124 (products per time unit). "Shortest process time" produces 19, 15 and 17 products of job type 1, 2 and 3, respectively. The production rate is 51/450 = 0.113 (products per time unit). "Shortest waiting time" produces 19, 18 and 19 products of job type 1, 2 and 3, respectively. The production rate is 56/450 = 0.124 (products per time unit).

Using the numbers of manufactured products as the lot sizes for the Petri net formulation, the problem is solved by the scheduling method presented in this chapter. The Petri net model of the system with one of the tested initial markings is shown in Figures 9-11. The number inside a place represents the number of tokens.

The heuristic function of (7) is used. w, A_1 and E_1 are estimated using the method discussed earlier, and are 0.25, 7.3 and 560, respectively. Table 22 shows the scheduling results. The first column indicates the simulation case whose number of manufactured products is used as the lot size.

In all three cases, the method presented in this chapter generates schedules with a shorter makespan. The production rates are increased by 13.7%, 24.8% and 16.9%, respectively. The results of Table 22 show that the scheduling method presented in this chapter performs better than the three heuristic dispatch rules for this scheduling problem. This is expected because the scheduling method presented in this chapter seeks to optimize globally as opposed to the local optimization using heuristic dispatch rules.

6. CONCLUSIONS

The Petri net-based heuristic scheduling method presented in this chapter formulates a scheduling problem with a timed place Petri net model. Once a complete model of the system is constructed, the scheduling algorithm uses this Petri net model to search over the reachability graph. Depending on the heuristic function used, the scheduling algorithm seeks a globally optimal or near optimal feasible schedule in terms of a firing sequence of the transitions of the Petri net model.

The formulation explicitly and easily handles the important characteristics of flexible manufacturing, such as routing flexibility, shared resources, concurrency, and lot sizes, and also facilitates maintaining the consistency of problem models. By setting the initial and the final markings appropriately, partial scheduling can be handled without any modification of the model. By using appropriate functions or combinations of functions in the scheduling algorithm, other performance criteria or multiple performance criteria can be adapted.

Since a complete Petri net model of a system is obtained during the problem formulation, the generated schedule in terms of a firing sequence of the transitions can be directly used by employing Petri net controllers in the supervisory control of the modeled system. Once a schedule to reach the desired final state

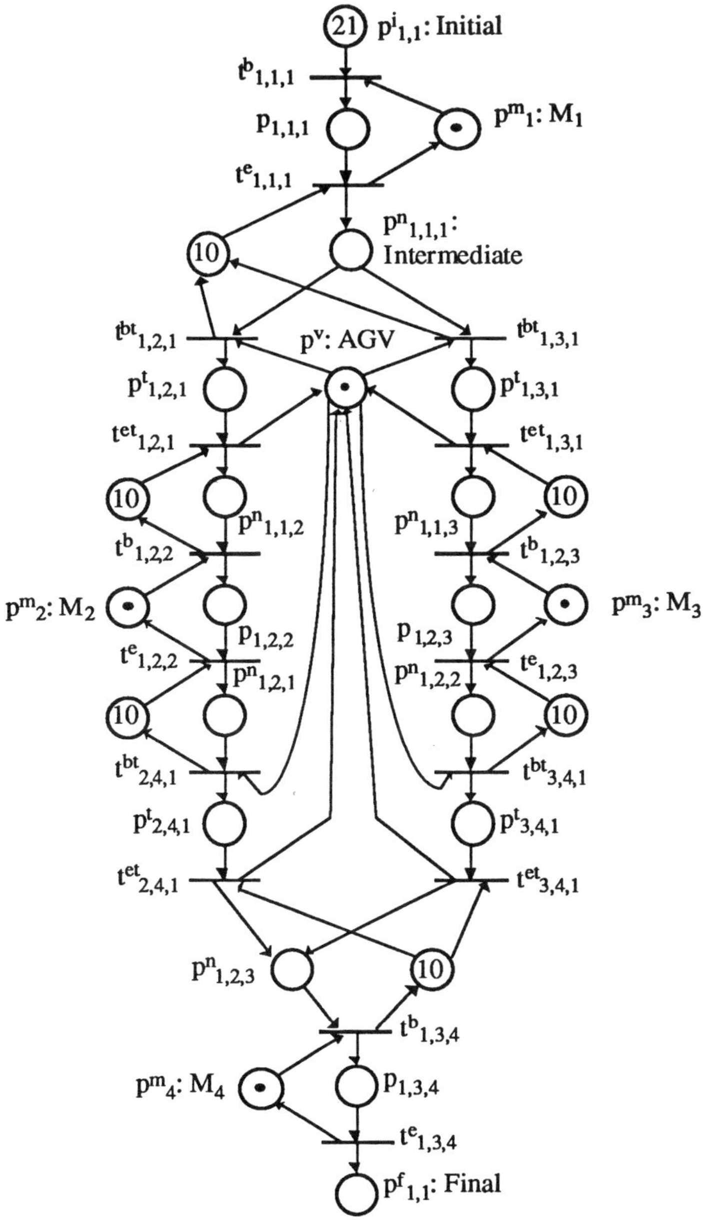

Figure 9: The Petri net sub-model for J_1 of Example 5.

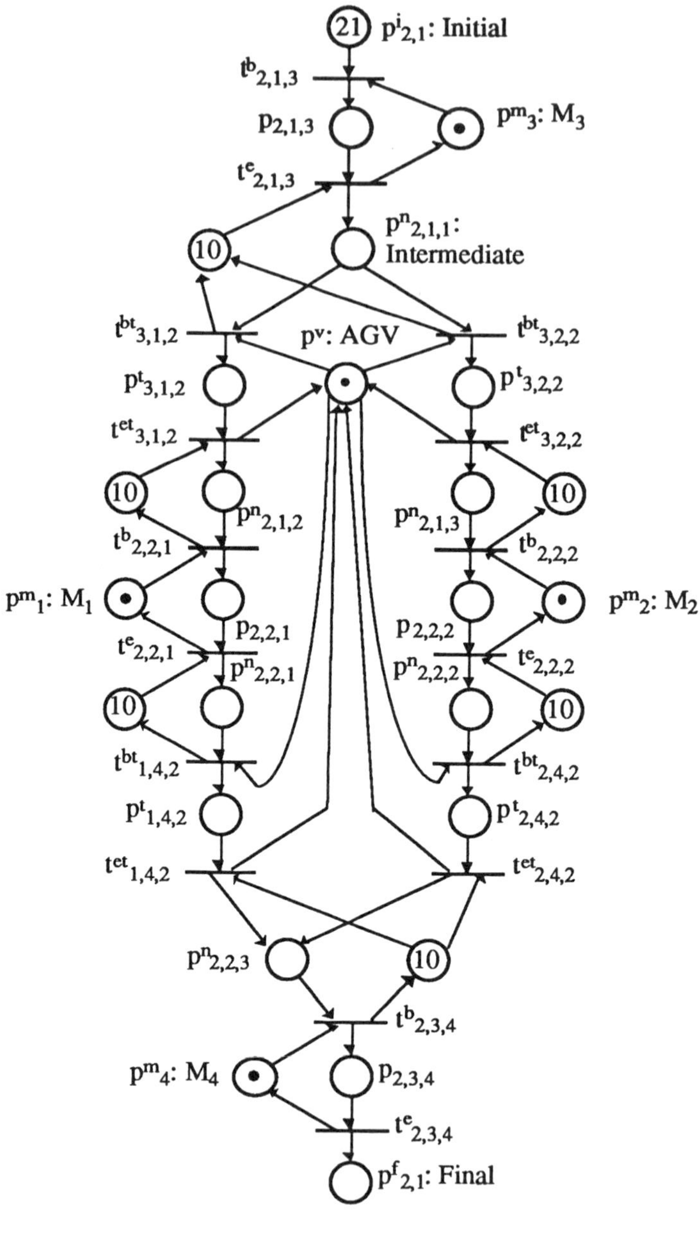

Figure 10: The Petri net sub-model for J_2 of Example 5.

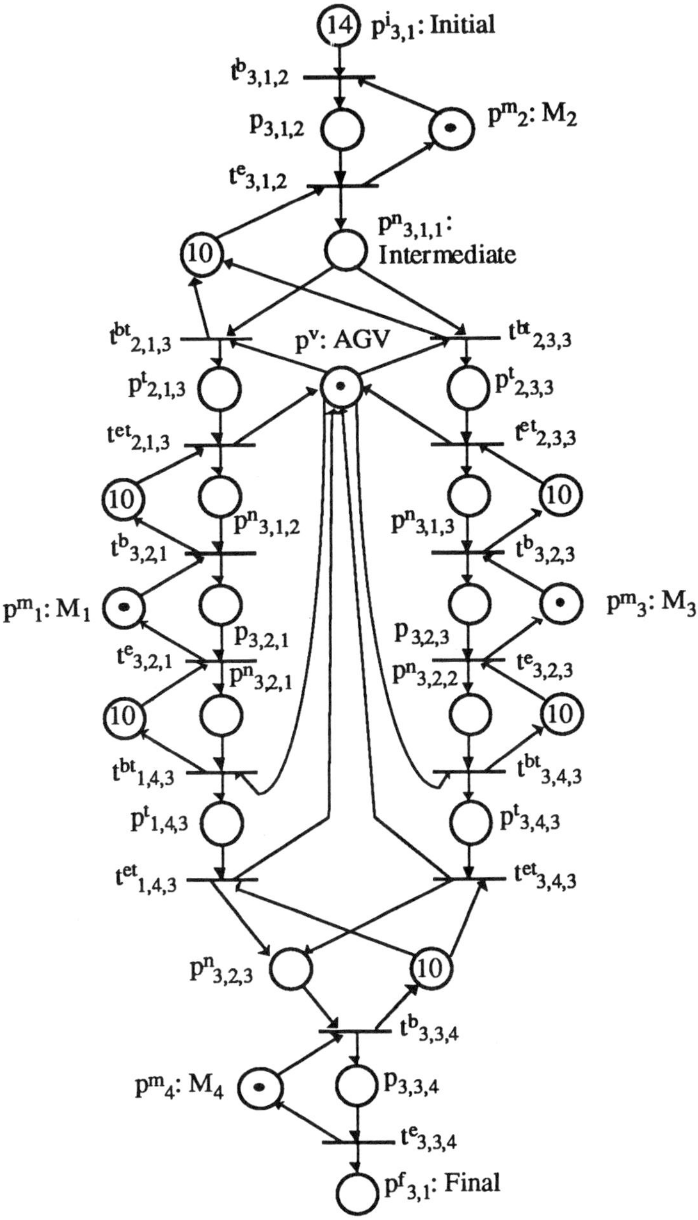

Figure 11: The Petri net sub-model for J_3 of Example 5.

Table 22: THE SCHEDULING RESULTS OF EXAMPLE 5.

Lot Size Used	Number of Iterations	Depth of Final Marking	Makespan of Solution	Production Rate (Petri Net Method)	Production Rate (Dispatch Rules)
From "shortest queue length"	727	560	396	56/396 = 0.141	0.124
From "shortest process time"	712	510	361	51/361 = 0.141	0.113
From "shortest waiting time"	723	560	385	56/385 = 0.145	0.124

is generated, potential deadlocks can be avoided by following the schedule since the generated schedule is always feasible. Hence the analytical overhead to guarantee the liveness of the model and the system is eliminated.

An experimental study of the efficiency of some heuristic functions is conducted. The function of (7) with a properly tuned w is a good heuristic function. The scheduling method is compared with results found in the literature. The method presented in this chapter compares favorably with the results simulated using heuristic dispatch rules.

Further research is under way to develop heuristic functions which satisfy the admissibility condition of the scheduling algorithm. A more comprehensive comparison of this method with other methods is also of research interest for the future.

REFERENCES

[1] S. Ahmad and B. Li, "Robot control computation in microprocessor systems with multiple arithmetic processors using a modified DF/IHS scheduling algorithm," *IEEE Transactions on Systems, Man, and Cybernetics,* vol. 19, no. 5, pp. 1167-1178, Sep/Oct 1989.

[2] M. Aicardi, A. D. Febbraro and R. Minciardi, "Analysis of deterministic discrete event systems via minimax algebra," *Proceedings of the 1991 IEEE International Conference on Systems, Man, and Cybernetics,* Charlottesville, VA, Oct 13-16, 1991, pp. 321-328.

[3] R. Y. Al-Jaar and A. A. Desrochers, "Performance evaluation of automated manufacturing systems using generalized stochastic Petri nets," *IEEE Transactions on Robotics and Automation,* vol. 6, no. 6, pp. 621-639, Dec 1990.

[4] K. R. Baker, *Introduction to Sequencing and Scheduling,* New York: John Wiley & Sons, 1974.

[5] E. Bowman, "The schedule-sequencing problem," *Operations Research,* vol. 7, no. pp. 621-624, 1959.

[6] P. J. Brucker, "Scheduling problems in connection with flexible production systems," *Proceedings of the 1991 IEEE International Conference on Robotics and Automation,* Sacramento, CA, April 1991, pp. 1778-1783.

[7] J. Carlier and P. Chretienne, "Timed Petri net schedules," in *Advances in Petri Nets,* Rozenberg (ed.), Berlin: Springer-Verlag, 1988, pp. 62-84.

[8] T. S. Chan and H. A. Pak, "Modelling of a controller for a flexible manufacturing cell," in *Simulation,* R. D. Hurrion (ed.), Kempston, UK: IFS (Publications) Ltd, 1986, pp. 105-118.

[9] C. L. P. Chen, "Time lower bound for manufacturing aggregate scheduling problems," *Proceedings of the 1991 IEEE International Conference on Robotics and Automation,* Sacramento, CA, April 1991, pp. 830-835.

[10] R. Conway, W. Maxwell and L. Miller, *Theory of Scheduling,* Reading, MA: Addison-Wesley, 1967.

[11] E. Falkenauer and S. Bouffouix, "A genetic algorithm for job shop," *Proceedings of the 1991 IEEE International Conference on Robotics and Automation,* Sacramento, CA, April 1991, pp. 824-829.
[12] S. French, *Sequencing and Scheduling: An Introduction to the Mathematics of the Job-Shop,* Ellis Horwood, 1982.
[13] S. B. Gershwin, R. R. Hildebrant, *et al.* "A control perspective on recent trends in manufacturing systems," *IEEE Control Systems Magazine,* vol. 6, no. 2, pp. 3-15, April 1986.
[14] H. P. Hillion and J. Proth, "Performance evaluation of job-shop systems using timed event-graphs," *IEEE Transactions on Automatic Control,* vol. 34, no. 1, pp. 3-9, Jan. 1989.
[15] D. J. Hoitomt, J. B. Perkins and P. B. Luh, "Distributed scheduling of job shops," *Proceedings of the 1991 IEEE International Conference on Robotics and Automation,* Sacramento, CA, April 1991, pp. 1067-1072.
[16] D. J. Hoitomt, P. B. Luh and K. R. Pattipati, "A Lagrangian relaxation approach to job shop scheduling problems," *Proceedings of the 1990 IEEE International Conference on Robotics and Automation,* Cincinnati, OH, May 1990, pp. 1944-1949.
[17] D. J. Hoitomt, P. B. Luh and K. R. Pattipati, "Job shop scheduling," *Proceedings of the First International Conference on Automation Technology,* Taipei, Taiwan, July 1990, pp. 565-574.
[18] M. A. Holliday and M. K. Vernon, "A generalized timed Petri net model for performance analysis," *Proceedings of the IEEE International Workshop on Timed Petri Nets,* Torino, Italy, July 1-3, 1985, pp. 181-190.
[19] L. E. Holloway and B. H. Krogh, "Synthesis of feedback control logic for a class of controlled Petri nets," *IEEE Transactions on Automatic Control,* vol. 35, no. 5, pp. 514-523, May 1990.
[20] E. Horowitz and S. Sahni, *Fundamentals of Computer Algorithms,* Rockville, MD: Computer Science Press, 1978.
[21] E. S. H. Hou and H. Y. Li, "Task scheduling for flexible manufacturing systems based on genetic algorithms," *Proceedings of the 1991 IEEE International Conference on Systems, Man, and Cybernetics,* 1991, pp. 397-402.
[22] C. Q. Jiang, M. G. Singh and K. S. Hindi, "Optimized routing in flexible manufacturing systems with blocking," *IEEE Transactions on Systems, Man, and Cybernetics,* vol. 21, no. 3, pp. 589-595, May/June 1991.
[23] P. Kapasouris, D. Serfaty, *et al.* "Resource allocation and performance evaluation in large human-machine organizations," *IEEE Transactions on Systems, Man, and Cybernetics,* vol. 21, no. 3, pp. 521-532, May/June 1991.
[24] R. Larson and A. Odoni, *Urban Operations Research,* Englewood Cliffs: Prentice-Hall, 1981, pp. 360-361.
[25] A. E. J. Lee, *Integrated Tooling and Scheduling of Flexible Machines: Theory and Algorithms,* Ph. D. Dissertation, Rensselaer Polytechnic Institute, Troy, NY, May 1989.

[26] D. Y. Lee and F. DiCesare, "Integrated scheduling of flexible manufacturing systems employing automated guided vehicles," *IEEE Transactions on Industrial Electronics,* to appear in December 1994.
[27] D. Y. Lee and F. DiCesare, "Scheduling flexible manufacturing systems using Petri nets and heuristic search," *IEEE Transactions on Robotics and Automation,* vol. 10, no. 2, pp. 123-132, April 1994.
[28] D. Y. Lee and F. DiCesare, "Scheduling flexible manufacturing systems with the consideration of setup times," *Proceedings of the 32nd IEEE Conference on Decision and Control,* San Antonio, TX, Dec. 15-17, 1993, pp. 3264-3269.
[29] D. Y. Lee and F. DiCesare, "Scheduling of flexible manufacturing systems employing automated guided vehicles," *Proceedings of the 9th International Conference on CAD/CAM, Robotics and Factories of the Future,* Newark, NJ, August 17-20, 1993, in press.
[30] D. Y. Lee and F. DiCesare, "Integrated models for scheduling flexible manufacturing systems," *Proceedings of the 1993 IEEE International Conference on Robotics and Automation,* Atlanta, GA, May 2-7, 1993, pp. 827-832.
[31] D. Y. Lee, *Scheduling and Supervisory Control of Flexible Manufacturing Systems Using Petri Nets and Heuristic Search,* Ph. D. Dissertation, Rensselaer Polytechnic Institute, Troy, NY, May 1993.
[32] D. Y. Lee and F. DiCesare, "Petri nets and heuristic search for periodic scheduling," *Proceedings of the 1992 IEEE International Conference on Systems, Man, and Cybernetics,* Chicago, IL, October 18-21, 1992, pp. 998-1003.
[33] D. Y. Lee and F. DiCesare, "Experimental study of a heuristic function for FMS scheduling," *Proceedings of the 1992 Japan-USA Symposium on Flexible Automation,* San Francisco, CA, July 13-15, 1992, pp. 1171-1177.
[34] D. Y. Lee and F. DiCesare, "FMS scheduling using Petri nets and heuristic search," *Proceedings of the 1992 IEEE International Conference on Robotics and Automation,* Nice, France, May 10-15, 1992, pp. 1057-1062.
[35] P. S. Liu and L. C. Fu, "Planning and scheduling in a flexible manufacturing system using a dynamic routing method for automated guided vehicles," *Proceedings of the 1989 IEEE International Conference on Robotics and Automation,* Scottsdale, AZ, 1989, pp. 1584-1589.
[36] Z. P. Lo and B. Bavarian, "Job scheduling on parallel machines using simulated annealing," *Proceedings of the 1991 IEEE International Conference on Systems, Man, and Cybernetics,* 1991, pp. 391-396.
[37] P. B. Luh, D. J. Hoitomt, *et al.* "Schedule generation and reconfiguration for parallel machines," *IEEE Transactions on Robotics and Automation,* vol. 6, no. 6, pp. 687-696, Dec. 1990.
[38] A. Manne, "On the job-shop scheduling problem," *Operations Research,* vol. 8, no. pp. 219-223, 1960.
[39] P. Mellor, "A review of job shop scheduling," *Operational Research Quarterly,* vol. 17, no. 2, pp. 161-171, June 1966.
[40] N. Nilsson, *Principles of Artificial Intelligence,* Palo Alto, CA: Tioga, 1980.

[41] J. O'Brien, *Scheduling Handbook,* New York: McGraw-Hill, 1969.
[42] K. Onaga, M. Silva and T. Watanabe, "On periodic schedules for deterministically timed Petri net systems," *Proceedings of the 4th International Workshop on Petri Nets and Performance Models,* Melbourne, Australia, Dec. 2-5, 1991, pp. 210-215.
[43] T. A. Owens and P. B. Luh, "A job completion time estimation method for work center scheduling," *Proceedings of the 1991 IEEE International Conference on Robotics and Automation,* Sacramento, CA, April 1991, pp. 110-115.
[44] J. Pearl, *Heuristics: Intelligent Search Strategies for Computer Problem Solving,* Reading, MA: Addison-Wesley, 1984.
[45] C. Ramamoorthy and G. Ho, "Performance evaluation of asynchronous concurrent systems using Petri nets," *IEEE Transactions on Software Engineering,* vol. SE-6, no. 5, pp. 440-449, Sep. 1980.
[46] J. Rickel, "Issues in the design of scheduling systems," in *Expert Systems and Intelligent Manufacturing,* Oliff (ed.), Elsevier, 1988, pp. 70-89.
[47] F. Rodammer and J. K. White, "A recent survey of production scheduling," *IEEE Transactions on Systems, Man, and Cybernetics,* vol. 18, no. 6, pp. 841-851, Nov.Dec. 1988.
[48] R. V. Rogers and K. P. White, "Algebraic, mathematical programming, and network models of the deterministic job-shop scheduling problem," *IEEE Transactions on Systems, Man, and Cybernetics,* vol. 21, no. 3, pp. 693-697, May/June 1991.
[49] R. Sengupta and S. Lafortune, "Optimal control of a class of discrete event systems," *Proceedings of the IFAC International Symposium on Distributed Intelligence Systems,* Arlington, VA, Aug 13-15, 1991, pp. 25-30.
[50] L. Shen, Q. Chen, *et al.* "Truncation of Petri net models of scheduling problems for optimum solutions," *Proceedings of the Japan/USA Symposium on Flexible Automation,* San Francisco, CA, July 13-15, 1992, pp. 1681-1688.
[51] H. Shih and T. Sekiguchi, "A timed Petri net and beam search based online FMS scheduling system with routing flexibility," *Proceedings of the 1991 IEEE International Conference on Robotics and Automation,* Sacramento, CA, April 1991, pp. 2548-2553.
[52] J. Sifakis, "Performance evaluation of systems using nets," in *Net Theory and Applications,* Brauer (ed.), Berlin: Springer-Verlag, 1980, pp. 307-319.
[53] M. Silva and R. Valette, "Petri nets and flexible manufacturing," in *Advances in Petri Nets,* Berlin: Springer-Verlag, 1989, pp. 374-417.
[54] A. S. Spachis and J. R. King, "Job-shop scheduling heuristics with local neighbourhood search," *International Journal of Production Research,* vol. 17, no. 6, pp. 507-526, Nov/Dec 1979.
[55] A. Vasquez and P. B. Mirchandani, "Concurrent resource allocation for production scheduling," *Proceedings of the 1991 IEEE International Conference on Robotics and Automation,* Sacramento, CA, April 1991, pp. 1060-1066.
[56] N. Viswanadham and Y. Narahari, "Stochastic Petri net models for per-

formance evaluation of automated manufacturing systems," *Information and Decision Technologies,* vol. 14, no. pp. 125-142, 1988.
[57] D. Zhang, "Planning using timed Pr/T nets," *Proceedings of the Japan/USA Symposium on Flexible Automation,* San Francisco, CA, July 13-15, 1992, pp. 1179-1183.
[58] Y. Zhang, "Solution to job-shop scheduling of FMS by neural networks," *Proceedings of the 1991 IFAC Workshop on Discrete Event System Theory and Applications in Manufacturing and Social Phenomena,* Shenyang, China, June 25-27, 1991, pp. 261-266.

7

Scheduling and Re-scheduling of AGVs for Flexible and Agile Manufacturing

Shifang Li Toshi Takamori Satoshi Tadokoro
Takamori Laboratory, Dept. Computer & System Engineering
Kobe University, Rokkodai 1-1, Nada, Kobe 657, Japan
Tel +81-78-803-1200
Fax +81-78-803-1217
E-mail lee@in.kobe-u.ac.jp.

Abstract

This chapter deals with a. modeling and scheduling flexible and agile manufacturing system (FAMS) using AGVs, and b. re-scheduling the system in the case of trouble, by using Timed Petri Net (TPN). The purpose is

1. *to provide a solution to the coordination problems of AGVs transferring a series of jobs to the appointed machining cells so as to minimize the idle time of each machining cell and total job processing time without collision of AGVs, and*

2. *to generate quickly another substituting effective schedule in the case of troubles of the FAMS, especially troubles of AGVs.*

TPN is applied to model the system in which a hierarchical PN is used for making the presentation and analysis easy. Two kinds of priority functions are defined to resolve conflicts arising among those active entities of the system. The priority functions express the degree of importance that the entities have depending on time so that precedence relationship among conflicting entities can be effectively determined. Therefore, quick scheduling and re-scheduling are possible even when troubles happen in the system. Meanwhile, Petri nets are checked to secure collision avoidance of AGVs. Some scheduling simulation results are shown in the chapter to verify the effectiveness of the method.

1 Introduction

In recent years, with the development of Factory Automation (FA), there is a growing tendency for practical production to construct a manufacturing system using robots, machining cells, transporting devices and multistory warehouses, so as to fit in with the needs of the multi-product and small-sized production [1]. It is necessary to generalize the control of those asynchronous and concurrent working equipments as well as synchronize exclusively them with each other. Especially, it is a very important theme when thinking about maximizing the operating rate of each machining cell and minimizing total job processing time or optimizing the schedule of FAMS, in which multiple kinds of jobs are machined by multiple kinds of machining cells according to their inherent orders. This is defined as a well-known optimization problem of large scale combination of complex discrete events (DEs). It is difficult to find strictly an optimal solution using branch-and-bound methods or dynamic programming because of huge amount of computing time. An efficient heuristic method is necessary to systematically work out a sub-optimal solution [2]. This method seems to become the mainstream in practical production.

Petri Nets (PNs) are a powerful tool to solve modeling and scheduling problems of FMS [3]. It is also applied effectively to the modeling and scheduling of AGVs system [4][5]. These papers apply PN to detect and resolve conflicts in the system, use simulation to provide performance information, and represent production processes of a job-shop manufacturing system where AGVs tranport workpieces. However, the computing time problem is not clearly specified, especially, little effort has been reported on the recovery of the schedule when the systems are in troubles.

In our research, Timed Petri Net (TPN) was applied to model FAMS using AGVs, and a (sub) optimal schedule of a repetitive job-shop type process was obtained under the restriction that the system secures collision avoidance of AGVs. This approach uses excellent faculties of PNs in representing synchronization, exclusion and concurrent action of sequence control demanded by the system and analysis of a numeral expression. Our research purpose is to solve the following problems:

- How AGVs cooperatively transport jobs to satisfy the (sub)optimal machining schedule without collision in limited space and with prescribed speed, and
- How rapidly a new efficient schedule is generated when some AGVs are in trouble.

The TPN model of the system is decomposed into a part of job processing on machining cells and part of job transporting with AGVs by using hierarchical PNs to make representation and analysis easy. In this chapter, according to analysis of simulation based on the TPN model, a (sub)optimal machining schedule (non-time delayed) of a repetitive job-shop type process is worked out. A process to assign AGVs to jobs and decide their transferring routes is then simulated so as to satisfy the schedule. In order to resolve the conflicts of jobs in machining cells and AGVs, priority functions of machining cells and AGVs are defined respectively to express the degree of importance that they have depending on time. The precedence relationships among machining orders and AGVs assignment can be immediately determined according to the functions. In addition, collision check nets are introduced to prevent two or more AGVs from entering the same transferring section simultaneously at any time. Our algorithm will guarantee a high operating ratio to machining system even in the case that the number of AGVs is not sufficient to satisfy the (sub)optimal machining schedule due to some troubles. Finally some simulation examples will be given to verify the effectiveness of the proposed method.

2 Scheduling Problems of FAMS Using AGVs

Let's consider the most general FAMS scheduling problem, or a repetitive job-shop type process. The total job processing time and operating ratio of a machining cell are evaluated to be as standard. The example illustrated

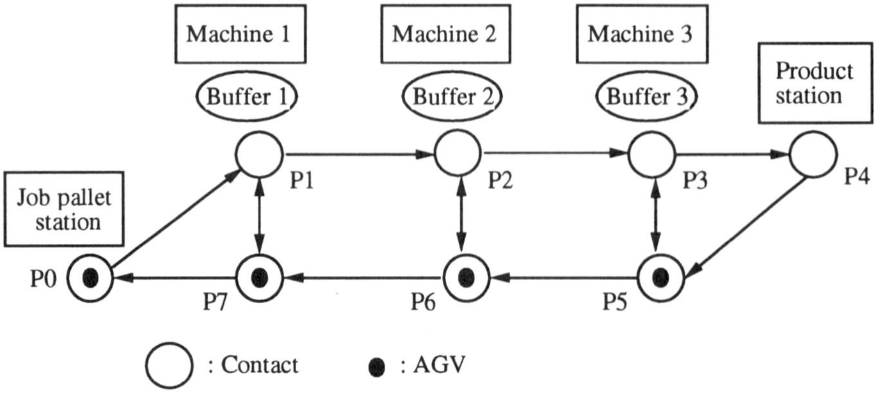

Figure 1: Flexible and Agile Manufacturing System Using AGVs

in Figure 1 shows a structure of FAMS using AGVs, where some sets of different jobs are transported by some AGVs and processed at different machines. Machining times which are standardized into integers and orders of the jobs are given, as those shown in Table 1. The numeral inside the parenthesis denotes the time units for the job to be processed at the machine. It is supposed that only one job can be processed at a machine at any time instant. Each pallet must be sent by an AGV to the appointed buffer beside the machines before processing to make the machines work with minimal idle time. Each buffer is divided into an input part and an output one to accommodate each job before and after machining respectively. A job in the

Orders	**1st**	**2nd**	**3rd**	
Job 1	Machine 1(2u)	Machine 2(3u)	Machine 3 (4u)	
Job 2	Machine 3(4u)	Machine 1(2u)	Machine 2(2u)	
Job 3	Machine 1(3u)	Machine 3(5u)	Machine 2(3u)	
Job 4	Machine 2(3u)	Machine 3(4u)	Machine 1(3u)	

Table 1: Machining Orders and Times(Unit Time)

input part is put on the machine automatically, it returns to the output part after machining. Then it waits to be transported to the next machine with an AGV. AGVs coordinately transport job pallets among the pallet stations, the buffers and the product stations according to Table 1. They move along one-way tracks in the manner as automobiles observe the traffic rule. In this example, AGVs move clockwise as shown by the arrows. The circles $P_p(p = 0 \sim 7)$ is called contact, and is to be used to detect an existence of an AGV at the end of the section. The control system must decide whether it should stop the AGV or let it continue to move in certain direction when an AGV arrives at a contact every time. So, no more than one AGVs can simultaneously enter a section at any time [6]. If some AGVs break down while machining processes are in progress, its efficient schedule will not enter suddenly disorder because some jobs are accumulated in the buffers, and the other AGVs will automatically share the remaining jobs and finish the work so as to minimize the idle time of machining system and total job processing time.

3 TPN model of the System

It is very complicated to present all the dynamic behaviors of machines and AGVs, and the flow of jobs by the model on the same level. Therefore, we break the system as shown in Figure 1 into two portions: machining and transferring events. Then Hierarchical TPN models is used to make the presentation and analysis easier.

3.1 Machining Events Model

Figure 2 is a TPN model corresponding to Figure 1 and Table 1. However, the transitions that express transporting events of AGVs are reduced by using the reduction techniques of PN [8]. The nets standing in a row such as P_{c12}, P_{c13}, and P_{c14} denote conflict relationship between Job 1 and the other jobs concerning Machine 1. They will also keep the next set of Job 1 from start processing until the current set of Job 1 finishes [7]. The place represented by P_{m12} shows a situation that the job was advanced in the defined machining order. Place P_{jps} and P_{ps} represent respectively the Job pallet station and the Product station. The token in place $P_{Mi}(i = 1,2,3)$ shows the condition whether Machine i is busy or not. The arrows at both ends of a arc that connects P_{Mi} and $T_{jo}(j$ = job number, o = machining order) symbolize a self loop PN. A dummy place P_{dummy} and a dummy

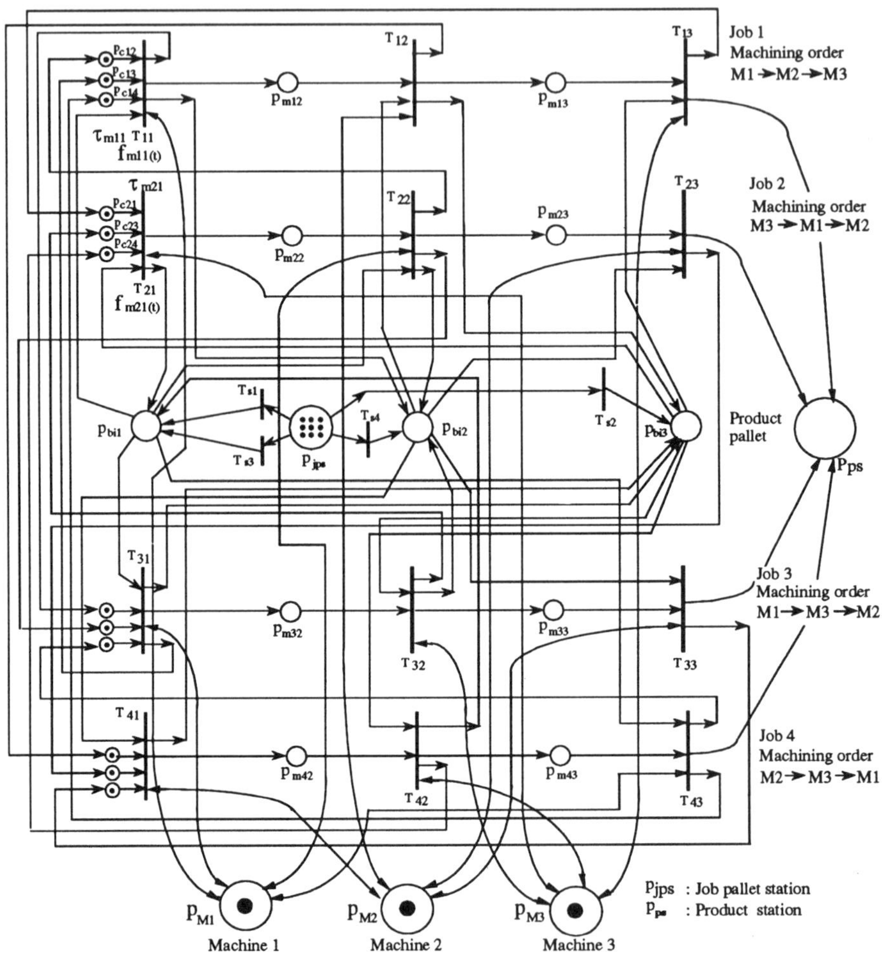

Figure 2: TPN Model of the Machining System

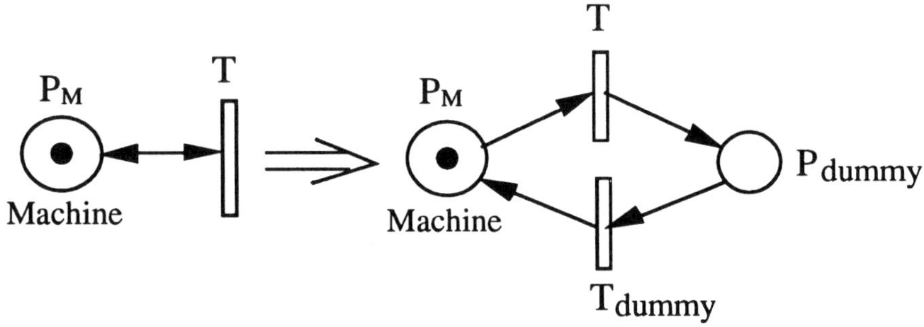

Figure 3: Transformation of a Self Loop PN

transition T_{dummy} are added as shown in Figure 3 for making computing possible. $P_{bim}(m = 1, 2, 3)$ is the reduced place which corresponds to the input part of the buffer that accommodates the jobs before processing on the machine beside it. Its detailed structure is developed as shown in Figure 4 (c). In Figure 4 (c), P_{bi11} represents space of Buffer 1 exclusively

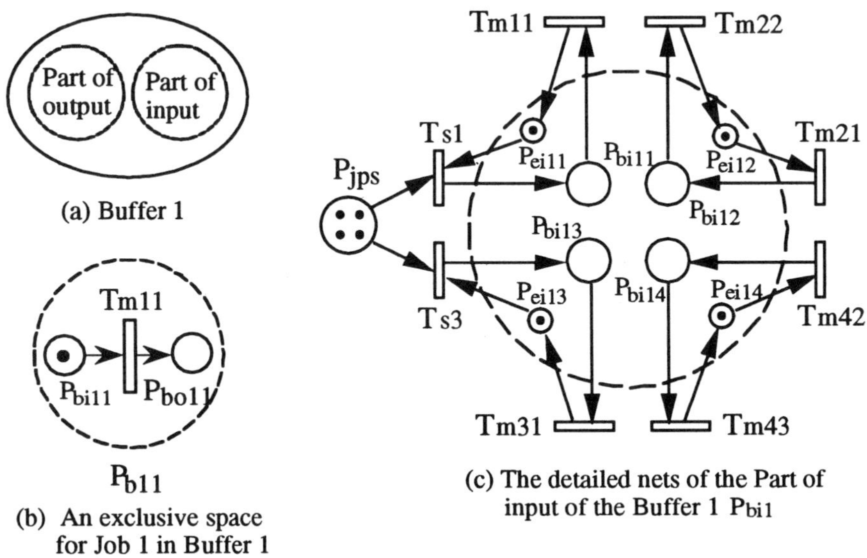

Figure 4: The Detailed Structure of Buffer 1

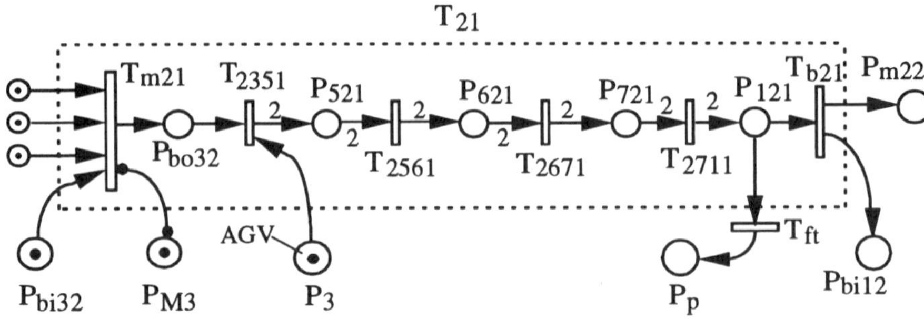

Figure 5: Detailed Transition of T_{21}

assigned to Job 1 before processing at Machine 1. Its capacity is limited by the number of tokens in P_{ei11}. Clearly, the spaces of the buffers are assigned to each job in the same proportion according to Table 1 to keep balance of total job progress. The structure of output part is omitted here for its similarity to the input part. You will find that every P_{bim} can accommodate one job before or after processing at any time instant. The transition T_{jo} in Figure 2 stands for the event in the machining order o of job j, in which the event of transporting process of an AGV is reduced by the rule of fusion of series transitions. The parameter τ_{mjo} indicates its firing period. In order to determine immediately precedence relationship of job processing, a machining priority function $f_{mjo}(t)$ is defined, which will be discussed in the next section. The initial marking corresponds to Figure 1. The jobs have entered Job pallet station, each machine is ready to process a job and there is no job in every buffer.

3.2 Transporting Events Model

Figure 5 shows the detailed transition of T_{21} in Figure 2. It develops the transporting process of an AGV which moves Job 2 from Buffer 3 to Buffer 1. P_{bi32} and P_{bo32} denote the places of input and output parts of Buffer 3 exclusively assigned to Job 2 respectively. The place P_{pjo} (p=contact number,j=job number, o=machining order), for example, P_{521} indicates whether Job 2 finished machining order 1 enters the contact P_5 shown in Figure 1 or not. The transition T_{2351} means that an AGV will transport Job 2 after machining order 1 from the place P_3 to P_5. The similar explanation applies to other transition T_{jflo}. All the detailed transition T_{jo} such as T_{21} consists of a job

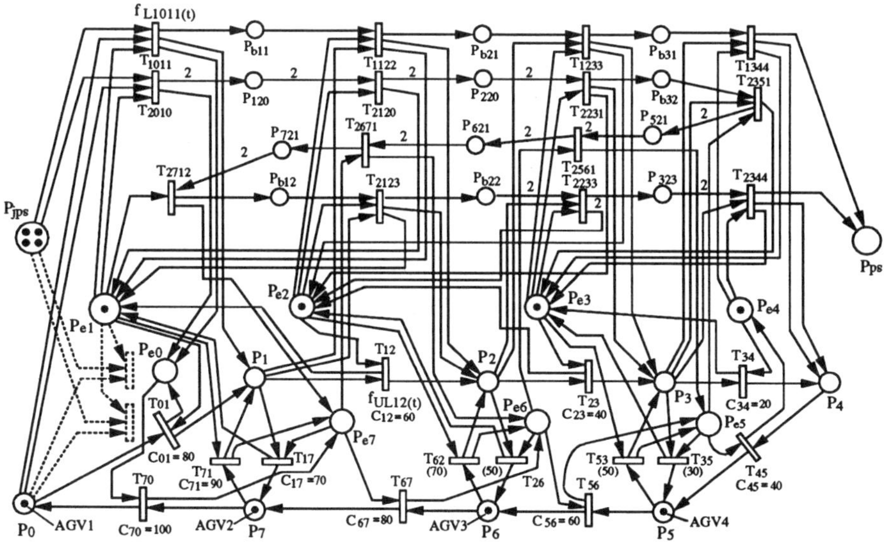

Figure 6: TPN Model of the AGVs System

transporting net of AGVs as shown in the upper part that is over the places P_{e1}, P_{e2}, P_{e3} and P_{e4} in Figure 6. However, the machining event transition such as T_{m21} in Figure 5 is reduced in this PN.

Figure 6 consists of three parts which are job loading, unloading, and collision check net of AGVs. Concerning the job loading net, for example, the series of transitions from T_{2010} to T_{2231} represent the process of transporting Job 2 from Job pallet station to Buffer 3 by an AGV. They also correspond to the transition T_{s2} in Figure 2. The transitions from T_{2351} to T_{2671} are those of T_{21} in Figure 5. Here, we should not overlook that those places such as P_{111}, P_{120} and P_{122} all denote the same place P_1 in order to distinguish behaviors of an AGV in transferring various jobs having different distances. It must be especially noticed that those transitions arranged in a row do not always continuously fire because they include the machining event transitions such as T_{m21} in Figure 5. The transitions with a constant C_{ft} consist of the unloading job net. The places $P_p (p = 0 \sim 7)$ correspond to the contacts in Figure 1 respectively. In order to rapidly determine the precedence relationship among assignments of AGVs to jobs at the same time, transporting priority function $f_{Ljfto(t)}$ or $f_{ULft(t)}$ is defined on each transition. In addition, the nets of Job 3 and Job 4 are simply drawn to

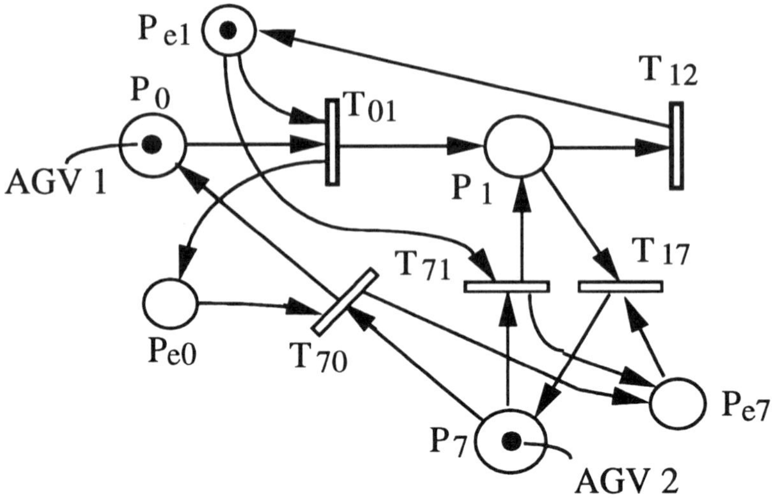

Figure 7: Collision Check Nets

be the part of those dotted lines because they are similar to the above nets. On the other hand, those places P_{ep} are used to check whether P_p is empty or not. The principle is shown in Figure 7. An AGV cannot enter P_p if corresponding P_{ep} has no token. It returns a token to P_{ep} after an AGV goes out of P_p. Therefore, it prohibits two or more AGVs to enter the same section at the same time so as to avoid collision.

A state change equation of the timed transition PN [10] can be established on the basis of the above TPN models.

$$M(t_{k+1}) = M(t_k) + (A(t_k)^+ - A^-)^T r(t_k) \tag{1}$$

$$M(t_k) \geq A^{-T} r(t_k) \tag{2}$$

where

$$\begin{cases} M(t_k): & \text{marking of the system at time } t_k \\ A(t_k)^+: & \text{output joint matrix} \\ A^-: & \text{input joint matrix} \\ r(t_k): & \text{firing enable vector at time } t_k \end{cases}$$

Equation (2) is used to work out the firing vector. It should be noticed that the output joint matrix $A(t_k)$ is also a time variable function because the token is bound in the transition so that it can not enter the output place during the firing time [10].

4 Definition of Priority Functions

A firing decision vector $d(t_k)$ should be efficiently determined to resolve conflicts in the firing vector $r(t_k)$. Machining and transporting priority functions are defined for this purpose.

4.1 Machining Priority Function

The machining priority function is defined as following.

$$f_{mjo}(t) = -\alpha_1 \lambda_{js} - \alpha_2 \frac{\Theta_i}{\Theta_{max}} \lambda_{jd}. \tag{3}$$

where

$$\begin{cases} \lambda_{js}, & \text{the set number of Job } j \\ \lambda_{jd}, & \text{waiting time until next process} \\ \Theta_i, & \text{operating period of Machine i in one set} \\ \Theta_{max}, & \text{maximum of } \Theta_i \\ \alpha_1, \alpha_2, & \text{positive constants, } \alpha_1 \gg \alpha_2 \end{cases}$$

Equation (3) immediately determines a precedence relationship of machining transitions in conflicts. We take $\alpha_1 \gg \alpha_2$ because the balance of machining progress of total jobs is important. The second term is the min-max standard which shortens the idle time of a machine having longer operating time.

4.2 Transporting Priority Function

Transporting priority functions of AGVs are defined by two formula. The first one deals with transitions of job loading. When equation (2) is satisfied with a transition,

$$f_{Ljfto}(t) = -\beta_{jft}(\Upsilon_{jis} - m\delta + (\lambda_{js} - 1)\Pi - t_k) \tag{4}$$

where

$$\begin{cases} \Upsilon_{jis}, & \text{firing start time of a machining transition determined} \\ & \text{by an algorithm for a desirable firing sequence} \\ & j\text{=job number, } i\text{=machine number, } s\text{=set number} \\ \Pi, & \text{a period for finishing one set of the jobs} \\ \delta, & \text{a necessary time for an AGV passing through one section} \\ m, & \text{the number of sections of Job j between the current} \\ & \text{position and the goal} \\ \beta_{jft}, & \text{positive constant, } \beta_{jft} \gg C_{ft} \text{ (inscribed in Figure 6)} \end{cases}$$

$f_{Ljfto}(t)$ is set to be zero when Equation (2) is not satisfied. Υ_{jis} can be obtained by an algorithm which will be described later. The priority function (4) assigns an AGV to a job whose time to fire is shortest considering its moving period.

The second function $f_{ULft}(t)$ is defined for transitions T_{ft} of unloading job as the following. The value of $f_{ULft}(t)$ is taken to be equal to the maximum value of the priority function $f_{Ljfto}(t)$ of those output transitions of the buffer. For example, $f_{UL12}(t) = MAX(f_{L1122}(t), f_{L2123}(t), \cdots)$. The reason is that an AGV unloading job may come close to a job which is calling for transportation instead of returning to the job pallet station. If all the values of those transporting priority functions are zero,

$$f_{ULft}(t) = C_{ft} \tag{5}$$

where C_{ft} is given to make AGVs move clockwise as shown in Figure 6.

The machining schedule and the AGVs schedule of the system are worked out respectively by the following algorithm.

1. A machining schedule of FAMS is solved according to Equation (1), (2) and the definded machining priority function (3), on the basis of the TPN model of the machining system illustrated in Figure 2 under the assumption that an AGV transportation time is zero. In the concrete, pay attention to the reduced transition T_{21} including one machining transition T_{m21} and transporting transitions T_{2351}, T_{2561}, T_{2671}, T_{2711} and T_{b21} (see Figure 5). Only the machining transition T_{m21} is treated as timed transition, and the others as immediate transitions, that is, their firing times are zero in this step. In general, a non-delayed machining schedule or the desirable firing sequence $\{\sigma_{jis}\}$ can be solved [7], and Υ_{jis} in Equation (4) is simply obtained from $\{\sigma_{jis}\}$. For example, it can be directly find that Υ_{412} and Υ_{413} are respectively 30 and 47 from the non-delayed time machining schedule shown in Figure 8.

2. Adding moving time of AGVs in computation and working out AGVs schedule according to Equation (1), (2) and the definded transporting priority function (4) or (5), on the basis of the TPN model of the AGVs system illustrated in Figure 6. Equation (4) gives AGVs an aim to transport jobs to the appointed buffers before Υ_{jis} so that the machining system performs to match the schedule of 1 if possible. It must be noticed that each P_{bij} in Figure 6 includs a machining transition T_{mij}, that is, the P_{bij} is treated as a timed place.

3. When firing machining transition can not match the firing sequence $\{\sigma_{jis}\}$ by circumstances of AGVs, $\{\sigma_{jis}\}$ will disorder, and the start time to perform its next machining order will depend on the condition of P_{mjo} (e.g. see P_{m22} shown in Figure 2 which is also corresponding to Figure 5). However, the Υ_{jis} is still as the aim to recovery σ_{jis}. Thus a practical machining schedule and an AGVs schedule are obtained.

4. Select one of the transitions to fire randomly if the priority functions are equal in the value to save computing time.

When the schedule is disordered by some AGV trouble, a new efficient schedule will be automatically generated by this algorithm because the priority functions guarantee AGVs to transport timely jobs so that the unfinished tasks can be automatically shared among the other AGVs.

5 Simulation Examples

Simulation was performed on a FAMS as shown in Figure 1 and Table 1 under conditions that it consists of n sets of 4 jobs, 3 machines, and AGVs need 1 unit time (u) to pass over one section.

5.1 A Non-Delayed Time Machining Schedule

At first, a desirable firing sequence of the machining system as illustrated in Figure 8 is worked out by the first step of the algorithm. The period of one set equals the maximum operating time 17u of Machine 3 or the schedule is a non-delayed time type. The result is obtained without branching because of the machining priority function. The schedule took a non-delayed time pattern from the first set. It became repetitive from the third set. The same kind of results are obtained in several simulation examples. Therefore, we think this modeling method and the definition of priority functions are effective.

A scheduling result in which moving time of AGVs is considered is presented in Figure 9 under conditions that the number of AGVs is 4, and that initial state of every buffer is empty.

Every job in the first and second sets has been ready for machining order 3 and 2 so as to run out the machining delayed time by transporting processes of AGVs from the first set. We can see that the repetitive non-delayed time machining schedule is kept, and the precedence relationship of

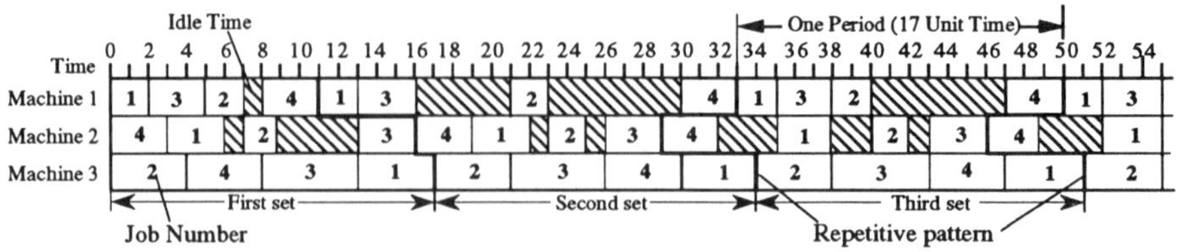

Figure 8: Non-Delayed Time Machining Schedule When the Moving Time of an AGV is assumed to be Zero

assigning AGVs to jobs are being rationally decided by using their value of the priority function. For example, this schedule matched the non-delayed time machining one in repetitive pattern from the start point. At time 3u, there are two choices that is whether AGV 4 in P_0 transports Job 3_3 (3_3 means the third set of Job 3, the index is the set number. The following is the same) to Buffer 1 or Job 1_2 to Buffer 2, both jobs should be machined at the same time 8u. The latter was selected because it has longer moving distance. In Figure 9 (a), at time 28u, all jobs are ready for the next machining order at corresponding buffers. In Figure 9 (c), at time 31u, all AGVs are ready for starting jobs of the next set though there's still a lot of time to the end of the second period.

5.2 Re-scheduling

The above schedule is obtained through the simulation that is performed to lead the number of AGVs to be enough to satisfy the non-delayed time machining schedule. However, it is more important how to generate an efficient machining schedule when the number of AGVs is insufficient, or

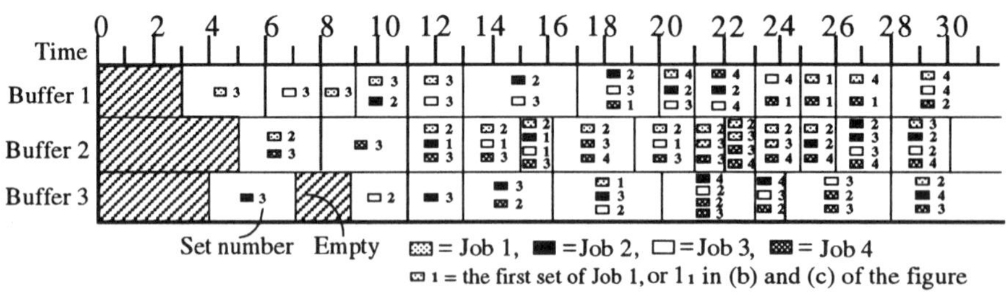

(a) State of the Buffers

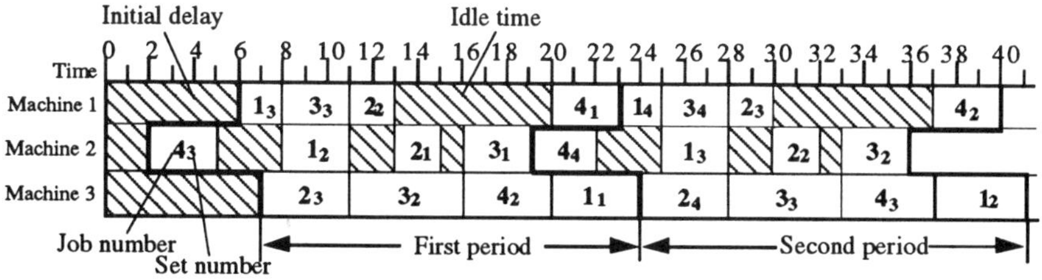

(b) Machining Schedule

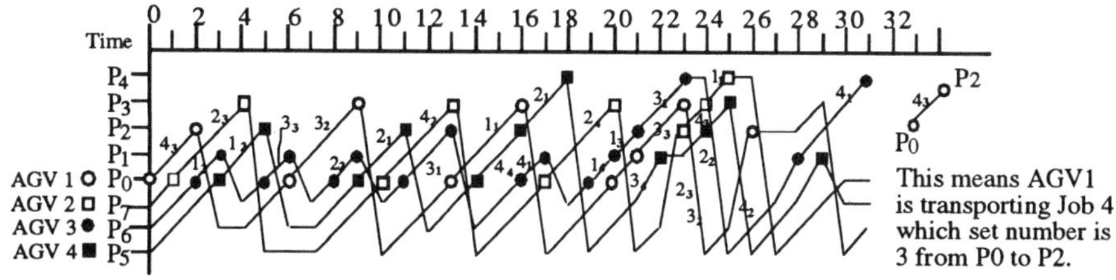

(c) AGVs Schedule

Figure 9: Obtained Scheduling Simulation Results When the Number of AGVs is 4

some one breaks down suddenly. Figure 10 illustrates the simulation result when AGV 4 is out of order at time 28u. We can see that machining schedule dose not change until the third period because of effects of the priority function and the buffers, and machines and the others AGVs are compelled to change the schedule due to insufficient number of working AGVs for the fourth period. However, unfinished processes are automatically assigned to the other AGVs, and the period increased very slightly. It means that this method is effective to solve scheduling problems even in the case that AGVs break down.

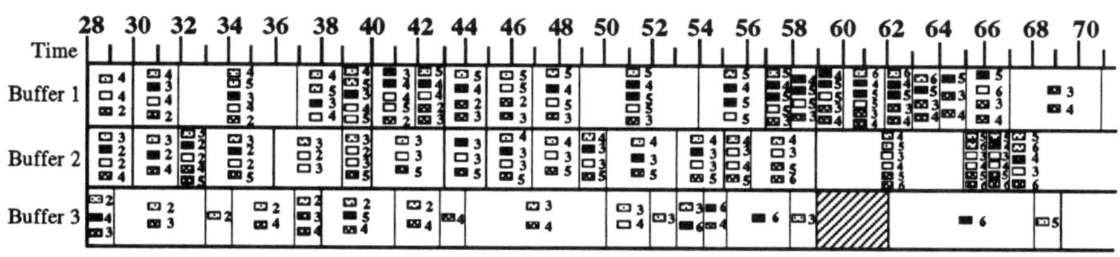

(a) State of the Buffers

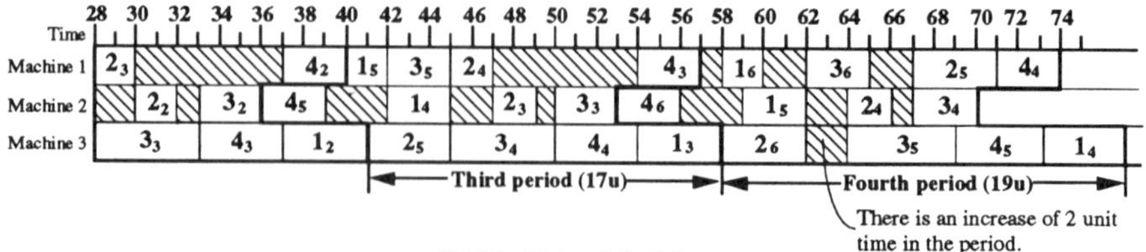

(b) Machining Schedule

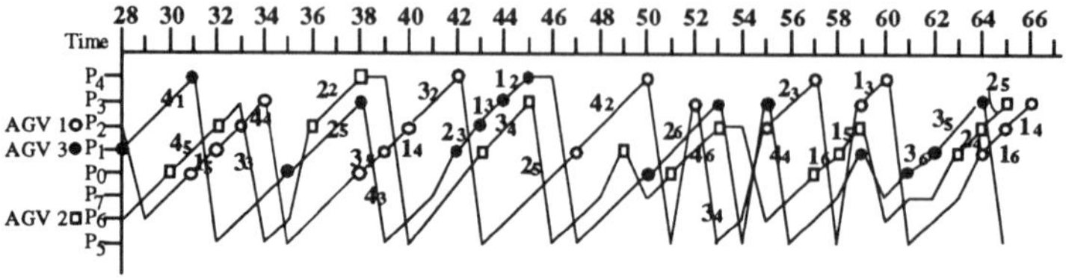

(c) AGVs Schedule

Figure 10: Obtained Scheduling Simulation Results When AGV 4 is Out of Order at Time 28u

6 Conclusion

This chapter presented an approach to modeling and scheduling of flexible and agile manufacturing system (FAMS) with input and output buffers using automatically guided vehicles (AGVs) on the basis of a Petri Net model. A large scale system can simply be represented by using hierarchical Petri Net (PN). Conflicts are resolved by definition of priority functions which efficiently determine a firing decision vector, and maintain high efficiency even if an AGV cannot operate in case of trouble. This method proves effective in simulation.

In industrially advanced nations, as a requisite for keeping up the manufacturing industry, the development of a manufacturing system with high flexibility and efficience is very important. It is necessary to model a manufacturing system based on a creative idea, simulate on various conditions to find problems, and spy out the optimal plan to design and manage the system. Our research should perform important duties for realization of the manufacturing system.

References

[1] J. Blazwicz, et al., "Scheduling Multiprocessor Tasks on Three Dedicated Processors," *Information Processing Letters 41*, pp. 429-442 (1992)

[2] C. L. Pape, "A Combination of Centralized and Distributed Methods for Multi-Agent Planning and Scheduling," *Proc. IEEE Int. Conf. on Robotics and Automation*, pp. 488-494 (1990)

[3] M. Silva, "Petri Nets and Flexible Manufacturing," *Advances in Petri Nets 1989, Lecture Notes in Computer Science 424*, G. Rozenberg(ed), Springer-Verlag, pp. 374-417 (1990)

[4] F. Archetti, et al., "Optimal Control Policies for Automated Storage/Retrieval System Using PN Models and Stochastic Optimization," *Proc. of the Intl. Workshop on Petri Nets & Performance Models*, Melbourne, Australia, pp. 258-267 (1991)

[5] J. Martinez, et al., "Petri Nets for the Specification of FMSs," *in Modeling and Design of Flexible Manufacturing Systems*, A. Kusiak(ed), Amsterdam: Elsevier, pp. 389-406 (1986)

[6] H. Shiizuka, "Practical Petri Nets–From Fundamentals to Software Tools," Corona Publishing Co., Ltd (1992)

[7] H. Arai, et al., "Applications of Timed Petri Net to Scheduling Problems of Repetitive Processes and Their Restoring Strategies in Emergency Stops," *Journal of Transactions of the Society of Instrument and Control Engineers*, Vol. 22, No. 9, pp. 955-961 (1984)

[8] T. Murata: "Static Analysis of Concurrent Programs by Petri Nets," *Journal of Information Processing Society of Japan*, Vol. 34, No. 6, pp. 701-709 (1993)

[9] J. L. Peterson, "Petri Net Theory and the Modeling of Systems," Prentice-Hall (1981)

[10] T. Hisamura, "Description and Fire of Petri-Net Taking Control Time into Consideration," *Proc. of the 23th SICE Annual Conference*, Tokyo, Japan, 2105, pp. 297-298 (1984)

8

8. Stochastic Petri Net Models of Communication and Flexible Systems

JIACUN WANG and SHENGBING JIANG

Dept. of Automatic Control
Nanjing University of Science and Technology
Nanjing 210014, P. R. China

8.1 Introduction

Petri net (PN) [1,2] has gained a widespread acceptance as a simple but powerful tool for the graphical description of the logical behaviour of discrete event systems. Indeed, PN has become a standard notation for the description of concurrency and synchronization in such fields as the modeling of protocols and computer system.

Classical PN includes no notion of timing, and thus allows only the logical description of the system behaviour, and the study of some logical properties of the system. For example, a PN model of a protocol can be used to verify the absence of deadlocks. On the other hand, it often happens that models of discrete event systems are developed in order to evaluate some time related performance indices, such as throughput and delay. For this reason, several different authors have proposed the use of augmented PN models in which means are provided for the specification of timing. We refer to these models as Timed PN (TPN).

From TPN it is possible to derive either analytical or simulation models of the system under investigation. The derivation of a simulation model from a TPN is a rather straightforward matter, and some packages exist for this purpose. The derivation of an analytical model from a TPN may instead be more complex and present subtle theoretical problems. In the discussion of TPN it is first necessary to separate the cases of deterministic and stochastic timing. Several authors have already considered either one of these environments. In this chapter we focus our attention on the case of stochastic timing. In this case TPN can be viewed as a graphical representation of a discrete time stochastic process, and the system behaviour can be studied using probabilistic arguments. In particular, the use of exponential distributions leads to an isomorphism between TPN and Markovian stochastic processes. This highly simplifies the model analysis since a wide gamut of results are readily available in this case.

The most natural way of introducing time into a PN refers to the interpretation of a PN in which, given a system condition (marking), some time must elapse before an event occurs (a transition fires). The event is the final result of some activity that is performed by the system when it is in the condition specified by the marking. Time is thus naturally associated with transitions, indicating that they can fire after some delay from the instant when they become enabled. The choice of associating time with transitions is the most frequent in the literature on TPN [3–10].

A different approach to introducing timing into a PN is to assign time to places, indicating that a condition becomes true only some time after an event has happened, because the new condition is the result of an activity triggered by the event. This is the approach followed by Sifakis [36].

In this chapter we restrict our attention to PN with stochastic timed transitions (stochastic timed transition PN or SPN) and their applications to communication and flexible systems. SPN can be used for several purposes in the process of modeling a discrete event system. First, SPN are very useful for a precise, unambiguous and compact specification of the system behaviour which is clearly understandable even by non specialists. Second, SPN can be very useful in the construction of system simulators. Third, in several cases SPN can be used as a high level interface for the description of a stochastic model of the system. The derivation of the stochastic model from the SPN description, as well as its solution, can be automated, and thus made completely transparent to the user, who need not be aware of the complex mathematical algorithms implemented.

8.2 Stochastic Petri Net Foundamentals

8.2.1 Stochastic Petri Net Models

The stochastic PNs (SPNs) described by Molloy [8, 9] are obtained by associating with each transition in a PN an exponentially distrbuted firing time. A formal definition of a SPN is thus the following:

$$\mathrm{SPN} = (P,\ T,\ I,\ O,\ m_0,\ \Lambda), \tag{8.1}$$

where $(P,\ T,\ I,\ O,\ m_0)$ is a PN and

$$\Lambda = \{\lambda_1,\ \lambda_2,\ \cdots,\ \lambda_m\} \tag{8.2}$$

is the set of firing rates (possibly marking–dependent) associated with the PN transitions.

Molloy [8, 9] has shown that SPNs are isomorphic to continuous time Markov chains (MCs) due to the memoryless property of the exponential distribution of firing times. SPN markings correspond to MC states. In particular, k–bounded PNs are isomorphic to finite MCs.

The sojourn time in each marking (state) is an exponentially distributed random variable with average

$$\left[\sum_{i \in H} \lambda_i\right]^{-1}$$

where H is the set of transitions that are enabled by the marking. The transition rate from marking m_i to marking m_j is obtained as

$$\sum_{k \in Hij} \lambda_k$$

where Hij is the set of transitions, which contains only one transition in most of the cases, enabled by marking m_i , and firing any one in Hij generates marking m_j . The infinitesimal generator Q of the MC isomorphic to the SPN can thus be easily obtained, and the steady state probability distribution of the MC, denoted by Π, is computed by solving the matrix equation:

$$\Pi Q = 0 \tag{8.3}$$

$$\sum_i \Pi_i = 1 \tag{8.4}$$

where Π_i is the steady state probability of state i.

For example, Fig.8.1 shows a SPN model with its reachable markings and its reachable graph. The line system of steady state probabilities is

$$(\pi_0, \pi_1, \pi_2, \pi_3, \pi_4) \begin{bmatrix} -\lambda_1 & \lambda_1 & 0 & 0 & 0 \\ 0 & -(\lambda_2 + \lambda_3) & \lambda_2 & \lambda_3 & 0 \\ 0 & 0 & -\lambda_3 & 0 & \lambda_3 \\ 0 & 0 & 0 & -\lambda_2 & \lambda_2 \\ \lambda_4 & 0 & 0 & 0 & -\lambda_4 \end{bmatrix} = 0$$

$$\pi_0 + \pi_1 + \pi_2 + \pi_3 + \pi_4 = 1$$

Let $\Lambda = (1, 1, 1, 1, 1)$ and solutions to this system are:

$$\pi_0 = \pi_4 = \frac{2}{7}, \pi_1 = \pi_2 = \pi_3 = \frac{1}{7}.$$

With the introduction of SPNs a link is established between two important classes of computer system models: the above–mentioned graph models, and

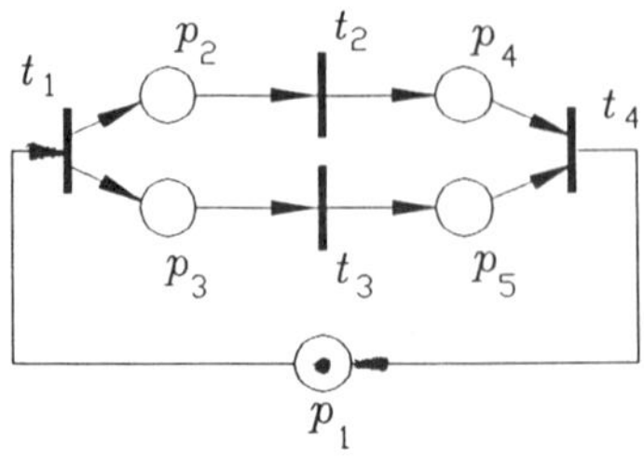

(a)

	p_1	p_2	p_3	p_4	p_5
m_0	1	0	0	0	0
m_1	0	1	1	0	0
m_2	0	0	1	1	0
m_3	0	1	0	1	0
m_4	0	0	0	1	1

(b)

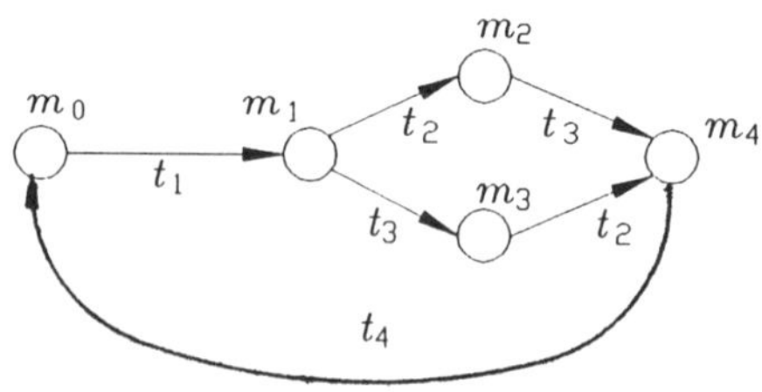

(c)

Fig.8.1 (a) A Simple SPN model (b) Reachable markings
(c) Reachable graph

the probabilistic models familiar to performance analysts. SPNs are a very useful tool for the analysis of computer systems since they allow the system operations to be precisely described by means of a graph that translates into a Markovian model useful for obtaining performance estimates. A very important feature of SPNs is that they can be understood by people who are not familiar with the probabilistic modeling approach, and that the derivation of the MC model and its solution can be made automatic, and thus transparent to the user.

8.2.2 Generalized Stochastic Petri Nets

The limitation of SPNs is that the graphical representation of systems rapidly becomes more difficult as system size and complexity increase. Moreover the number of states of the associated MC grows very fast as the dimensions of the graph increase. SPNs can thus be used to model only systems of limited size.

Often it is not desirable to associate a random time with each transition, since one would rather associate times only with the events believed to have the largest impact on system performance. A typical example is the case in which the operating sequence of a system comprises activities whose durations differ by orders of magnitude. It is then conceivable to model the brief activities only from the logical point of view whereas time is associated with the longer lasting ones. This choice becomes particularly convenient if by doing so the number of states of the associated MC model is reduced, thereby reducing the solution complexity.

GSPNs[10,44] are obtained by allowing transitions to belong to two different classes: immediate and timed. Immediate transitions fire in zero time once they are enabled. Timed ones fire after a random, exponentially distributed enabling time. Usually we use the convention of drawing timed transitions as thick bars, and immediate transitions as thin bars. Firing rates are obviously associated only with timed transitions, and they may depend on the GSPN marking. A formal definition of a GSPN is thus as in (8.1), where the array Λ now contains only m' elements, m' being the number of timed transitions in the GSPN. For example, for the SPN model shown in Fig.8.1, if we let transition t_4 be immediate, then the model is changed to a GSPN one with $\Lambda = \{\lambda_1, \lambda_2, \lambda_3\}$.

Several transitions may be simultaneously enabled by a marking. If the set of enabled transition H comprises only timed transitions, then the enabled timed transition t_i $(i \in H)$ fires with probability

$$\frac{r_i}{\sum_{k \in H} r_k}$$

exactly as with SPNs. If H comprises both immediate and timed transitions, then only immediate ones can fire. If H comprises zero or more timed transitions and only one immediate transition, then this immediate one fires. When H comprises several immediate transitions it is necessary to specify a probability density function on the set of enabled immediate transitions according to which the firing transition is selected. The subset of H comprising all enabled immediate transitions together with the associated probability distribution is called a random switch. This associated probability distribution will be called a switching distribution. Different markings may originate a single random switch whenever they enable the same set of immediate transitions upon which a single (possibly marking–dependent) switching distribution can be difined. If the probability associated with an enabled immediate transition is zero, the transition cannot fire, and thus behaves as if it were not enabled.

This condition, in some cases, can be modeled in a simpler way by means of inhibitor arcs. An inhibitor arc connects a place to a transition and is represented by a line terminating with a circle rather than with an arrowhead at the transition. The firing condition of a transition can thus be generalized by saying that a transition is enabled when all of its "normal" input places contain at least one token and no token is contained in its "inhibitor inputs". The firing of the transition removes a token from each of its normal input places only. Inhibitor arcs do not expand the modeling power of GSPNs, but in some cases, they allow a simpler description of the system operations, since the use of inhibitor arcs may reduce the number of random switches to be defined in the GSPN.

By examining GSPN behavior as a function of time we can easily realize that it is equivalent to the time behavior of a stochastic point process (SPP) $\{X(t),\ t>0\}$ with a finite state space. A one–to–one correspondence exists between GSPN markings and SPP states. Sample functions representing a possible behavior sequence of the SPP, may present "multiple discontinuities" due to the sequential firing of one or more immediate transitions. The process is observed to spend a nonnegative amount of time in markings enabling timed transitions only, while it transits in zero time through markings enabling immedite transitions. We call a state (or marking) of the former type tangible and a state (or marking) of the latter type vanishing.

The GSPN application purposes allow the following assumptions to be made.

(1) The reachability set is finite.
(2) Firing rates do not depend on time.
(3) The initial marking is reachable with a nonzero probability from any marking in the reachability set. No marking (or group of markings) exists that "absorbs" the process.

These assumptions further specify the nature of the SPP that can thus be classified as a finite state space, and a stationary (homogeneous), irreducible, and continuous–time stochastic point process.

The authors in [10] developed a method of analysis, based on identifying an embedded MC within the SPP, for the evaluation of the steady–state probability distribution of the GSPN markings. They assumed, however, that the SPP is irreducible, and no relatively infinite number of events in a finite–time interval can occur, which may happen when the duration of activities in the system differ by several orders of magnitude.

Ammer et al.[11,12] proposed a more general method of analysis for GSPN's. This method is based on characterizing the above process as a stochastically discontinuous Markov process (SDMP), the analysis of which is based on the concept of state aggregation, and can be divided into two steps. First, a discrete parameter MC is obtained by considering only immediate transitions in the reachability graph of the GSPN. This MC determines a partition of the state space in terms of ergodic classes and transient states. Each ergodic class consists of one marking or several instantaneous markings. From the analysis of this MC we can determine the trapping probabilities from the transient states to the ergodic classes as well as the steady–state probabilities of the states within each ergodic class. Secondly, an aggregated continuous parameter MC is obtained, the state space of which is defined on the set of ergodic classes. The steady–state probabilities of the ergodic classes can then be evaluated by analyzing this aggregated MC.

8.2.3 Extended Stochastic Petri Net Models

An extended SPN (ESPN) [13] is such a SPN whose timed transitions can be classified into three classes: exclusive transitions–for all markings that enable an exclusive transition t_i, this is the only enabled transition; competitive cransitions–for all markings enabling a competitive (non–exclusive) transition t_i, all enabled transitions are in conflict with t_i ; concurrent transitions–for some marking enabling a concurrent (non–exclusive) transition t_i some enabled transition is not in conflict with t_i. It is then possible to show that the SPP underlying an ESPN is a semi–Markov process provided that with all concurrent transitions are associated exponentially distributed firing times, and that the firing policy of all competitive transitions is such that, when they become enabled, the new firing delay is resampled from the distribution. Exclusive and competitive transitions can be associated with arbitrarily distributed firing delays. The steady state distribution of the semi–Markov process underlying an ESPN is then obtained with standard techniques[14,15].

Let $\{X_t,\ t>0\}$ be a semi–Markov process, X_n be the n–th state visited,then $\{X_n,\ n>0\}$ is called the embedded Markov chain of this semi–Markiv process. A semi–Markov process $\{X_t,\ t>0\}$ is said to be irreducible if its embedded Markov chain $\{X_n,\ n>0\}$ is irreducible. Let T_{ii} denote the time between successive transition into state i and let $\mu_{ii}=E[T_{ii}]$. Let H_i denote the distribution of time that the semi–Markov process spends in state i before making a transition, and let μ_i denote its mean value. We have

Theorem 8.1[15] If the semi–Markov process is irreducible and if T_{ii} has a non–lattice distribution with finite mean, then

$$P_i = \lim_{i\to\infty} P\{X_i = i \mid X_0 = J\} \tag{8.5}$$

exists and is independent of the initial state. Futhermore,

$$P_i = \frac{\mu_i}{\mu_{ii}} \tag{8.6}$$

Theorem 8.2[15] Suppose that the conditions of Theorem 8.1 hold and the embedded Markov chain $\{X_n,\ n>0\}$ is positive recurrent. Then

$$P_i = \frac{\pi_i \mu_i}{\sum_j \pi_j \mu_j} \tag{8.7}$$

where π_i is the steady probability of state of the Markov chain $\{X_n,\ n>0\}$.

Wang et al.[15,43] also discussed the time delay analysis method of descrete event systems represented by ESPN models. In Section 8.3.2 we will give an example of its application.

8.2.4 Deterministic Stochastic Petri Net Models

Deterministic stochastic Petri nets(DSPNs) [16,17,18] consider a different extension which allows firing delays of timed transition to be either constant, or exponentially distributed random variables. The reason for considering such a model is the following. Normally users are rather happy with exponential distributions whenever the duration of the activity that is modeled with a transition has a coeffinient of variation significantly greater than zero. They tend instead to be dissatisfied with an exponential distribution assumption when the activity to be modeled has a constant (or almost constant) duration. Providing the two alternatives of deterministic and exponential firing times will significantly improve the acceptance of SPN as a modeling tool. In fact, the introduction of deterministic timing appears to be even more important to the naive user than the availability of phase type distributions with which they may not be familiar.

Of course, in order to keep the underlying stochastic model reasonably simple, some restrictions must be observed in the use of deterministic firing times. In the cases of exclusive and competitive transitions a constant firing delay can be used freely, provided that the SPN execution policy is race with enabling memory. In the case of concurrent transitions, the SPN execution policy can be race with age or enabling memory, but all markings must enable at most one deterministic transition.

Under the above restrictions, it is possible to map the DSPN onto a semi–Markov process, using an approach quite similar to that of the "embedded Markov chain" (EMC) discussed in [19] for the analysis of M / G / 1 queueing systems; a discrete–time Markov chain can be defined at the instants of deterministic transition firings. When either no deterministic or an exclusive deterministic transition is enabled in a given marking m_i, m_i trivially maps onto a state of the EMC. In the case of markings enabling several competitive transitions, among the deterministic ones only that with the shortest firing time can fire, in race with the possibly enabled exponential transitions; the DSPN marking maps onto a state of the EMC. Both the transition probabilities to other states and the mean sojourn time in the state can be computed. When a deterministic transition t_i is enabled concurrently to exponential transitions, the next state of the EMC is sampled only at the instant of firing of t_i , i.e., the EMC state evolution does not account for the marking changes due to the firings of exponential transitions during the enabling interval t_d , but "delays" those possible state changes to the instant of firing of the deterministic transition. The state transition probability of the EMC is computed using the Chapman–Kolmogorov equation as:

$$exp\ (\tau_d\ Q')\ D\ (t_d)$$

where Q' is the infinitesimal generator of the exponentially distributed part of the process, and $D(t_d)$ is the state change matrix corresponding to the firing of transition t_d. From the steady state solution of the EMC, it is possible to compute the steady state probability distribution of the DSPN, by weighting the state probabilities with the mean sojourn times, and converting the probabilities of states enabling concurrent deterministic transitions with the conversion matrix:

$$C(t_d) = \int_0^{\tau_d} \theta \; exp \; (\theta \; Q') \; d\theta \qquad (8.8)$$

which accounts for correspondence between marking sojourn times in the DSPN and state sojourn times in the EMC.

Both the EMC transition probability matrix and the probability conversion matrix can be efficiently computed using a truncated Taylor series expansion of matrix exponential. The algorithm outlined above has been implemented as part of a software tool for the automated analysis of GSPN [20]. The extension of the algorithm to the case of a single deterministic transition in concurrence with phase distributed transitions is conceptually trivial.

8.2.5 Arbitrary Stochastic Petri Nets

An arbitrary stochastic Petri net (ASPN) [21] is also an extention of a SPN. An ASPN is a six–tuple (P, T, I, O, m_0, f) where (P, T, I, O, m_0) is a Petri net and $f : T \rightarrow R$ is a firing delay function of arbitrary distributions.

Moment generating function (MGF) based methcds [21] can be used to find a lower and upper bound of an ASPN model. Exact solutions can be obtained for those ASPNs in which transitions with nonexponential distributions are converted into subnets in which each transition has a firing delay of exponential distribution by using the existing technigues [31].

Moment generating functions are derived from Laplace transform [22]. They offer a solid mathematical foundation for GERT analysis and the approach discussed in this chapter. Note that Pr, p_1, p_2, and p will be used as probability parameters only in ths section.

A moment generating function (MGF) of a random variable is defined as follows [22, 23]:

$$M(s) = \int_{-\infty}^{\infty} e^{st} f(t) dt \qquad (t \text{ is continuous}) \qquad (8.9)$$

or

$$M(s) = E\;(e^{st}) \qquad (8.10)$$

where s is an arbitrary parameter and $f(t)$ is a probability density function of random variable t. Note that $M(0) = 1$.

Suppose that t represents the time duration for event or activity E whose probability to occur at a certain state is $p(0 < p < 1)$. Then a transfer function of E is defined as $pM(s)$. The following property lays down a theoretical foundation for moment generating function approaches to evaluating stochastic Petri nets.

Property[21] Suppose that t_1 and t_2 are two transitions whose firing time delays are X_1 and X_2, respectively. X_1 and X_2 are independent. For $i=1$ and 2, X_i has MGF $M_i(s)$, t_i fires at probability p_i and has transfer function $W_i(s)$, and $0 < p_i < 1$. t is a transition with time delay Y with MGF $M(s)$, firing probability p, and has transfer function $W(s)$.

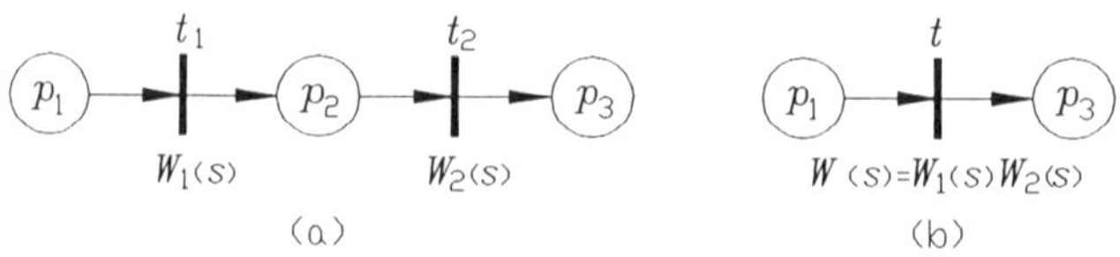

Fig.8.2 Sequence structure and its equivalent

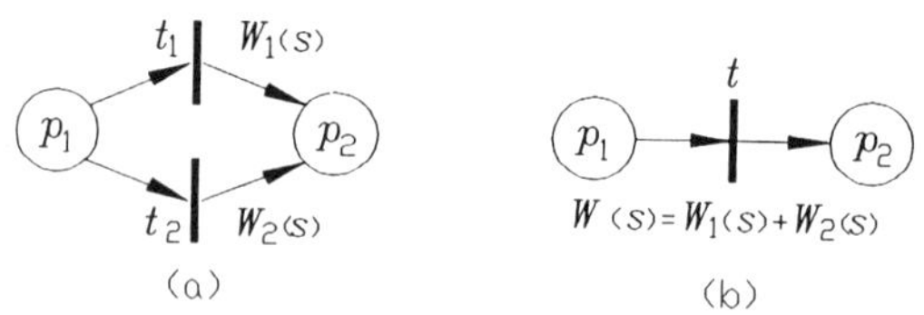

Fig.8.3 Parallel choice structure and its equivalent

Fig.8.4 Loop structure and its equivalent

P1. For a sequence Petri net structure (Figure 8.2a) and the equivalent transition t:

$$M(s) = M_1(s)\, M_2(s) \text{ and } W(s) = W_1(s)\, W_2(s).$$

P2. For a parallel choice Petri net structure (Figure 8.3a) and the equivalent transition t:

$$M(s) = \frac{p_1 M_1(s) + p_2 M_2(s)}{p_1 + p_2},$$

$$W(s) = W_1(s) + W_2(s).$$

P3. For a loop Petri net structure (Figure 8.4a) and the equivalent transition t:

$$M(s) = \frac{(1-p_2)\, M_1\,(s)}{1-p_2\, M_2\,(s)} \text{ and } W(s) = \frac{W_1\,(s)}{1-W_2\,(s)} \text{ if } p_2 \neq 1.$$

There are five steps when the MGF–based method is applied to a practical DEDS:

(1). ASPN Modeling. Using Petri net design methodologies such as bottom–up [24], topdown [25, 26], and hybrid approaches [27–29], we can synthesize an ordinary Petri net model for a system based on its operations and relationship among these operations. After we get such an ordinary Petri net, time requirements for various operations result in an ASPN model, in which every transition is associated with an appropriate time delay that is either constant or random.

(2). Generation of Reachability Graph.

(3). Generation of State Machine Petri Net. A state machine Petri net is generated based on the derived reachability graph and information on firing delays of transitions. This state machine Petri net is an ASPN with a particular structure, i.e., each transition has exactly one input place and one output place. A place in the net can have multiple input and output transitions. The place with more than one output transition is called a choice place.

(4). Derivation of Transfer Functions. For the above state machine Petri net, the transfer functions of interesting indices can be derived in two steps. The first step is to identify the places or states of interest in the state machine Petri net and create a new one as a sink place such that their output arcs and transitions will be directed to that sink place. The sink will absorb any number of tokens like a sink transition [25]. Next, all Petri net structures of a sequence, parallel choice, or loop form are stepwise reduced to a transition with an equivalent transfer function being stepwise calculated [30]. The final equivalent transition is the one between the source place and the sink one. It is noted that Mason's formula can be directly used in such a reduction process.

(5). Evaluation of Performance Measures. To obtain the i–th moments, we simply take the i–th ($i \geq 1$) derivative of a transfer function of a performance index. Means and derivations of certain measures can be obtained. The analytical results may be obtained by inverting their transfer functions.

Except the MGF based method, Huang et al. developed a hybrid state analysis (HSA) method for the performance evaluation of ASPN models [32,33]. The basic idea is to make the extended state, hybrid state, which is composed of marking and the enabling times of enabled transitions under this marking, propagates as a Markov process by the inclusion of supplementary varibles. By the HSA method we can obtain the exact expressions for the probability distribution density functions of time delays of ASPN models with no restrictions to transitions′ firing policy and firing time distributions.

8.2.6. Aggregation for Complex Stochastic Petri Nets

Although SPNs are highly effective in modeling concurrency, synchronization, and communication, they suffer from two major drawbacks. First, there is the explosion of the state space of its reachability graph when suffcient details are introduced in the model. Second, the presence of timed transitions having orders of magnitude difference in their firing rates leads to an underlying stiff Markov chain, which results in numerical instability in their solution technique. These have been a major hindrance in using SPNs for modeling practical systems with a large number of resources.

The technique of time scale decomposition (TSD) of SPNs developed by Ammar et al. [12,34,35] offers a step in the right direction for a solution to the above fundamental problems.

The essence of TSD technique is described as follows:

1) It is applicable to a system containing activities whose duration differs by orders of magnitude. This is to insure a good accuracy for the solution technique.

2) The given system can be decomposed into a hierarchical sequence of aggregated subnets each of which is valid at a certain time scale.

3) The aggregation at each level is done by assuming that the transitions included in the lower level are immediate transitions.

4) At each level of the hierarchy, the current marking of an aggregated subnet determines the number of tokens in the subnet at the lower level, which are then analyzed to determine the rate of transitions in the aggregated subnet.

Let us consider the GSPN = $(P, T, I, O, m_0, \Lambda)$, and that the set of T is composed of the set of slow transitions T_s and the set of fast transitions T_f .

The aggregated GSPN which can replace the GSPN is denoted by AGSPN = $(P_a, T_{as}, I_a, O_a, m_0, \Lambda_a)$. The steps of the time scale decomposition (TSD) algorithm are as follows:

Step 1: Represent the system using a GSPN = $(P, T, I, O, m_0, \Lambda)$, identify the sets T_s and T_f and determine the presence of recurrent and nonrecurrent subnets which will be aggregated into places $p_{ui} \in P_u$ at the slow time scale.

Step 2: Find the set T_{as}, P_a and the initial marking m^1 of the AGSPN. Let RG be the reachability set of the AGSPN.

Step 3: Let the current marking be $m^k = (m_1^k, m_2^k, \cdots, m^k_{|P_a|})$. Insert this into the reachabiliy set RG of the AGSPN.

Step 4: The initial marking m_i^k of the subnet i determines the marking dependent firing rates $\lambda_{a1}, \lambda_{a2}, \cdots, \lambda_{a|T_{as}|}$.

Step 5: Starting with the current marking m^j of the AGSPN we reach the current marking m^k of the AGSPN through firing of some transition $t_{ks} \in T_{as}$ with nonzero firing rate given by R_u.

Step 6: The new marking m^k defines some new initial marking m_i^k of subnet i, where $i = 1, \cdots, |P_a|$.

Step 7: Repeat from Step 3 to 6 while a new AGSPN marking m^k results from firing of some transition t_{ks} from some current marking m^j.

Step 8: The AGSPN can now be solved for the steady state solution of the markings in the reachability set RG.

Step 9: The token probability distribution of some place $p_i \in P$ of the given GSPN (i.e., GSPN of Step 1) can now be determined as follows:

$$Prob[p_i \text{ contains } m \text{ tokens}] = \sum_{k=1}^{n} Prob\,[m_i^k]\, d_{pi}\,(m)$$
$$= \sum_{k=1}^{n} Prob\,[m^k]\, d_{pi}\,(m)$$

where m_i^k is the initial marking of subnet i and place p_i belongs to subnet i. In addition, $d_{pi}(m)$ is the token probability distribution of place p_i with initial marking m_i^k.

It can be easily shown that the decomposition of a GSPN into fast subnets is equivalent to the block diagonalization of the nearly decomposable stochastic matrix of a stiff MC. Courtois [46] has shown that the error due to such decomposition is in o(ε), where ε is the maximum degree of coupling between aggregates, which are the fast subnets in our case.

Valette [36] and Suzuki et al. [37] presented a method for expanding or reducing a Petri net representation to the desired level of detail using step by step refinement of transitions and places (or abstraction of subnets to transitions). Wang et al. [38,39] transplanted the abstraction (reduction) technique mentioned above to propose hierarchical models for time delays of complex ASPN modelling systems. The basic idea is to substitute subnets (blocks) by transitions such that these transitions' firing time distribution only depend on the subnets substituted, and the probability nature of time delay of the model preserves invariable after such substitutions. Thus the time delay analysis of a complex ASPN models can be done by solving several smaller models.

8.3 Performance Analysis of Communication and Flexible Systems

8.3.1 SPN Model of A Semi–duplex Protocol

The SPN shown in Fig. 8.5 models a semi–duplex protocol where a dropped message will not be retransmitted. The model comprises 8 places and 8 transitions. The process that a communication system using such a protocol transmits a frame of message can be described as follows. The system is initially in

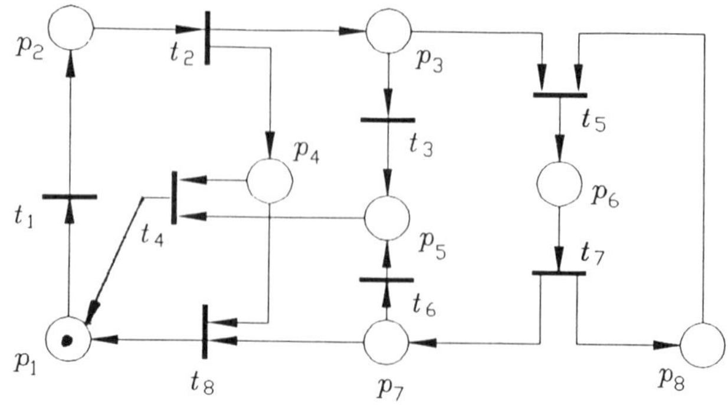

Fig.8.5 SPN model of a semi–duplex protocol

the state that the buffer which stores the messages to be transmitted is available (a token in p_1) and the receiver is ready to receive (a token in p_8). After a random time delay a frame of message arrives (t_1 fires and a token moves into p_2). Then the frame is transmitted (t_2 fires and both of p_3 and p_4 are placed a token). If the frame is received by the receiver and the error detection routine drops the message (t_3 fires and a token moves into p_5), then after a fixed time delay the system returns to its initial state and is ready to receive and transmit a new frame of message (t_4 fires, and a token moves into p_1). If the frame is received and the error detection routine okays (t_5 fires and a token moves into p_6), then the receiver sends a ACK back (t_7 fires, and both p_7 and p_8 are placed a token). If the ACK is received by the sender of the system and the error detection routine drops the ACK (t_6 fires and a token moves into p_5), then after a fixed time delay the system returns to its initial state (t_4 fires and a token moves into p_1). Otherwise, if the error detection routine okeys, the system returns to its initial state at ones (t_8 fires, and a token moves into p_1).

The normal assumptions of exponential message length and interarrival times make the distributions of transitions t_1, t_2 and t_3 exponential. Transitions t_3 and t_4 occur when the message is received and the error detection routine either okays or drops the message. If the CRC check is not embedded in the line drive routine, this calculation would take a constant time for a particular message length and therefore would also have an exponentially distributed execution time. If the CRC check is embedded in the line driver, the net can be further reduced by collapsing transitions t_3 and t_5 onto transition t_2. Similarly transitions t_6 and t_8 have exponentially distributed execution times.

Transition t_4 in the real system is not exponentially and a fixed timeout value is used. Notice that t_4 enabled only after the message was transmitted and the CRC checked. That means that the probability density function of random variable representing the time until t_4 fires would be Erlangian instead of an impulse at the fixed time value. In this analysis the time until t_4 fires assumed to be exponentially distributed. That implies that the time value is actually random.

Solving for the reachability set we find 6 markings. Notice that probability π of no token being in place p_1 is the probability that the subsystem is busy and cannot accept new messages. Solving for the steady state probabilites of markings, we can then find the probability of the subsystem being busy. In this moment messages arrive to the system at a Poisson rate λ and are dropped if they arrive when no buffers are available. Therefore, the actual throughput of the system is $\lambda(1-\pi)$.

Consider a system using this protocol which has a 2400 baud rates with a 3 percent error probability and 400 bit packets and a fixed timeout value of 1 second. Then we can obtain some transitions' firing rates as: $\lambda_2 = \lambda_7 = 6$, $\lambda_3 = \lambda_6 = 3.093$, $\lambda_5 = \lambda_8 = 100$, and $\lambda_4 = 1$.

Using these values and varying the rate of the firing rate λ_1 of transition t_1 we can plot the throughput S versus offered load. The plot is shown Fig.8.6.

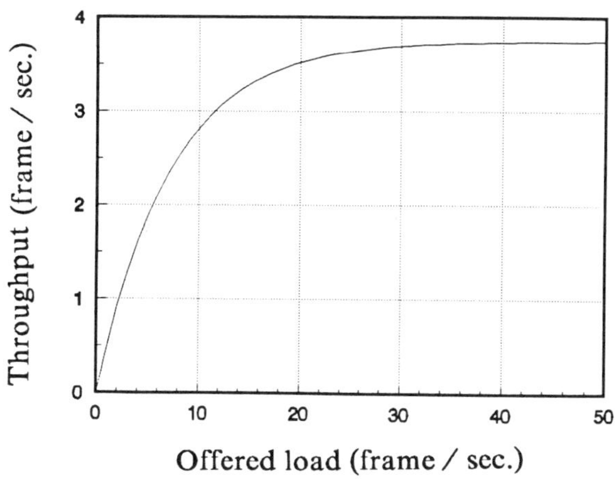

Fig.8.6 Throughput versus offered load for the protocol

8.3.2 ESPN Model of CSMA / CD Protocol

Carrier sense multiple access (CSMA) protocols are quite common in today's local area computer networks (LAN). In particular, the CSMA / CD protocol (CSMA with collision detection) [40] is used in Ethernet [41,42], as well as in many other LAN, and it is being standardized by IEEE with project 802. A precise and detailed specification of the CSMA / CD Ethernet protocol with DSPN is possible, and it has the advantage of being useful for the verification of the protocol correctness as well as for its performance analysis. The same modeling tool thus serves the three main areas of protocol modeling: specification, validation and analysis.

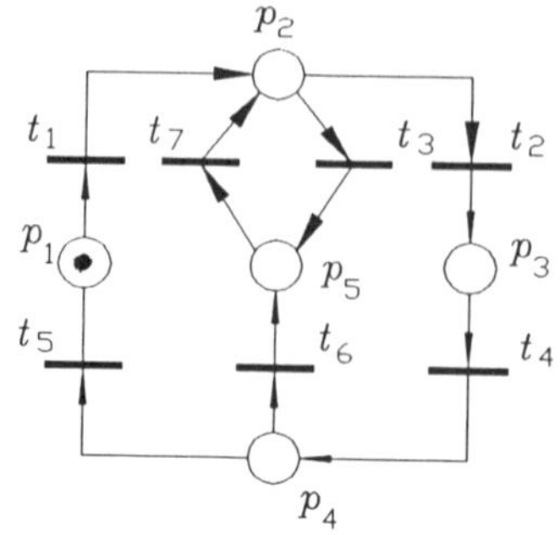

Fig.8.7 SPN model of a reduced CSMA / CD protocol

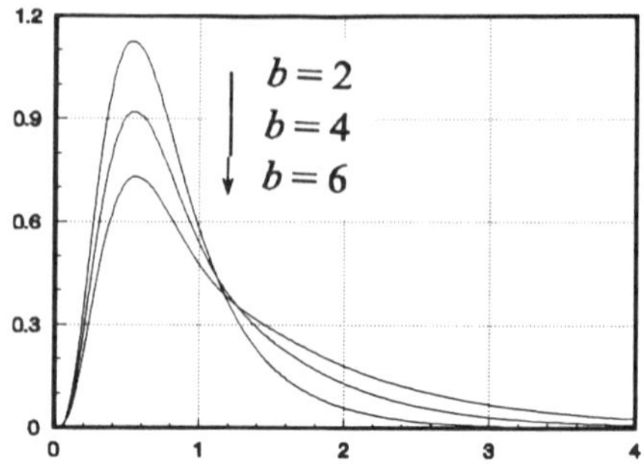

Fig.8.8 Time delay density function of a packet transmission

In this section we present in Fig.8.7 a reduced SPN model of the CSMA / CD protocol to provide an example of the use of our techniqu[43]. The model comprises 5 places and 7 transitions. The local area computer network (LAN) operations under the CSMA / CD protocol can be discribed as follows. A LAN station is initially in the state (place p_1) which models the situation in which the station is modulation a message to be transmitted into packet. After an exponentially distributed time delay the station generates a packet (transition t_1 fires and a token moves to p_2). The packet transmission then can begin (t_2 fires) only if the station senses the channel idle. Otherwise, if the station senses the channel busy (t_3 fires), it resubmits the packet transmission request after a time delay (t_7 fires). Once the packet transmission has begun (token in p_3), the station will go on to sence the channel for a while (t_4 fires, a token moves to p_4). If it doesn't sense conflict, the packet transmission will continue to complete. Otherwise, if it senses conflict (t_6 fires), it will resubmit the packet transmission request after a time delay (t_6 fires).
Denote by α the probability that a conflict (including collision and channel busy) happens. Denote by b the number of the station in a LAN. Suppose $\alpha = (b-1) / 20$, $b < 10$. The waiting time to avoid conflict is assumed to be uniformly distributed, Let

$$\psi_1(\tau) = 5e^{-5\tau}, \qquad \psi_2(\tau) = 10(1-\alpha)e^{-10(1-\alpha)\tau},$$
$$\psi_3(\tau) = 10\alpha e^{-10\alpha\tau}, \qquad \psi_4(\tau) = 5e^{-5\tau},$$
$$\psi_5(\tau) = 5(1-\alpha)e^{-5(1-\alpha)\tau}, \qquad \psi_6(\tau) = 5\alpha e^{5\alpha\tau},$$
$$\psi_7(\tau) = 1, \tau \in (0,1].$$

Then the ESPN model is semi–Markovian with 5 reachabe markings. According to the results developed in [15,43], if we disregard the influence that each execution path whose path probability is less then 0.02 has on $f(\tau)$, we can show the time delay density function of a packet transmission in Fig.8.8 approximatly.

8.3.3 GSPN Model of A Flexible Manufacturing System

An FMS is composed of machines, buffers, a transport system, and a control system. Machines are usually able to process different types of parts. Buffers are needed because different machines have different processing speeds. The transport system can move parts from one point to another within the FMS cell, following the commands issued by the control system. The control system decides the routing of parts with the objective of attaining the highest possible throughput and system utilization with the lowest possible inventory. Usually a part to be processed needs a support (pallet) on which it is loaded at beginning of its processing and from which it is unloaded after the completion of processing.

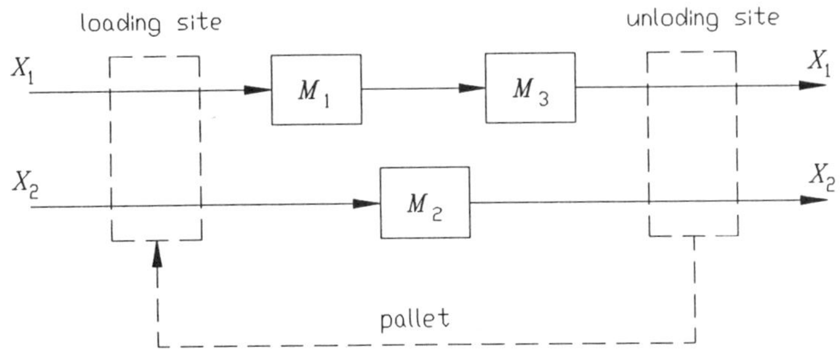

Fig.8.9 A flexiable manufacturing system

Consider as an example the FMS whose functional structure is depicted in Figure 8.9. It is composed of three different machines M_1, M_2 and M_3. Two types of parts (X_1 and X_2) can be processed by the three machines. Parts of type X_1 need to be processed by M_1 and M_3 in sequence, and parts of type X_2 need to be processed by M_2.

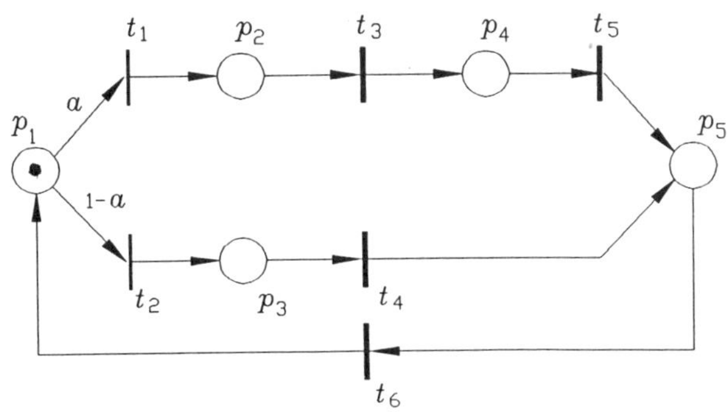

Fig.8.10 GSPN model of the FMS

Tab.8.1 Description of the GSPN model of Fig.8.10

p_1	parts available
p_2	part of type X_1 waiting for M_1
p_3	part of type X_2 waiting for M_2
p_4	part of type X_1 waiting for M_3
p_5	pallet available
t_1	M_1 loads part of type X_1
t_2	M_2 loads part of type X_2
t_3	M_1 processes
t_4	M_2 processes
t_5	M_3 processes
t_6	pallet moves with a part

The GSPN model of the FMS is shown in Fig.8.10, and the description on this model is given in Tab.8.1. There are two immediate transitions: t_1 and t_2, and four timed transitions: t_3, t_4, t_5 and t_6. The two immediate transitions t_1 and t_2 are both enabled under the initial marking $(1,0,0,0,0)^T$, thus set $\{t_1,t_2\}$ is a random switch, and a switching distribution should be given for the performance analysis of the model. We assume that the firing probabilities of t_1 and t_2 are a and $1-a$ as shown in Fig.8.10. Futher assume that $\lambda_3=\lambda_4=\lambda_5=\lambda_6=1$. From the above data and varying a we can plot the usages of muchines M_1, M_2 and M_3, which correspond to stead state probability of markings $(0,1,0,0,0)^T$, $(0,0,1,0,0)^T$ and $(0,0,0,1,0)$, versu a, which is shown in Fig.8.11.

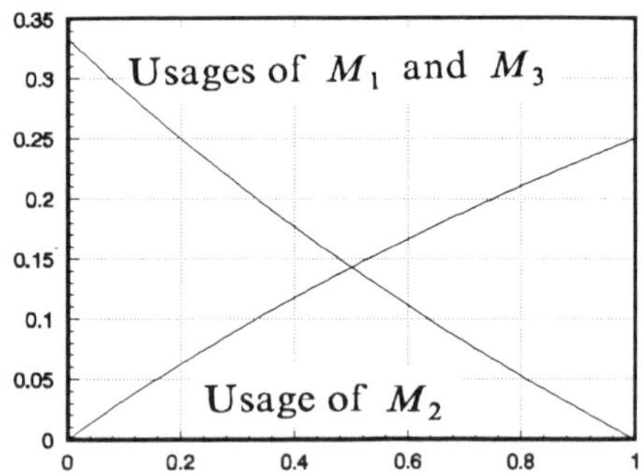

Fig.8.11 Usages of machines M_1, M_2 and M_3 versus a

8.3.4 ASPN Models of An Assembly System

A. System Description

In this section, we consider a robotic assembly system consists of two robot arms, R_1 and R_2, and two workstations, WS_1 and WS_2, as shown in Fig.8.12. The specification of this system is as follows:

Fig.8.12 The layout of a two–robot and two–workstation assembly system

(1). When either of WS_1 and WS_2 is ready to complete the assembly task, it requires its right robot and acquire it if it is available. This operation of WS_1 (WS_2) takes an exponentially distributed random time whose firing rate is λ_1 (μ_1).

(2). After a workstation acquires its right robot, it requests its left robot and acquires it if it is available. This operation of WS_1 (WS_2) needs an exponentially distributed random time whose firing rate is λ_2 (μ_2).

(3). When a workstation starts an assembly task, it cannot be interrupted until it completes. It takes an exponentially distributed random time to finish the assembly in WS_1 (WS_2), with firing rate λ_2 (μ_2).

(4). When WS_1 (WS_2) completes, it releases two robots in a constant time c_1 (c_2).

B. Petri Net Modeling

For this system, its Petri net model can be easily designed as shown in Fig.8.13 with neither dotted arcs nor transition t. The discription of the elements of set P and T is shown in Table 8.2.

Notice that firing transitions t_1 and t_5 leads to a system deadlock. To resolve such a deadlock, we can model deadlock detection and resolution as a transition t with the dotted arcs in Fig. 8.13. In particular, the function of t is to release two robots and two requests from both workstations when each workstation takes its right robot. Its time delay is a random variable distributed uniformly over $[a, b]$. It is remarked that similar ordinary Petri net model has been constructed and discussed for a resource–sharing problem between two processes [2]. The main difference lies in that this model is stochastic and includes a deadlock detection and resolution transition. In addition, two more places and three more transitions are used to model the system considered. The firing time delays of transitions t_1, t_2, t_3, t_5, t_6, and t_7 are exponentially distributed random variables with rates λ_1, λ_2, λ_3, μ_1, μ_2, and μ_3, respectively. The firing delays of transitions t_4 and t_8 are constants, c_1 and c_2, respectively. Firing delay of t has the uniform distribution over $[a, b]$. The following discussion focuses on the derivation of closed–form analytical performance results.

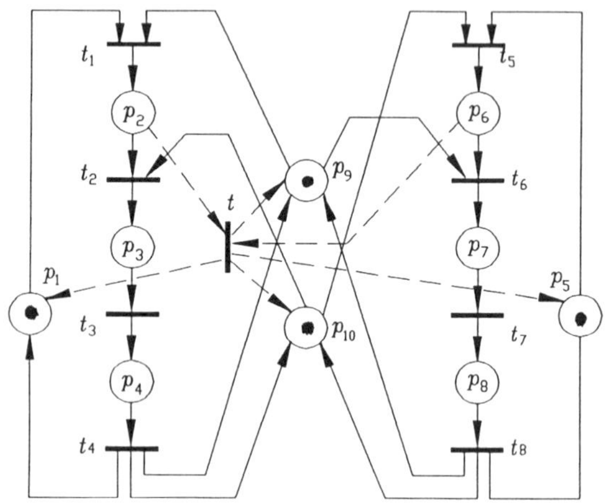

Fig.8.13 ASPN model for a two–robot and two–workstation assembly system

Tab.8.2 Description of the elements of Fig.8.13

p_1	WS_1 requests R_1
p_2	WS_1 requests R_2
p_3	WS_1 has R_1 and R_2
p_4	WS_1 ready to release robots
p_5	WS_2 requests R_2
p_6	WS_2 requests R_1
p_7	WS_2 has R_1 and R_2
p_8	WS_2 ready to release robots
p_9	R_1 available
p_{10}	R_2 available
t_1	WS_1 acquires R_1
t_2	WS_1 acquires R_2
t_3	Assembling at WS_1
t_4	Releasing robots from WS_1
t_5	WS_2 acquires R_2
t_6	WS_2 acquires R_1
t_7	Assembling at WS_2
t_8	Releasing robots from WS_2
t	Deadlock resolution

C. Reachability Graph

The reachability graph as shown in Fig.8.15 is very easily derived from the above Petri net model as shown in Fig.8.14. There are totally eight states:

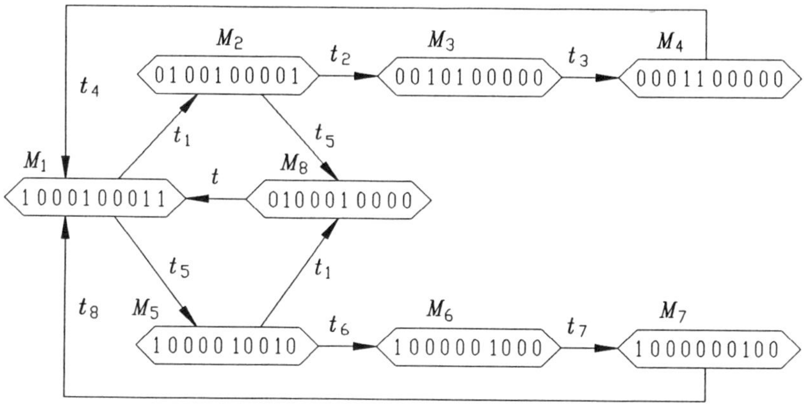

Fig.8.14 The reachability graph of the ASPN in Figure 8.13

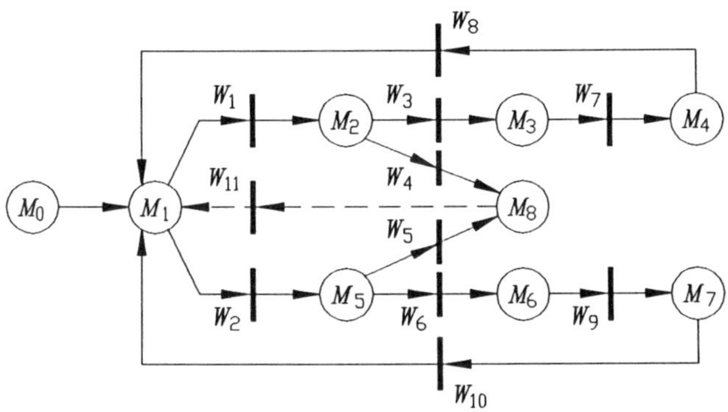

Fig.8.15 The state machine Petri net

M_1: Initial state representing both stations are requesting their robots and robots are ready.

M_2: WS_1 acquires R_1 and requtsts R_2
M_3: WS_1 acquires R_2.
M_4: Two robots are performing assembly at WS_1.
M_5: WS_2 acquires R_2 and requests R_1.
M_6: WS_2 acquires R_1.
M_7: Two robots are performing assembly at WS_2.
M_8: Deadlock state (WS_1 acquires R_1 and WS_2 acquires R_2).

The transitions whose firings lead the system to change from one state to another are also included in Fig.8.14.

D. State Machine Petri Nets

Corresponding the reachabitity graph as shown in Fig.8.14 its state machine Petri net follows as shown in Fig.8.15. There are eight places represented by M_i ($1 < i < 8$) and eleven transitions. For simplicity, W_i will be used to represent transition i and is actually the transfer function of the corresponding transition in the following discussion and figures. M_0 is a source place that functions as a source of tokens for this net. It can be regarded as a source transition that can produce an infinite number of tokens. M_0, as well as M_0' which is a sink place to absorb any number of tokens, is used to derive equivalent transfer functions.

The transfer functions of the transitions are derived by using the results in section 8.2.5 as follows:

$$W_1 = \frac{\lambda_1}{\lambda_1+\mu_1}\,\frac{\lambda_1+\mu_1}{\lambda_1+\mu_1-s}; \qquad W_2 = \frac{\mu_1}{\lambda_1+\mu_1}\,\frac{\lambda_1+\mu_1}{\lambda_1+\mu_1-s};$$

$$W_3 = \frac{\lambda_2}{\lambda_2+\mu_1}\,\frac{\lambda_2+\mu_1}{\lambda_2+\mu_1-s}; \qquad W_4 = \frac{\mu_1}{\lambda_2+\mu_1}\,\frac{\lambda_2+\mu_1}{\lambda_2+\mu_1-s};$$

$$W_5 = \frac{\lambda_1}{\lambda_1+\mu_2}\,\frac{\lambda_1+\mu_2}{\lambda_1+\mu_2-s}; \qquad W_6 = \frac{\mu_2}{\lambda_1+\mu_2}\,\frac{\lambda_1+\mu_2}{\lambda_1+\mu_2-s};$$

$$W_7 = \frac{\lambda_3}{\lambda_3-s}; \qquad W_8 = e^{c_1 s};$$

$$W_9 = \frac{\mu_3}{\mu_3-s}; \qquad W_{10} = e^{c_2 s};$$

$$W_{11} = \frac{e^{bs}-e^{as}}{b-a}.$$

E. Transfer Functions for Performance Measures

In order to find deadlock rate, we cut the flow transition W_{11} to M_1 and add a sink place M_0' to obtain the net in Fig.8.16. M_o' is used to absorb all tokens generated by its input transitions. Then stepwise reduction is applied to this net shown in Fig.8.16.

Let $\lambda_1=0.8$, $\lambda_2=1$, $\lambda_3=1$, $\mu_1=1$, $\mu_2=1$, $\mu_3=1$, $c_1=2$ and $c_2=2$, and suppose that a and b are two changing parameters. Then, the reduction proce–

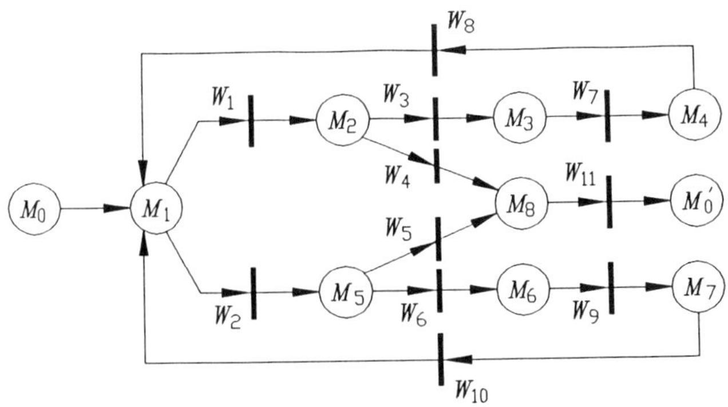

Fig.8.16 A net for solution of deadlock rate

dure and solution method developed in [27–30] lead to the following deadlock rate of the closed–form:

$$R_d = \frac{a + b}{11.444 + a + b}.$$

The exactly same reduction procedure or more directly Mason's rule [45], can lead to the following formulas, which describe the relationship between the production rates of the workstations and the deadlock resolution time.

$$PR_1 = \frac{1}{6.041 + 0.522(a + b)},$$

$$PR_2 = \frac{1}{4.299 + 0.387(a + b)}.$$

From these formulas and by noting that $(a+b)/2$ is the mean deadlock resolution time we can get Fig.8.17 and conclude clearly that

1. When $(a+b)/2$ increases, R_d becomes larger and approaches 1 while throughputs PR_1 and PR_2 become smaller and approach zero.

2. When $b \to \infty$, $R_d = 1$, $PR_1 = 0$, and $PR_2 = 0$. This means the system will be dead locked and the deadlock cannot be resolved: thus the production rates are zero.

3. When $a = b = 0$, $R_d = 0$, $PR_1 = 0.1$, and $PR_2 = 0.233$. This corresponds to the case that a deadlock is immediately resolved. Note that $PR_1 < PR_2$ because the firing rate of transition t_1 is less than that of t_5 in Fig.8.16 though t_2, t_3, t_6, and t_7 have the same firing rate.

The throughput of W_2 is greater than W_1 under the condition $\lambda_2 = \mu_2 = \lambda_3 = \mu_3 = 1$, $c_l = c_2 = 2$, $\lambda_1 = 0.8$, and $\mu_1 = 1$. Difference between two assembly stations lies only in their speeds to obtain their right robots. Since W_2 has the faster speed than W_1 its throughput is greater. This fact can be confirmed both graphically and theoretically, i.e.,

$$PR_2 - PR_1 = \frac{1.542 + 0.135(a+b)}{(4.299 + 0.387(a+b)(6.041 + 0.522(a+b))} > 0.$$

More results of a closed form can be derived when we allow some parameters to change and others to be fixed.

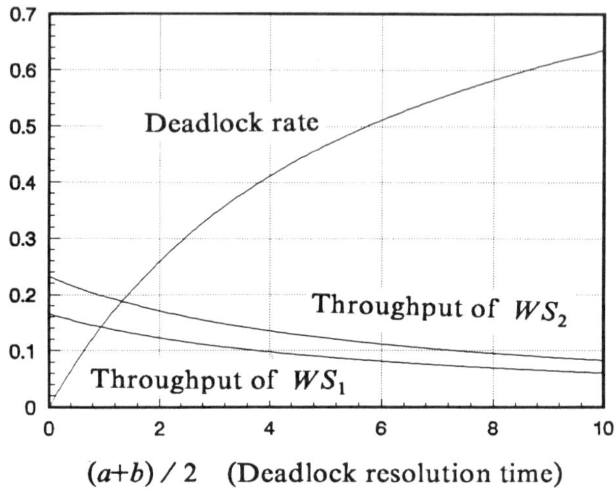

Fig.8.17 Deadlock rate, throughputs of Workstations 1 and 2 versus $(a+b)\,/\,2$

8.3.5 Hierarchical Models of A Buffer Multiprocessor Subsystem

Consider a buffer–multiprocessor subsystem consisting of n processors and a pool of buffers, as shown in Fig.8.18 [35]. Tasks are arriving as a Poisson process with the rate λ . If the buffer pool is empty they are rejected; otherwise they are placed in buffers to be served by the first available processor. As soon as a task residing in a buffer gets a processor, the buffer is released and can be used to hold new tasks. We also assume that the processing times are independent and are exponentially distributed with each processor having a processing rate μ.

Faults occur in both buffers and processors as a Poisson process with rates f_b and f_p respectively. For simplicity and to show the general technique, we assume that faults in both the buffer and processor occur only when they are waiting for a task. Repairs are performed on both buffers and processors (Poisson process with rates r_b and r_p) one at a time basis (single server).

Fig.8.19 shows the GSPN representation of the system GSPN $= (P, T, I, O, \Lambda)$ under study, where

$$P = \{p_1, p_2, p_3, p_4, p_5, p_6\}$$

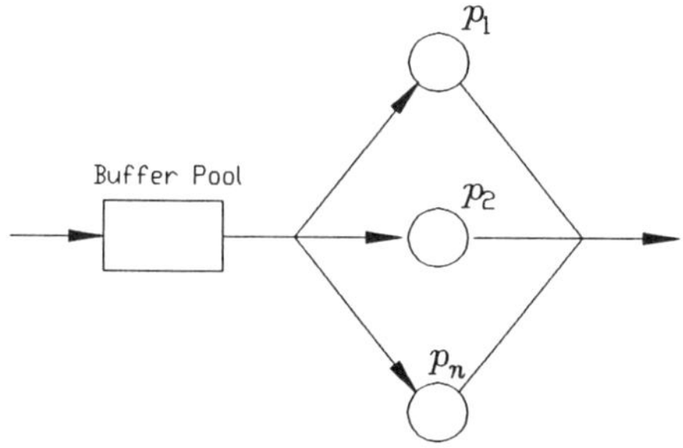

Fig.8.18 Buffer–multiprocessor subsystem

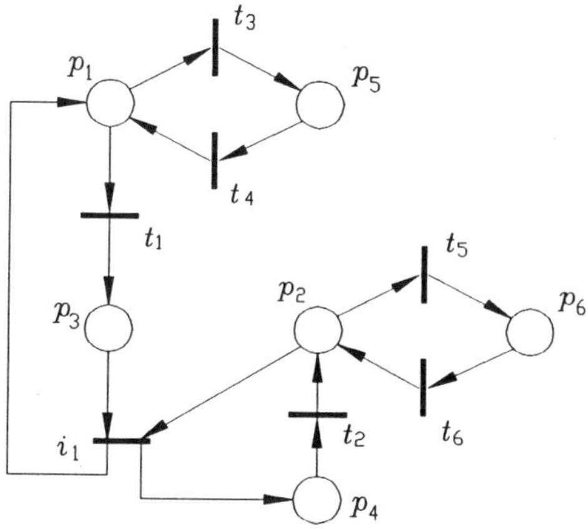

Fig.8.19 GSPN representation of buffer–multiprocessor subsystem

$$T = \{t_1, t_2, t_3, t_4, t_5, t_6, i_1\}$$

$$\Lambda = \{\lambda_1, \lambda_2(m), \lambda_3(m), \lambda_4, \lambda_5(m), \lambda_6\}$$

$$\lambda_2(m) = m(p_4)\lambda_2,$$

$$\lambda_3(m) = m(p_1)\lambda_3,$$

$$\lambda_5(m) = m(p_2)\lambda_5,$$

$$\lambda_1 = \lambda, \qquad \lambda_4 = r_b,$$

$$\lambda_2 = \mu, \qquad \lambda_5 = f_p,$$

$$\lambda_3 = f_b, \qquad \lambda_6 = r_p.$$

A description of the elements of the set P and T is given in Table 8. 3.

Table 8.3 Description of the GSPN Modeling Buffer–Multiprocessor Subsystem

p_1	available buffers
p_2	available processors
p_3	task in a buffer waiting for a free processor
p_4	task executing in a processor
p_5	failed buffers
p_6	failed processors
t_1	task arrival with rate λ
t_2	completion of task in processor with rate $m(p_2)\ \mu$
t_3	available buffer failure with rate $m(p_1)\ f_b$
t_4	buffer repair with rate r_b
t_5	available processor failure with rate $m_2 f_p$
t_6	processor repair with rate r_p
i_1	if a processor is available and a task is waiting in the buffer then start processing of that task and free the buffer

When modeling a system using GSPN's, transitions correspond to activities taking place in the system. Therefore activities such as job arrival, job completion, failure and repair of processor and memory each will be represented by a timed transition. Since the failure and repair rates are normally orders of magnitude smaller than the rate of the activities such as job arrival and completion, the sets of transitions T_s and T_f can be easily defined.

In order to solve the given GSPN model, we suppose that:

$$\mu = 1, \qquad f_b = 10^{-4},$$

$$\rho = 0.8, \qquad r_p = 10 f_p,$$

$$f_p = 10^{-4}, \qquad r_b = 10 f_b.$$

where $\rho = \lambda / \mu$.

From the above numerical values, clearly $T_s = \{t_3, t_4, t_5, t_6\}$ and $T_f = \{t_1, t_2, i_1\}$. At the fast time scale the failure and repair activities are so far apart in time that the system can be assumed fault free. The detail of the process of solving the problem using TSD method is omitted here, which is available in [35].

Tables 8.4–8.5 show the mean throughput for various processor buffer combinations. The exact thoughput is computed by solving the whole network and finding the token probability distribution in place $p_4 \in P$. From these

Table 8.4 Variation of Mean Throughput with One Processor

No.of Processors	No.of Buffers	Exact mean T–put	Mean T–put using TSD	Complexity		
				exact no.of states	TSD no.of states	TSD No.of state in subnet
1	1	0.5277	0.5277	8	4	(1,1)=3
1	3	0.6708	0.6710	24	8	(2,1)=4 (3,1)=5
1	6	0.7329	0.7329	63	14	(4,1)=6 (5,1)=7 (6,1)=8
1	9	0.7536	0.7600	120	19	(7,1)=9 (8,1)=10 (9,1)=11

Table 8.5 Variation of Mean Throughput with Two Processors

No.of Processors	No.of Buffers	Exact mean T–put	Mean T–put using TSD	Complexity		
				exact no.of states	TSD no.of states	TSD No.of state in subnet
2	1	0.9914	0.9922	15	6	(1,1)=3 (1,2)=4
2	3	1.1655	1.1600	42	12	(2,1)=4 (2,2)=4 (3,1)=5 (3,2)=6
2	6	1.1951	1.1969	105	21	(4,1)=6 (4,2)=7 (5,1)=7 (5,2)=8 (6,1)=8 (6,2)=9
2	9	1.1968	1.2008	195	29	(7,1)=9 (7,2)=10 (8,1)=10 (8,2)=11 (9,1)=11 (9,2)=12

tables, it is clear that the TSD algorithm is quite accurate and the complexity is significantly less in terms of the size of the reachability graph. The complexity columm in the tables compares the complexity of the exact and the TSD method in tems of the size of the reachability graph that we have to solve.

To find the throughput with the addition of a new buffer or processor to the system, it is not necessary to solve the whole problem. Only the solution of the additional part is required which arises from the addition of the new resources. Similarly, to find the throughput after removal of a buffer or processor, it is not necessary to solve the system again. All that is needed is to rearrange previous solutions. This is because the set of initial markings for which the system must be solved is a superset of the set of initial markings when solving with fewer resources. Therefore, in Tables 8.4–8.5, only the set of additional initial markings are shown. Initial markings are $(m_1,m_2) = ns_1$, where ns_1 is the size of the reachability set for this initial marking.

8.4 Concluding Remark

Stochastic Petri nets, that were introduced by assciating stochastic time with transitions of classic Petri net, enhanced the modelling and analysis power of Petri nets. As a concluding remark, let us frist recall the definitions of various forms of SPNS and the differences among them.

Orignal SPN models associates a firing rate with each transition in the net. An isomorphism exists between SPN models and continuous time Markov chain(MC). The infinitesimal generator of the MC isomorphic to the SPN can thus be easily obtained, and the steady state probability distribution over the MC states can be computed. The limitation of SPN is that the graphical representation of systems becomes rapidly difficult when system size and complexity increase, and the number of states of associated MC grows very fast with the dimensions of the graph. Basic SPN can thus be used to model limited size systems only.

GSPN were introduced to reduce the problem associated with the explosion of the number of MC states by defining two types of transitions: time transitions and immediate transitions. Immediate ones model the activities of a relatively short delay time. It was shown that GSPN are still equivalent to Markovian models, and that their solution can be obtained with less effort than it is necessary for the same size SPN. This is due to the fact that, since immediate transitions have priority over timed transitions, the number of reachable marking in a GSPN is subset of the reachability set of the associated PN.

However, the solution of GSPN model still becomes a formidable task when sufficient details are introduced into such models. Additionally, the presence of orders of magnitude different in the duration of activities represented by timed transitions in a model leads to numerical instability. The technique of time scale decomposition (TSD) of GSPN's developed by Ammar et al. offered a step in the right direction for a solution to the above fundamental problems. The decomposition is equivalent to the hierarcical aggregation of the underlying Markov process.

Although in GSPN models exponentially distributed firing times coexist with firing delay that are deterministically equal to zero, this does not represent a significant extension of the type of firing time distributions available in the model. In order to further enhance the analysis power of SPNs, some extensions of SPN to arbitrarily distributed firing times were developed. In a ESPN, a transitions are classified into three classes: exclusive, competitive and concurrent. When with all concurrent transitions is associated an exponentially distributed firing time, and with all exclusive and competitive transitions are associated with arbitrarily distributed firing delays, the SPP underlying a ESPN is a semi–Markov process. DSPN allows firing delays of timed transiton to be either constant or exponentially distributed random variables. In the cases of exclusive and competitive transitions a constant firing delay can be used freely, provided that the SPN execution policy is race with enabling memory. In the case of concurrent transition, the execution policy can be race with age with age or enabling memory, but all markings must enable at most one deterministic transition. Under these restrictons, the DSPN aslo can be mapped into a semi–Markor process.

Moment generating function (MGF) based method approach the performance analysis of SPN in a different way from the above method. Instead of solving the resulting Markov (or semi–Markov) models, MGF–based methods derive the MGF of interesting performance measures. For an arbitrary stochasti Petri net (ASPN), in which arbitrary distributions are incorporated into SPNs, the method can be used to find a lower and an upper bound for some time–related indices. One limitation of this approach is that the explosion of states may lead to very complicated computation.

In order to show the application of orignal SPN and its extension forms to communication and flexible systems, examples are given in this chapter. From these examples we can see the powerful modeling and analysis ability of the tools. Such merits exceed the scope that can be reached by other analysis tools of discrete event systems, such as queueing theory, max–min algebra method, perturbation analysis method, and so on.

It should be noted that the research field on stochastic timed Petri net, promoted by the need of applications, is very extensive. The results given in this chapter are a small but most splendid fraction of the whole.

REFERENCES

〔1〕 Petri, C. A., *Communication with Automata*, Ph. D Thesis, Technical Report RADC–TR–65–377, New York, January 1966.

〔2〕 Peterson, J. L., *Petri Net Theory and the Modeling of Systems,* Prentice–Hall, 1981.

〔3〕 Noe J. D., Nutt G.j., Macro E–Nets Representation of Parallel Systems, *IEEE Trans. on Computers,* August 1973.

〔4〕 Ramamoorthy, C. V., Ho, G. S., Performance Evaluation of Asynchronous Concurrent Systems Using Petri Nets, *IEEE Transactions on Software Engineering,* 6(5) 1980.

〔5〕 Zubberek, W. M., Timed Petri Nets and Preliminary Performance Evaluation, *Proc. of the 7–th Annual Symposium on Computer Architecture,* 1980.

〔6〕 Symons, F. J. W., Introduction to Numerical Petri Nets, a General Graphical Model of Concurrent Processing Systems, *A.T.R.*, 14(1), January 1980.

〔7〕 Natkin, S., *Reseaux de Petri Stochastiques* ,These de Doctiur–Ingegneur, CNAM–Paris, June 1980.

〔8〕 Molloy, M. K., *On the Integration of Delay and Throughput Measures in Distributed Processing Models,* Ph.D Thesis, University of California, Los Angeles, 1981.

〔9〕 Molloy, M. K., Performance Analysis Using Stochastic Petri Nets, *IEEE Trans. Comput.,* 31(9), 1982, 913–917.

〔10〕 Ajmone Marsan, M., Balbo, G., Conte, G., A Class of Generalized Stochastic Petri Nets for the Performance Evaluation of Multiprocessor Systems, *ACM Transactions on Computer Systms,* 2(2), 1984, 93–122.

〔11〕 Ammar, H. H. and Liu, R. W., Analysis of the Generalized Stochastic Petri Nets by State Aggregation, *Proc. Int. Workshop Timed Petri Nets,* Torino, Italy, July 1985, 88–95.

〔12〕 Ammar, H. H., Huang, Y. F. and Liu, R. W., Hierarchical Models for Systems Reliability Maintainability, and Availability, *IEEE Trans. Circuits Syst.*, 34(6), 1987, 629–638.

〔13〕 Dugan, J. B., Trivedi, K. S., Geist, R. M., Nicola, V.F., Extended Stochastic Petri Nets: Applications and Analysis, *Performance 84,* Paris, France, 1984, 507–519.

〔14〕 Sheldon, M.R., *Stochastic Processes,* John Wiliey Sons, 1983.

〔15〕 Wang, J. C. and Huang, Z. T., Time Delay Analysis on Assembly Lines, *Proc. of the 11th Int. Conf. on Production Res.*, China, 1991, 1151–1154.

〔16〕 Marsan, M. A. and Chiola, G., On Petri Nets with Deterministic and Exponential Transition Firing Times, *Proc.7th Workshop on Application and Theory of Petri Nets,* Oxford, England, June 1986.

〔17〕 __, On Petri Nets with Deterministic and Exponential Distributel Firing Times, *Advances on Petri Nets'87*, 1987, 132–145.

〔18〕 __, Modeling Discrete Event Systems with Stochastic Petri Nets, *Proceedings of Int. Symp. of Computer and Systems'85*, 739–742.

〔19〕 Kleinrock, L., *Queueing Systems, Vol.1 Theory*, John Wiley, 1974.

〔20〕 Chiola, G., A Software Package for the Analysis of Generalized Stochastic Petri Net Models, proc. of *International Workshop on Timed Petri Nets*, Torino Italy, July 1985.

〔21〕 Zhou, M. C., Guo, L., Dicesare, F., Integration of Pertri Nets and Moment Generating Function Approaches for System Performance Evaluation, *Journal of Systems Integration,* 3(1) 1993, 43–62.

〔22〕 Whitehouse, G. E., *System Analysis and Design Using Network Techniques,* Prentice Hall: Englewood Cliffs, NJ, 1973.

〔23〕 Guo, D. L., DiCesare, F. and Zhou, M. C., A Moment Generating Function Based Approach for Evaluating Extended Stochastic Petri Nets, *IEEE Transactions on Automatic Control*, 37(12), 1992.

〔24〕 Agerwala, T. and Choed–Amphai, Y., A Synthesis Rule for Concurrent System, *Proc. of 15th Design Automation Conference,* Las Vegas, NV, 1978, 305–311.

〔25〕 Murata, T., Petri Nets: Properties, Analysis and Application, *Proc of the IEEE,* 77(4), 1989, 541–579.

〔26〕 Valette, R., Analysis of Petri Nets by Stepwise Refinements, *J. of Compt. and Syst. Sci.,* 18, 1979, 35–46.

〔27〕 Zhou, M. C. and DiCesare, F., *Petri Net Synthesis for Discrete Event Control of Menufacturing Systems*, Kluwer Academic Publishers, Boston, MA, 1993.

〔28〕 Zhou, M. C., DiCesare F. and Desrochers A. A., A Hybrid Methodlogy for Synthesis of Petri Nets for Manufacturing Systems, *IEEE Trans. on Robotics and Automation*, 8(3), 1992, 350–361.

〔29〕 Zhou, M. C. and DiCesare, F., Hybrid Synthesis of Petri Nets for Manufacturing Systems: A case study, *3rd Int. Symp. on Robotics & Manufacturing*, Vancouver, Canada, July 1990, 667–672.

〔30〕 Guo, D. L., DiCesare, F. and Zhou, M. C., Moment Generating Function Apptoach to Performance Analysis of Extended Stochastic Petri Nets, , *Proc. of IEEE Int. Conf. on Robotics and Automation*, Sacramento, CA, 1991, 1309–1314.

〔31〕 Chen, P., Bruell, S. C. and Balbo, G., Alternative Methods for Incorporating Nonexponential Distributions into Stochastic Timed Petri Nets, *Proc. the Third IEEE Int. Workshop on Timed Petri Nets and Performance Models*, Tokyo, 1989, 187–197.

〔32〕 Huang, Z. T., Zhu, X. and Wang, J. C., A New Method for Time Delay Analysis of Stochastic Petri Nets Models, Proc. of the 1991 IFAC Workshop on DES Theory and Applications, China, June 1991, 80–85.

〔33〕 Huang, Z. T., Wang, J. C. and Zhou, M. C., Hybrid State Analysis

Method in Stochastic Petri Nets, *Proc. of IEEE Symposium on SMC*, USA, Oct. 1992.

〔34〕 Ammar, H. H., Huang, Y. F. and Liu, R. W., Analysis by Aggregation of the Generalized Stochastic Petri Nets with Application to Reliability/ Maintainability and Fault Diagnosis, *Proc. Int. Symp. Circuits and Systems*, Kyoto, Japan, June 1985.

〔35〕 Ammar, H. H. and Islan, M. R., Time Scale Decomposition of a Class of Generalized Stochastic Petri Net Models, *IEEE Trans. on Software Engineering*, 15(6), 1989, 809–820.

〔36〕 Sifakis, J., Use of Petri Nets for Performance Evaluation in Measuring, *Modelling and Evaluating Computer Systems*, North Holland, 1977.

〔37〕 Suzuki, I. and Murata, T., A Methode for Stepwise Refinements and Abstractions of Petri Nets, *J. of Computer and System Science*, Vol. 27, 1983, 51–76.

〔38〕 Wang, J. C., A New Method for the Aggregation of Complex Stochastic Petri Net Models Based Upon Time Delay, *Proc. of the 1991 IFAC Workshop on DES Theory and Applications*, China, June 1991, 148–154.

〔39〕 Wang, J. C. , Wang, Z. G. , Huang, Z. T. and Xiong, H. X., Hierarchical Models for Time Delays on C2 Systems, *Preprints of the IFAC Symp. on LSS*, China, 1992, 541–546.

〔40〕 Tobagi, F. A., Hunt, V. B., Performance Analysis of Carrier Sense Multiple Access with Collision Detection, *Computer Networks*, 4(5), 1980.

〔41〕 *The Ethernet, a Local Area Network.* Date Link Layer and Physical Layer Specification. Version 2.0 , Digital, Intel, Xerox, November 1982.

〔42〕 Metcalfe, R. M., Boggs, D. R., Ethernet: Distributed Packet Swithing Techniques for Computer Networks, *Communication of the ACM*, 19(7), 1976.

〔43〕 Wang, J. C., Huang, Z. T. and Wang, Z. G., Time Delay Analysis of Semi–Markov Stochastic Petri Net Models, *Robotics and Flexible Manufacturing Systems,* Elsevier Science Publishers B. V., North–Holland, 1992, 339–348.

〔44〕 Marsan, M. A., An Introduction to Generalized Stochastic Petri Nets, *Microelectron. Reliab.*, 31(4), 1991, 699–725.

〔45〕 Mason, S. J. and Zimmermann, H. J., *Electronic Circuits, Signal, and Systems*, John Wiley & Sons: NY, 1960.

〔46〕 Courtois, P. J. and Semal, P., Computable Bounds for Conditional Steady–state Probabilities in Large Markov Chains and Queueing Models, *IEEE J. Select. Areas Commun.*, 4(6), Sept. 1986.

9

Deadlock Avoidance Policy for Flexible Manufacturing Systems

Keyi Xing[1,2], **Baosheng Hu**[2] and **Haoxun Chen**[2]
[1] Department of Applied Mathematics, Xidian University
[2] Systems Engineering Institute, Xian Jiaotong University
Xi'an, 710049, P. R. China

Abstract: Multiple products through a flexible manufacturing system (FMS) with limited resources can lead to deadlock situation. In this chapter, we study the problem of deadlock avoidance by using the Petri net (PN) model for FMS and introducing the concept of deadlock structure. The necessary and sufficient conditions to prevent deadlock are characterized. We use a state feedback restriction policy which prevents some enabled transitions from firing for avoiding the deadlock in the system. When the number of any key kind of resources is greater than 1, this policy is minimally restrictive and maximizes the resource usage. We present the Petri net realization of these restriction policies. The Petri net controlled by the policies may be used as the Petri net model for an FMS in which deadlock can never occur. The restriction policies can be easily implemented. Several examples are provided for illustration.

1. Introduction

In recent years there has been considerable interest in the development of control theory and methods for design, scheduling, operation and performance analysis of flexible manufacturing systems (FMS's). Given a model of FMS and a specification of the desired behavior for the controlled system, the basic objective is to synthesize the appropriate controller to realize the specified closed—loop behavior.

In the general FMS, raw products of various types enter the system at discrete points of times and are processed concurrently. Sharing a limited number of resources such as machines, robots and buffers, each product has a particular operation routing that determines the order in which resources must be assigned to the product. The concurrent flow of multiple products in a system, which all compete for a finite set of resources, can lead to a deadlock situation. A deadlock is a state under which a set of products are in circular waiting for resources, each product in the set is waiting for a resource being held by another product. The deadlock could occur, for example, in the system which consists of two machines M_1, M_2 and one robot R which can pick up and deliver products between machines. Suppose that a kind of products is processed through machines M_1, M_2, M_1 in order, if a new product enters the system and is processed in M_1, a product waiting in M_2 to return to machine M_1 for the second process would not be able to proceed. This two products are in deadlock. In such resource — shared system, deadlocks constitute a major issue to be addressed at the design and operation phases [2,3,4].

In the recent literatures, several efforts have focused on the problem of deadlocks in FMS [2, 3, 4]. Studies on deadlock avoidance using PN — based models are presented. A deadlock avoidance algorithm (DAA) for FMS with concurrently competing process flow is proposed by Banaszak and Krogh [4]. The proposed feedback policy allows a process step to start only when certain resource conditions are satisfied under the current system state. This control policy is sufficient for avoiding deadlocks, but not necessary. It is not a minimal restriction policy either. The existing deadlock avoidance schemes are not the most effective for controlling an FMS. A good deadlock avoidance policy should allow the maximal use of resources to optimize the throughput of the FMS.

In this chapter, we address the important issues of deadlock avoidance in FMS using PN—based techniques. We first give the conditions in which the system is in deadlock situation and introduce the deadlock structure concept. Then we construct a restriction policy to guarantee that deadlock can never occur in the closed—loop system under the proposed control policy. The control policy proposed is a feedback policy that inhibits requests for resources only when they will lead to deadlocks. Thus the policy is minimally restrictive and maximizes the resource usage when the number of any key kind of resources is greater than 1. This policy has its simple Petri net realization for a class of FMS. So the closed-loop system may be modeled by a live Petri net.

In the next section, we introduce definitions and notations of Petri net model of the concurrent system. The structure and conditions of deadlock in PN model is discussed in Section 3. The restriction policy for the problem of deadlock avoidance is solved in Section 4. In Section 5 we present the Petri net realization of the restriction policy. Section 6 gives an example for illustration. Section 7 is the conclusion.

2. Petri Net Models of Concurrent Process Flows

We use Petri nets (PN's) to model the flow of products in an FMS. PN provides a convenient representation of concurrent contention for resources by multiple products in the system [2, 3, 4]. A PN Z is a five tuple: $Z=(P, T, I, O, m_0)$ where P is a finite set of places, T is a finite set of transitions, $I \subseteq P \times T$ and $O \subseteq T \times P$ are the sets of directed arcs connecting state places and transitions, and $m_0: P \to N$ is the initial marking of Z with N as the set of nonnegative integers, the marking m of a PN Z indicates the number of tokens in each place, which is the current state of the system. The set of all markings is denoted as M. A transition $t \in T$ is enabled iff $m(p) > 0$ when $(p, t) \in I$ for any $p \in P$. Let $T_e(m)$ denote the set of enabled transitions under a marking m. An enabled transition t fires at a marking m' yielding the new marking m,

$$m(p) = m'(p) + \#(t,p) - \#(p,t) \qquad \text{for } p \in P$$

where $\#(a,b) = \begin{cases} 1 & (a,b) \in I \cup O \\ 0 & (a,b) \notin I \cup O. \end{cases}$

The marking m is said to be reachable from a marking m' and denoted by $m'[t>m$. The marking m is reachable from the initial marking m_0 if there are a transition sequence $\sigma = t_0 t_1 \ldots t_k$ and a sequence of markings $m_0, \ldots, m_{k+1} = m$ such that $m_i[t_i > m_{i+1}$, $i = 0, 1, \ldots, k$. The reachable set is the set of all markings reachable from m_0 with different sequences of transition firings and denoted by $RM(Z, m_0)$.

Let $Z^i = (P^i, T^i, I^i, O^i, m_0^i)$, $i=1,2$, be two Petri nets, the composition of Z^1 and Z^2, denoted by $Z^1 \parallel Z^2$, is a Petri net $Z^1 \parallel Z^2 = (P^1 \cup P^2, T^1 \cup T^2, I^1 \cup I^2, O^1 \cup O^2, m^0)$ where

$$m^0(p) = \begin{cases} m_0^1(p) & p \in P^1 \\ m_0^2(p) & p \in P^2. \end{cases}$$

and initial markings m_0^1 and m_0^2 must be consistent, i.e. $m_0^1(p) = m_0^2(p)$ for

$p \in P^1 \cap P^2$.

Let us consider an FMS with m types of resources, denoted by r_1, $r_2, \ldots, r_m$ and n different types of products, denoted by $q_1, q_2, \ldots, q_n$. The manufacturing process of each product is supposed to be uniquely defined as a routing through the system (i. e. a sequence of resources to be used). We will denote by $\sigma(q_i)$ the routing corresponding to product—type q_i.

In this paper we adopt the Petri nets as models of the FMS's. We illustrate the modeling methodology using the following example model.

Example 1. Consider the FMS illustrated in Fig. 1. In which two machining cells W_1, W_2 serviced by a robot S, which can pick up and deliver products between cells. There are two types of products q_1 and q_2 to be produced. Each cell has two same kind machines. The machine kind in work cell W_j is denoted by M_j, $j=1,2$. Then the set of resources is $R=\{M_1, M_2\}$. The capacity of each resource kind is 2.

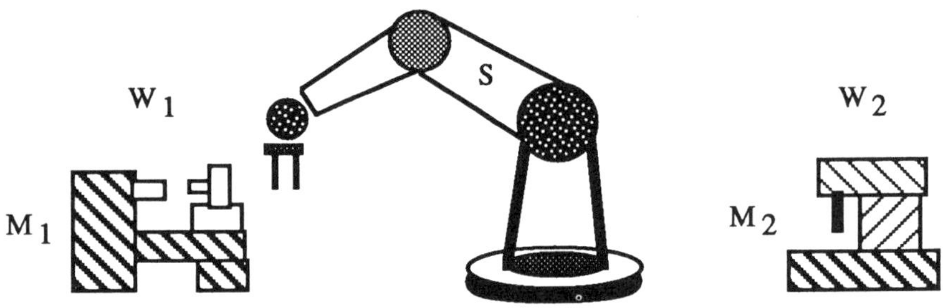

Fig. 1. A simple manufacturing system with two machining cells

Suppose q_1, q_2 are produced by sequencing machines M_1, M_2, M_1 and M_1, M_2 in order, respectively. Hence the process routing is specified as follows:

$$\sigma(q_1) = (M_1, M_2, M_1)$$
$$\sigma(q_2) = (M_1, M_2)$$

Let us now develop the Petri net model of the FMS in two steps:

The first step is to model the routing of each type of products as a line of transitions and places, as shown in Fig. 2 for Example 1. Each transition firing corresponds to one process step completing and the next process step starting. A token in a line represents a product as it flows through the system and tokens in place P_0^i represent the number of products of type q_i to be initiated and tokens in place $P_{l_i+1}^i$ represent the number of completed products of q_i. L_i is length of production for q_i, places p_1^i, p_2^i, ..., $p_{L_i}^i$ correspond to process steps. We assume that each process step requires only one resource. The initial marking of p_j^i is defined as $m_o(p_j^i)=0$.

The second step is to model the usage of resources, we assign a resource place to each type of resources r_i, denoted by a_{r_i}, tokens in place a_{r_i} indicate available resources of type r_i. The initial marking is defined as $m_0(a_{r_i})=C_{r_i}$, where C_{r_i} is available capacity of type r_i. The process step p_j^i using resource r, denoted by $R(p_j^i)=r$, is represented by introducing arc from a_r to t_j^i and arc from t_{j+1}^i to a_r, $i=1,2,\ldots,n$, $j=1,\ldots,L_i$.

The complete PN model for the FMS $Z=(P,T,I,O,m_0)$ is obtained as shown in Fig. 3 for Example 1. The complete PN model for a general FMS is followed.

$$Z=(P,T,I,O,M_0)$$

where $P=P_p\cup P_r$,
where $P_p=\{p_j^i,\ i=1,2,\ldots,n,\ j=0,1,\ldots,L_i+1\}$
and $P_r=\{a_{r_i},\ i=1,2,\ldots,m\}$

$$T=\{t_j^i,\ i=1,2,\ldots,n,\ j=0,1,\ldots,L_i+1\}$$
$$I=\{(p_j^i,\ t_{j+1}^i),\ i=1,\ldots,n,\ j=0,1,\ldots,L_i\}\cup \{(a_{R(p_j^i)},t_j^i),\ i=1,2,\ldots,n,\ j=0,\ldots,L_i\}$$
$$O=\{(t_j^i,\ p_j^i),\ i=1,2,\ldots,n,\ j=0,1,\ldots,L_i+1\}\cup \{(t_{j+1}^i,\ a_{R(p_j^i)}),\ i=1,\ldots,n,\ j=1,\ldots,L_i\}$$
$$m_0(p)=\begin{cases} C_r & p\in\{a_{r_i},\quad i=1,2,\cdots,n\} \\ 0 & \text{otherwise.}\end{cases}$$

This model is called production Petri net (PPN). We restrict our discussion on this class of PPNs in this chapter.

Note that in our PPN, each transition has at most two input places and two output places. One input place may be due to the product routing requirement and the other due to the resource requirement. For a given transition $t\in T$, we let ${}^o t$ and ${}^r t$ denote an input process place in P_p and input resource place in P_r for t, respectively. Similarly, t^o, t^r denote an output process place and resource place for t, respectively. For example, ${}^o t_1^1=p_0^1$, ${}^r t_1^1=a_{M_1}$ in Fig. 3. For a given set of transitions A, let ${}^o A$ denote the union set of ${}^o t$ for all $t\in A$, i.e. ${}^o A=\{{}^o t\,|\,t\in A\}$. Similarly for A^o. A transition $t\in T$ is process enabled at the marking m iff $m({}^o t)\geqslant 1$ and resource enabled iff $m({}^r t)\geqslant 1$.

Let $\overline{T}=\{t_j^i\,|\,i=1,\ldots,n,\ j=1,\ldots,L_i\}$. The firing of each transition in $\overline{T}$ requires some resource.

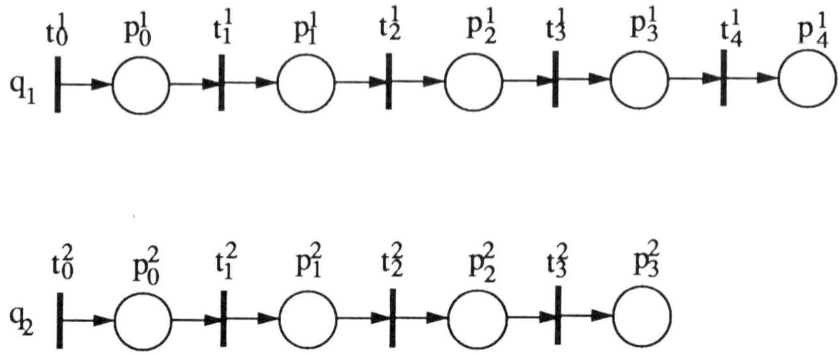

Fig. 2. Petri net model for product routings in Fig. 1

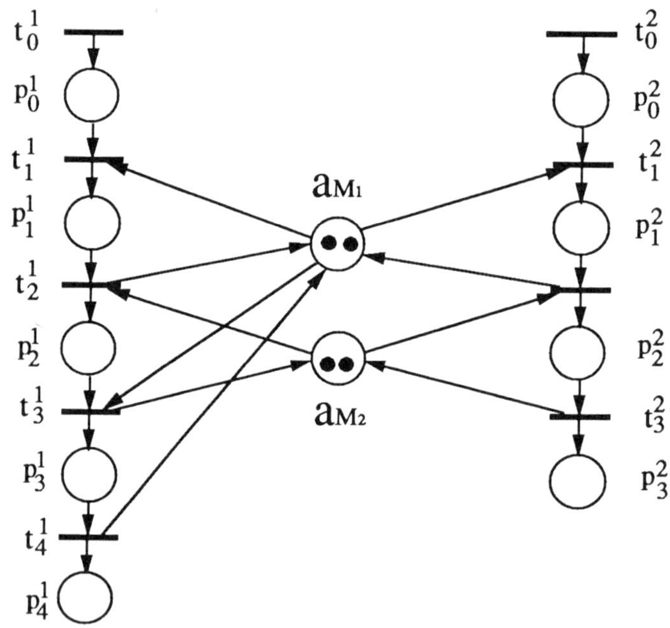

Fig. 3. A complete Petri net model for the FMS in Fig. 1

3. Deadlock Structures and Deadlock Conditions

When one or more kinds of products enter an FMS, it is possible to lead to a deadlock situation. For example, if the marking m of PN model in Fig. 3 in previous section is

$$m(p) = \begin{cases} 2 & p \in \{p_1^1, p_2^1\} \\ 0 & \text{otherwise} \end{cases}$$

then transitions t_2^1, t_3^1 can never fire in future, which leads the other transitions, i.e. t_3^1, t_4^1, t_1^2, t_2^2, t_3^2 to deadlock.

A formal definition of deadlock in an FMS is as follows [4]:

Given sets of resources R, products Q, and a PPN Z, a set of transitions D_0 is said to be *in deadlock* for a marking $m \in RM(Z, m_0)$ if 1) all transitions in D_0 are process enabled under the marking m, and 2) no transition in D_0 is resource enabled for any marking $m' \in RM(Z, m)$.

From the definition of deadlock for a PPN Z, if D_1 and D_2 are in deadlock for the marking $m \in RM(Z, m_0)$, then $D_1 \cup D_2$ is in deadlock for the marking m, hence for the marking $m \in RM(Z, m_0)$, there is a uniquely maximal set, denoted by $D(m)$, which is in deadlock in the sense of inclusion of sets.

Lemma 1: *If $D(m)$ is not empty for some marking $m \in RM(Z, m_0)$ then it has the following properties:*

1). $m(^\circ t) > 0$, *for any* $t \in D(m)$,

2). $R(^\circ D(m)) \supseteq R(D(m)^\circ)$,

3). *For any* $r \in R(D(m)^\circ)$, $\sum_{t \in D(m)} m(^\circ t, r) = C_r$,

4). $D(m) \subseteq \overline{T}$,

where $R(^\circ D(m)) = \{R(^\circ t) \mid t \in D(m)\}$,

$R(D(m)^\circ) = \{R(t^\circ) \mid t \in D(m)\}$,

and

$$m(^\circ t, r) = \begin{cases} m(^\circ t) & R(^\circ t) = r \\ 0 & \textit{otherwise.} \end{cases}$$

Proof: The first and last properties are trivial. If $r \in R(D(m)^\circ)$ and $r \in R(t^\circ)$ for some transition $t \in D(m)$, then $m(a_r) = 0$ and there exist transitions $t_1, t_2, \ldots, t_k \in T$ such that $m(^\circ t_i) > 0$, $R(^\circ t_i) = r$, $i = 1, \ldots, k$ and $\sum_{i=1}^{k} m(^\circ t_i) = C_r$. If some transition t_i can be fired under a marking $m' \in RM(Z, m)$ and $m'[t_i > m''$, then $m''(a_r) = 1$. Hence t is resource and process enabled for the marking $m'' \in RM(Z, m)$. This contradiction implies that $t_1, \ldots, t_k$ are in deadlock for the marking m and hence $t_1, \ldots, t_k \in D(m)$, $r \in R(^\circ D(m))$. Therefore 2) and 3) hold. ●

The properties in Lemma 1 can be explained as follows:

1) t is process—enabled,

2) The resources required for the firings of transitions in $D(m)$ are also required for the process steps in ${}^0D(m)$.

3) All resources required for the firing of every transition in $D(m)$ are being used by the process steps in ${}^0D(m)$ under the marking m. Thus $D(m)$ is in deadlock.

4) The firing of every transition in $D(m)$ requires some resource.

Lemma 2: *Given sets of resources R, products Q, a PPN Z, and a marking $m \in RM(Z, m_0)$, if a nonempty transition set $T_1 \subseteq \overline{T}$ satisfies that*

1) $m({}^o t) > 0, \qquad t \in T_1$,

2) $R({}^o T_1) = R(T_1^o)$,

3) *For any* $r \in R({}^o T_1), \qquad \sum_{t \in T_1} m({}^o t, r) = C_r$

then T_1 is in deadlock for the marking m.

Proof: For every transition $t \in T_1$, by 1) t is process enabled and let $R(t^o) = r$, by 2) and 3), $m(a_r) = 0$, hence t is not resource enabled in the marking m. For any $t_1 \notin T_1$ if $R(t_1^o) \notin R(T_1^o)$, then the firing of t_1 does not change the number of tokens in ${}^o T_1$, ${}^r T_1$, T_1^o and T_1^r. But if $R(t_1^o) \in R(T_1^o)$, then t_1 is not resource enabled for the marking m, therefore T_1 is in deadlock for the marking m. ●

The conditions 2) and 3) in Lemma 2 are that the resources required for the firings of transitions in T_1 are being used under the marking m. Thus T_1 is in deadlock. Lemma 2 establishes a sufficient condition which leads to deadlock situation.

Theorem 1: *For any maximal nonempty deadlock set $D(m) \subseteq \overline{T}$, there exists a nonempty transition set $D \subseteq D(m) \backslash \{t_1^i, i = 1, 2, \ldots, n\}$ such that*

1). $m({}^o t) > 0 \qquad t \in D$,

2). $R({}^o D) = R(D^o)$,

3). *For any* $r \in R({}^o D), \qquad \sum_{t \in D} m({}^o t, r) = C_r$.

Proof: We first note that $T_1 = D(m) \backslash \{t_1^i, i = 1, 2, \ldots, n\}$ is nonempty. If $R({}^o T_1) = R(T_1^o)$, then $D = T_1$ satisfies the conditions in Theorem 1. Suppose now $R({}^o T_1) \neq R(T_1^o)$. For any resource $r \in R({}^o T_1) \backslash R(T_1^o)$, forming $T_2 = T_1 \backslash \{t \in T_1 \mid R({}^o t) = r\}$, then T_2 is nonempty. Repeat the above process, since the types of resources are finite, there exist nonempty sets $T_k \subset T_{k-1} \ldots \subset T_2 \subset T_1 \subset D(m)$ such that T_k is nonempty and $R({}^o T_k) = R(T_k^o)$. For any resource $r \in R(D(m)^o)$, since $D(m)$ is maximal, we have $\sum_{t \in D(m)} m({}^o t, r) = C_r$. If $r \in R({}^o T_k) = R(T_k^o)$ and $t \in D(m)$, $R({}^o t) = r$, then t

$\in T_k$, hence $\sum_{t \in T_k} m({}^o t, r) = C_r$ and $D = T_k$ satisfies the conditions in Theorem 1. ●

We call a nonempty transition set $D \subseteq \overline{T} \setminus \{t_1^i,\ i=1,2,\ldots,n\}$ satisfying $R({}^0 D) = R(D^0)$ as a *deadlock structure* and let $\mathscr{D} = \{D \mid D \subseteq T$ is a deadlock structure$\}$ and for simplicity let $R(D)$ denote $R({}^o D) = R(D^o)$ for $D \in \mathscr{D}$.

Theorem 1 states that if deadlock occurs in a PPN under some reachable marking m, then there exist deadlock structures which are in deadlock. Thus if there is no deadlock structure in a PPN or every deadlock structure is not in deadlock, then deadlock will not occur.

If $D = \{t_{j_1}^{i_1}, t_{j_1+1}^{i_1}, \ldots, t_{j_1+k_1}^{i_1}, t_{j_2}^{i_2}, t_{j_2+1}^{i_2}, \ldots, t_{j_2+k_2}^{i_2}, \ldots, t_{j_s}^{i_s}, t_{j_s+1}^{i_s}, \ldots, t_{j_s+k_s}^{i_s}\} \subseteq \overline{T} \setminus \{t_1^i,\ i=1,\ldots,n\}$ is a deadlock structure with the property: $R(\{t_{j_r+k_r}^{i_r}\}^0) = R({}^0 t_{j_{r+1}}^{i_{r+1}})$, $r=1,\ldots,s-1$, and $R(\{t_{j_s+k_s}^{i_s}\}^0) = R({}^0 t_{j_1}^{i_1})$, then D is called a *basic deadlock structure*. Let $\mathscr{B} = \{D \mid D$ is a basic deadlock structure$\}$. Now we construct the set $\mathscr{C}$ from $\mathscr{B}$ as follows:

(1) $\mathscr{B} \subseteq \mathscr{C}$,

(2) Any $D' \in \mathscr{C} \setminus \mathscr{B}$ can be constructed by repeating the following two steps:

(a) If $D \in \mathscr{C}$, then $D' = D \cup \{t_j^i, t_{j+1}^i, \ldots, t_{j+k}^i \mid R({}^0 t_j^i), R((t_{j+k}^i)^0) \in R(D)\} \in \mathscr{C}$,

(b) If $D_1, D_2 \in \mathscr{C}$, then $D' = D_1 \cup D_2 \in \mathscr{C}$.

Then $\mathscr{C}$ is the set of deadlock structures and $\mathscr{C} \subseteq \mathscr{D}$. In the following, we shall prove that $\mathscr{C} = \mathscr{D}$ and point out a method for computing the deadlock structures.

Let $D \in \mathscr{D}$. If there exist $t_j^i, t_{j+1}^i, \ldots, t_{j+k}^i \in D$ with $R({}^0 t_j^i)$, $R((t_{j+k}^i)^0) \in R(D)$ such that $D_0 = D \setminus \{t_j^i, t_{j+1}^i, \ldots, t_{j+k}^i\}$ is a nonempty deadlock structure, then $D_0 \in \mathscr{C}$ implies that $D \in \mathscr{C}$. So we suppose that there exist no $t_j^i, t_{j+1}^i, \ldots, t_{j+k}^i$ with $R({}^0 t_j^i)$, $R((t_{j+k}^i)^0) \in R(D)$ such that D_0 is a nonempty deadlock structure. For proving $D \in \mathscr{C}$, we only need to prove that there is a basic deadlock structure in D. For any $t_j^i, t_{j+1}^i \ldots, t_{j+k}^i \in D$, $t_{j-1}^i, t_{j+k+1}^i \notin D$, if $R({}^0 t_j^i) \in R(\{t_j^i, t_{j+1}^i, \ldots, t_{j+k}^i\}^0)$, then $R({}^0 t_j^i) = R(\{t_{j+l}^i\}^0)$ for some t_{j+l}^i, $0 \leqslant l \leqslant k$ and $\{t_j^i, t_{j+1}^i, \ldots, t_{j+l}^i\}$ is a basic deadlock structure. Thus it is assumed that $R({}^0 t_j^i) \notin R(\{t_j^i, t_{j+1}^i, \ldots, t_{j+k}^i\}^0)$. Since $R({}^0 D) = R(D^0)$, there exist $t_{j_1}^{i_1}, t_{j_1+1}^{i_1}, \ldots, t_{j_1+k_1}^{i_1} \in D$, $t_{j_1-1}^{i_1}, t_{j_1+k_1+1}^{i_1} \notin D$

such that $R({}^0t_j^i)=R(\{t_{j_1+l_1}^{i_1}\}^0)$ for some $0\leqslant l_1\leqslant k_1$. Similarly for $t_{j_1}^{i_1}$ there exist $t_{j_2}^{i_2}, t_{j_2+1}^{i_2}, \ldots, t_{j_2+k_2}^{i_2} \in D$, $t_{j_2-1}^{i_2}, t_{j_2+k_2+1}^{i_2} \notin D$ such that $R({}^0t_{j_1}^{i_1}) = R((t_{j_2+l_2}^{i_2})^0)$ for $0\leqslant l_2\leqslant k_2$. Repeating the above process, we can get t_j^i, $t_{j_1}^{i_1}, \ldots, t_{j_k}^{i_k}, \ldots$, and $t_{j_1+1}^{i_1}, \ldots, t_{j_k+l_k}^{i_k}, \ldots$, such that $R({}^0t_{j_s}^{i_s}) = R((t_{j_{s+1}+l_{s+1}}^{i_{s+1}})^0)$ for $s=1,2,\ldots$. Since $\{t_{j_1}^{i_1}, \ldots, t_{j_s}^{i_s}, \ldots\}\subseteq\overline{T}$ is finite, there exist $t_{j_s}^{i_s}$ and $t_{j_n}^{i_n}$ such that $t_{j_s}^{i_s}=t_{j_n}^{i_n}$. For simplicity, suppose that $t_{j_1}^{i_1}=t_{j_k}^{i_k}$ for $k>0$. Then $\{t_{j_1}^{i_1}, t_{j_1+1}^{i_1}, \ldots, t_{j_1+l_1}^{i_1}, \ldots, t_{j_k}^{i_k}, t_{j_k+1}^{i_k}, \ldots, t_{j_k+l_k}^{i_k}\}$ is a basic deadlock structure.

The number of deadlock structures in the PPN Z is related to the numbers of times and order in which resources are used. For example, if only one type of products is produced, the number of deadlock structures is increased exponentially with the number of repeated utilization times of the same kind of resources. In most manufacturing practice, the number of the times for which a kind of shared resources is used to produce an individual product from its raw form to final is two or three. So far, we do not encounter any case in which a shared resource is repeatedly used over four times for an individual product.

The deadlock structures in PN shown in Fig. 3 are $D_1=\{t_2^1, t_3^1\}$, $D_2=\{t_3^1, t_2^2\}$ and $D_3=D_1\cup D_2$.

Theorem 2: *For any deadlock structure $D_0\in\mathscr{D}$, there exist a nonempty subset D' of D_0 and a reachable marking $m'\in RM(Z,m_0)$ such that D' is in deadlock for the marking m'.*

Proof: We construct D' and m' recurrently from D_0 and m_0.

First, let $D'_0=\varnothing$ and $k=0$.

Repeat the following procedure till $D_k=\varnothing$.

Let $t_j^i\in D_k$ be a transition such that if $i_1<i$ or $j_1>j$, then $t_{j_1}^{i_1}$ is not in D_k and let $R({}^\circ t_j^i)=r$ and $D_{k+1}=D_k\setminus\{t_j^i\}$. Now we consider the following two possible cases:

Case 1. $r\notin R({}^0 D_{k+1})$

In this case, we consider the sequence of transitions $\sigma=t_0^it_1^i\ldots t_{j-1}^i$. Note that $m_0(a_r)=C_r$, so $\sigma^{C_r}=\sigma\sigma\ldots\sigma$ is a valid sequence for m_0 and let $m_k[\sigma^{C_r}>m_{k+1}$, then $m_{k+1}({}^\circ t_j^i)=C_r$ and $m_{k+1}(a_r)=0$.

Let $D'_{k+1}=D'_k\cup\{t_j^i\}$ and $k:=k+1$.

Case 2. $r\in R({}^\circ D_{k+1})$.

Let $D'_{k+1}:=D'_k$, $m_{k+1}:=m_k$ and $k:=k+1$.

Finally we get the chains of transition sets:

$$D_0, D_1, D_2, \ldots, D_k=\varnothing$$

and

$$D'_0, D'_1, D'_2, \ldots, D'_k$$

For any transition $t \in D'_k$, 1) $m_k({}^{\circ}t) > 0$, 2) $R({}^{\circ}D'_k) = R(D'^{\,0}_k) = R({}^{\circ}D_0)$ and 3) for any $r \in R({}^{\circ}D'_k)$, since $t^i_j \in D'_k$ such that $m_k({}^{\circ}t^i_j) = C_r$, hence $\sum_{t \in D'_k} m_k({}^{\circ}t, r) = C_r$, by Lemma 2, $D' = D'_k$ is in deadlock for $m' = m_k$. ●

Theorem 2 implies that if the system contains some deadlock structures, then it may reach some marking m from its initial marking m_0 such that $D(m) \neq \emptyset$. So in order to avoid deadlock, it is necessary to restrict the number of tokens in every deadlock structure D, that is, the number of tokens in ${}^{\circ}D$ is not greater than $\sum_{r \in R(D)} C_r - 1$.

Now we give the following definitions and notations used in the sequel.

Given a deadlock structure D, a transition t^i_j of T is called a *precedence transition* of D if $t^i_j \notin D$ and $t^i_{j+1} \in D$. Let $F(D)$ denote the set of all precedence transitions of D. A transition t^i_j of D is called a *rear transition* of D if $t^i_{j+1} \notin D$. Let $B(D)$ denote the set of all rear transitions of D.

A cyclic deadlock structure chain in $\mathscr{D}$ is a set of deadlock structures $V = \{D_1, D_2, \ldots, D_l\}$ in which, for any $D_k \in V$, there exist deadlock structures D_i, $D_j \in V$ and transitions t_i, t_j such that $t_i \in D_i \cap F(D_k)$ and $t_j \in F(D_j) \cap D_k$.

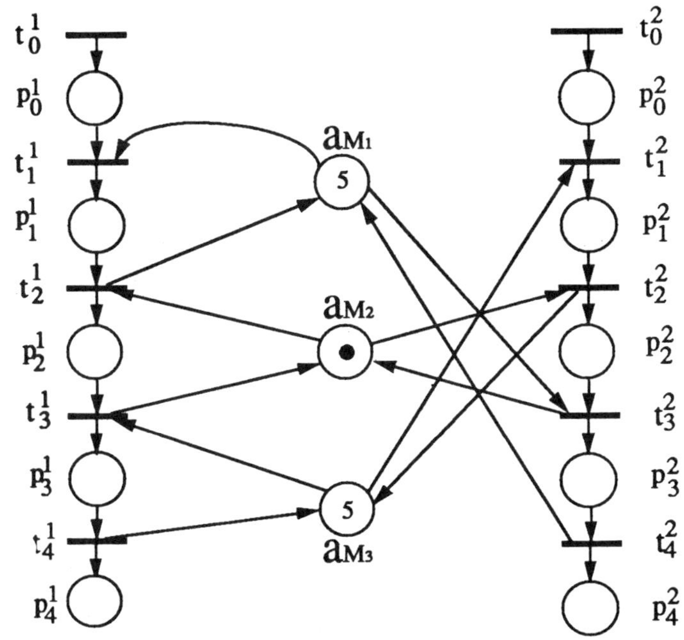

Fig. 4. A Petri net model for the FMS in Example 2 with cyclic deadlock structure chains

Example 2. Consider the FMS in which there are three different kinds of machines M_1, M_2, M_3 and a robot S, which can pick up and deliver products between machines. Suppose that two kinds of products q_1, q_2 are produced in sequencing machines M_1, M_2, M_3 and M_3, M_2, M_1 in order, respectively. The capacities of machine kinds M_1, M_2 and M_3 are 5, 1 and 5, respectively. Then the complete Petri net model for the FMS is depicted in Fig. 4. $D_1=\{t_3^1, t_2^2\}$, $D_2=\{t_2^1, t_3^2\}$ and $D_3=D_1\bigcup D_2$ are deadlock structures. $F(D_1)=\{t_2^1, t_1^2\}$, $F(D_2)=\{t_1^1, t_2^2\}$, $t_2^2\in D_1\bigcap F(D_2)$, $t_2^1\in F(D_1)\bigcap D_2$. $\{D_1, D_2\}$ is a cyclic deadlock structure chain.

4. Maximally Permissive Feedback Policies for the Problem of Deadlock Avoidance

Since a transition may not fire even if it is enabled, the set of transitions which can fire at the marking m is not unique. So in this paper we are interested in nondeterministic state feedback policies of the form as follows [8]:

$$\rho: M \rightarrow 2^T.$$

A feedback policy ρ maps each marking $m\in M$ into a collection of subsets of T. A feedback policy ρ_1 is more permissive or less restrictive than a feedback policy ρ_2, denoted by $\rho_1>\rho_2$, if for all marking $m\in M$, $\rho_2(m)\subseteq\rho_1(m)$ and for some $m'\in M$, $\rho_1(m')\neq\rho_2(m')$.

A set of transitions B is control enabled under the marking m and control policy ρ, if there exists set $B'\in\rho(m)$ such that $B\subseteq B'$.

Example 3. In the PPN shown in Fig. 4 of Example 2, D_1, D_2 and D_3 are deadlock structures. By Theorem 2 to avoid deaddlock, it is necessary to guarantee that the number of tokens in ${}^0D_i$ for every deadlock structure D_i is not greater than $\sum_{r\in R(D_i)} C_r - 1$. Thus the deadlock avoidance policy ρ for the PPN in Fig. 4 must guarantee that the numbers of tokens in ${}^0D_1$ and ${}^0D_2$ are not greater than 5, respectively and the number of tokens in ${}^0D_3$ is not greater than 10. If ρ satisfies only these conditions, then a transition set B is control enabled under the control policy ρ and a marking m iff

$$|B\bigcap F(D_i)| + \sum_{r\in R(D_i)}\sum_{t\in D_i} m({}^0t, r) \leqslant \sum_{r\in R(D_i)} C_r - 1$$

for each $D_i, i=1,2,3$.

Under the initial marking m_0 shown in Fig. 4, every transition set is control enabled. But under the marking m where $m(p_1^1)=5$, $m(p_2^1)=1$, m

$(p_0^2)=1$, $m(p_1^2)=4$, $m(a_{M_3})=1$ and $m(p)=0$ for other places, every transition set containing t_1^2 is not control enabled. Since otherwise, after only t_1^2 fires, the number of tokens in ${}^0D_1$ is 6, greater than the number 5 permitted by ρ.

When the feedback policy ρ is applied to a PPN Z, the closed—loop system, denoted by $\rho \| Z$, permits only the control, process and resource enabled transitions to fire. The set of reachable markings from a marking m under a feedback policy ρ will be denoted as $RM(\rho \| Z, m)$. It is easily shown that if $\rho_1 > \rho_2$, then $RM(\rho_2 \| Z, m) \subseteq RM(\rho_1 \| Z, m)$.

In this chapter, our objective is to find a feedback policy that restricts resource allocation decision to guarantee that deadlock cannot occur in the closed—loop system and to maximize resource usage. To avoid deadlock in the closed—loop system $\rho \| Z$, we note that it is necessary to guarantee that $RM(\rho \| Z, m_0)$ does not contain any marking m in $RM(Z, m_0)$ such that $D(m) \neq \emptyset$, but this is not sufficient, since in the closed—loop system, a process and resource enabled transition may not be allowed to fire due to the restriction of feedback policy ρ.

For the deadlock in the closed—loop system, we adopt the following definition similar to [4].

Given sets of resources R, products Q, a PPN Z for the production sequence and a feedback policy ρ, a set of transitions D is said to be in ρ—*deadlock* under a marking $m \in RM(\rho \| Z, m_0)$ if the following conditions hold:

1) all transitions in D are process enabled under the marking m,

2) $D \cap T_e(m') \cap B = \emptyset$ for any transition set $B \in \rho(m')$ and all marking $m' \in RM(\rho \| Z, m)$.

Example 4. Reconsider the PPN Z shown in Fig. 4 and the control policy for the PPN Z defined in Example 3. It is easy to see that the marking m is reachable under the control policy ρ, that is, $m \in RM(\rho \| Z, m_0)$

where $$m(p) = \begin{cases} 5 & p \in \{p_1^1, p_1^2\} \\ 1 & p = a_{M_2} \\ 0 & \text{otherwise} \end{cases}$$

Under the marking m, t_2^1 and t_2^2 are process and resource enabled. If only t_2^1 fires under the marking m, then the number of tokens in ${}^0D_2$ is 6. If only t_2^2 fires, then the number of tokens in ${}^0D_1$ is 6. But Those are not permitted by the control policy ρ. Thus t_2^1 and t_2^2 can never fire in $\rho \| Z$ from the marking m. $\{t_2^1, t_2^2\}$ is in ρ—deadlock under the marking m.

By the definition, we can get a uniquely maximal set of transitions, denoted by $D_\rho(m)$, which is in ρ—deadlock under the marking $m \in RM(\rho \parallel Z, m_0)$.

We consider the following two cases:

Case I: 1) $\mathscr{D}$ has no cyclic deadlock structure chain or 2) for any cyclic deadlock structure chain $V = \{D_1, D_2, \ldots, D_k\}$, $\Phi := R(F(D_1)^0)) \cap R(F(D_2)^0)) \cap \ldots \cap R(F(D_k)^0)) = \emptyset$ or when $\Phi \neq \emptyset$, then $C_r \geqslant 2$, $r \in \Phi$. In this case, $r \in \Phi$ is called the key resource.

For example, in the PPN Z shown in Fig. 4 $\{D_1, D_2\}$ is a cyclic deadlock structure chain. $\Phi = R(f(D_1)^0) \cap R(F(D_2)^0) = \{M_2, M_3\} \cap \{M_1, M_2\} = \{M_2\}$. M_2 is a key resource.

Deadlock Avoidance Policy ρ_1 (DAP ρ_1):

Given sets of resources R, products Q, and a PPN Z for production sequence, ρ_1 is a restriction policy defined as follows for each marking $m \in RM(Z, m_0)$. A set of transitions $B \subseteq T$ is ρ_1—control enabled iff for each $D \in \mathscr{D}$

$$|B \cap F(D)| + \sum_{r \in R(D)} \sum_{t \in D} m({}^\circ t, r) \leqslant \sum_{r \in R(D)} C_r - 1$$

In other words, ρ_1 is the restriction policy which guarantees

$$\sum_{r \in R(D)} \sum_{t \in D} m({}^\circ t, r) \leqslant \sum_{r \in R(D)} C_r - 1$$

for any marking $m \in RM(\rho_1 \parallel Z, m_0)$ and any deadlock structure D.

Let $D_1, D_2 \in \mathscr{D}$. If $R(D_1) \cap R(D_2) = \emptyset$, then

$$\sum_{r \in R(D_1)} \sum_{t \in D_1} m({}^\circ t, r) \leqslant \sum_{r \in R(D_1)} C_r - 1$$

and

$$\sum_{r \in R(D_2)} \sum_{t \in D_2} m({}^\circ t, r) \leqslant \sum_{r \in R(D_2)} C_r - 1$$

implying that

$$\sum_{r \in R(D_3)} \sum_{t \in D_3} m({}^\circ t, r) \leqslant \sum_{r \in R(D_3)} C_r - 1$$

where $D_3 = D_1 \cup D_2$.

Let $\mathscr{D}_1 = \{D_1 \cup D_2 \mid D_1, D_2 \in \mathscr{D}, R(D_1) \cap R(D_2) = \emptyset\}$.

If $D_1 \subset D_2$ and $R(D_1) = R(D_2)$, then

$$\sum_{r \in R(D_2)} \sum_{t \in D_2} m({}^\circ t, r) \leqslant \sum_{r \in R(D_2)} C_r - 1$$

implying that

$$\sum_{r \in R(D_1)} \sum_{t \in D_1} m({}^o t, r) \leqslant \sum_{r \in R(D_1)} C_r - 1$$

Let $\mathscr{D}_2 = \{D_1 \in \mathscr{D} \mid D_1 \subset D_2 \in \mathscr{D}$ and $R(D_1) = R(D_2)\}$ and let $\mathscr{D}' = \mathscr{D} \backslash \{\mathscr{D}_1 \cup \mathscr{D}_2\}$. Then we need only to consider the restriction of ρ_1 on $D \in \mathscr{D}'$ and in Case I we will do so.

Now we can demonstrate that when the policy ρ_1 is used for Z, then in the closed—loop system, ρ_1—deadlock will not occur and ρ_1 is the least restrictive deadlock avoidance policy in Case I.

Theorem 3: *Given sets of resources R and products Q and a PPN Z for the production sequences, if the DAP ρ_1 is applied to Z in Case I, then for any marking $m \in RM(\rho_1 \parallel Z, m_0)$, the set of transitions in ρ_1-deadlock is empty, that is, $\rho_1 \parallel Z$ is live, and ρ_1 is the least restrictive deadlock avoidance policy.*

Proof: We first note that if for some deadlock structure D_i and a marking m,

$$\sum_{r \in m(D_i)} \sum_{t \in D_i} m({}^0 t, r) = \sum_{r \in R(D_i)} C_r$$

then $D_i \subseteq D(m)$ and deadlock occurs. Hence the condition

$$\sum_{r \in R(D_i)} \sum_{t \in D_i} m({}^0 t, r) \leqslant \sum_{r \in R(D_i)} C_r - 1$$

is necessary to guarantee that no deadlock occurs in the closed—loop system. Since ρ_1 only restricts the number of tokens of ${}^0 D_i$ for each deadlock structure $D_i \in DS$. If ρ_1 can prevent deadlock from occuring in the closed—loop system, then ρ_1 is the maximally permissive deadlock avoidance policy. Thus we only need to prove that ρ_1 - deadlock can never occur in $\rho_1 \parallel Z$.

. If $D_{\rho_1}(m)$ is nonempty for some marking $m \in RM(\rho_1 \parallel Z, m_o)$, then we can prove that there exists a cyclic deadlock structure chain $V = \{D_1, D_2, \cdots, D_k\}$ such that the sets of resources $R(D_i)$ are same and the sets of process enabled transitions in D_i under the marking m are same for all deadlock structure $D_i, i = 1, 2, \cdots, k$. But this is not impossible. Thus $\rho_1 \parallel Z$ is live. See [9] for detail. ●

The deadlock structures in PN shown in Fig. 3 are $D_1 = \{t_2^1, t_3^1\}$, $D_2 = \{t_3^1, t_2^2\}$ and $D_3 = D_1 \cup D_2$. Since $D_1 \subset D_3$, $D_2 \subset D_3$ and $R(D_1) = R(D_2) = R(D_3)$, hence $\mathscr{D}' = \{D_3\}$. The least restrictive deadlock avoidance policy requires that the number of tokens in ${}^0 D_3$ is not greater than 3.

Case II: Otherwise is considered in this case.

In this case we adopt the following policy.

Deadlock Avoidance Policy ρ_2 (DAP ρ_2):

Given sets of resources R and products Q and a PPN Z for the production sequences, the DAP ρ_2 is defined on $\mathscr{D}$ recurrently as follows:

Initially let $\mathscr{C} = \emptyset$.

(1). Let deadlock structure $D_1 \in \mathscr{D} \setminus \mathscr{C}$,

If D_1 is not union set of any two deadlock structures in $\mathscr{D}$ and there exists no deadlock structure D_2 such that $R(D_1) = R(D_2)$, then the number of tokens in $^{\circ}D_1$ is restricted by the policy ρ_2 within $L(D_1)-1$, where $L(D_1) = \sum_{r \in R(D_1)} C_r$

Add D_1 to $\mathscr{C}$.

If D_1 is not union set of any two deadlock structures in $\mathscr{D}$, but there exist deadlock structures $D_2, \ldots, D_k$ such that $R(D_1) = R(D_i)$, $i = 2, \ldots, k$, then the number of tokens in $^{\circ}D$ is restricted by ρ_2 within $L(D)-1$, where $L(D) = \sum_{r \in R(D)} C_r$ and $D = D_1 \cup D_2 \cup \ldots \cup D_k$.

Add the various union sets of $D_1, D_2, \ldots, D_k$ to $\mathscr{C}$. Thus for each union set D', the number of tokens in $^{\circ}D'$ is not greater than $L(D)-1$.

(2). Let $D = \bigcup_{i=1}^{k} D_i$, $D_i \in \mathscr{C}$, $i = 1, 2, \ldots, k$.

If $\{D_1, D_2, \ldots, D_k\}$ is not cyclic deadlock structure chain, then the restriction of ρ_2 for D is equivalent to that of ρ_2 for D_i, $i = 1, \ldots, k$. That is, the number of tokens in $^{\circ}D$ is not greater than $L(D)-1$, where $L(D) = \sum_{t \in D} \overline{m}(^{\circ}t) + 1$ and $\overline{m}$ is a marking of Z such that $\sum_{t \in D_i} \overline{m}(^{\circ}t) = L(D_i) - 1$. $i = 1, \ldots, k$.

Add D to $\mathscr{C}$

If $\{D_1, D_2, \ldots, D_k\}$ is a cyclic deadlock structure chain, then the number of tokens in $^{\circ}D$ is restricted by ρ_2 within $L(D)-1$, where $L(D) = \sum_{t \in D} \overline{m}(^{\circ}t)$ and $\overline{m}$ is a marking of Z such that $\sum_{t \in D_i} \overline{m}(^{\circ}t) = L(D_i) - 1$. $i = 1, \ldots, k$.

Add D to $\mathscr{C}$

Repeat the above procedure until $\mathscr{C} = \mathscr{D}$.

That is, a set of transitions $B \subseteq T$ is control enabled under the marking m if for each $D \in \mathscr{D}$

$$|B \cap F(D)| + \sum_{r \in R(D)} \sum_{t \in D} m(^{\circ}t, r) \leqslant L(D) - 1.$$

Theorem 4: *Given sets of resources R, products Q and a PPN Z for the production sequences, if the DAP ρ_2 is applied to Z in Case* II *, then for any marking $m \in RM(\rho_2|Z, m_0)$, the set of transitions in ρ_2-deadlock is empty, that is, $\rho_2 \parallel Z$ is live.*

Proof: Similar to the proof of theorem 3 we can prove that $\rho_2 \parallel Z$ is live. See [9] for detail. ●

We can see that ρ_2 has very little restriction. That is, if the numbers of tokens in deadlock structures are larger than that permitted by ρ_2, the closed loop system may be in deadlock. But it is not least restrictive. For this we reconsider Example 2 in Section 3 for illustration. Since there are three deadlock structures D_1, D_2 and D_3. $\{D_1, D_2\}$ is a cyclic deadlock structure chain. M_2 is a key resource. Thus the policy ρ_2 restricts the number of tokens in $^{\circ}D_1 = \{p_2^1, p_1^2\}$ within $C_{M_2} + C_{M_3} - 1 = 5$, the number of tokens in $^{\circ}D_2 = \{p_1^1, p_2^2\}$ within $C_{M_1} + C_{M_2} - 1 = 5$ and the number of tokens in $^{\circ}D_3 = {}^{\circ}(D_1 \cup D_2) = \{p_1^1, p_2^1, p_1^2, p_2^2\}$ within $L(D_3) - 1 = 9$ where $L(D_3) = \sum_{t \in D_3} \overline{m}(^{\circ}t) = 10$. But for the reachable marking $m \in RM(\rho \parallel Z, m_o)$ where $m(p_1^1) = 5$, $m(p_2^1) = 1$, $m(p_1^2) = 4$ and $m(p) = 0$ for other place p, the number of tokens in $^{0}D_3$ is 10, great than 9, t_3^1 can fire. Therefore the set of transitions in deadlock is empty. That is, $D_{\rho_2}(m) = \emptyset$. ρ_2 is not least restrictive since ρ_2 permits only 9 tokens to enter the places in $\{p_1^1, p_2^1, p_1^2\}$.

5. The Petri Net Realization of Deadlock Avoidance Policies

In this section we consider the Petri net realization of deadlock avoidance policies given in Section 4.

1). The Petri net realization of DAP ρ_1

For every given deadlock structure $D_i \in \mathscr{D}'$, we construct a Petri net $X^i = (P^i, F(D_i) \cup B(D_i), I^i, O^i, m_0^i)$ where P^i has only one place, denoted by cp_i,

$$I^i = P^i \times F(D_i),$$

$$O^i = B(D_i) \times P^i,$$

$$m_0^i(cp_i) = \sum_{r \in R(D_i)} C_r - 1$$

Let $X = X^1 \| X^2 \| \ldots \| X^k$ where $k = |\mathscr{D}'|$, then X is a Petri net and $X = (P_X, T_X, I_X, O_X, m_{X0})$ where

$$P_X = \bigcup_{i=1}^{k} P^i,$$

$$T_X = \bigcup_{i=1}^{k} (F(D_i) \cup B(D_i)),$$

$$I_X = \bigcup_{i=1}^{k} (P^i \times F(D_i)),$$

$$O_X = \bigcup_{i=1}^{k} (B(D_i) \times P^i),$$

$$m_{X0}(cp_i) = m_0^i(cp_i),\ i = 1, \ldots, k,$$

We can easily see that the restriction of X on Z is equivalent to that of ρ_1 on Z. Hence X is a Petri net realization of ρ_1 and is called Petri net controller. The closed—loop system for Z with Petri net controller X is also a Petri net $X \| Z = (P \cup P_X, T, I \cup I_X, O \cup O_X, m_0')$ where

$$m_0'(p) = \begin{cases} m_0(p) & p \in P \\ m_{X0}(p) & p \in P_X. \end{cases}$$

Consider the Petri net in Fig. 3, $\mathscr{D}' = \{D_3\}$ where $D_3 = \{t_2^1, t_3^1, t_2^2\}$. So $X^3 = (P^3, F(D_3) \cup B(D_3), I^3, O^3, m_0^3)$ where $P^3 = \{cp_3\}$, $F(D_3) = \{t_1^1, t_1^2\}$, $B(D_3) = \{t_3^1, t_2^2\}$, $I^3 = \{(cp_3, t_1^1), (cp_3, t_1^2)\}$, $O^3 = \{(t_3^1, cp_3), (t_2^2, cp_3)\}$ and $m_0^3(cp_3) = 3$. Then $X = X^3$ shown in Fig. 5 is the Petri net realization of the least restrictive deadlock avoidance policy for the FMS in Example 1.

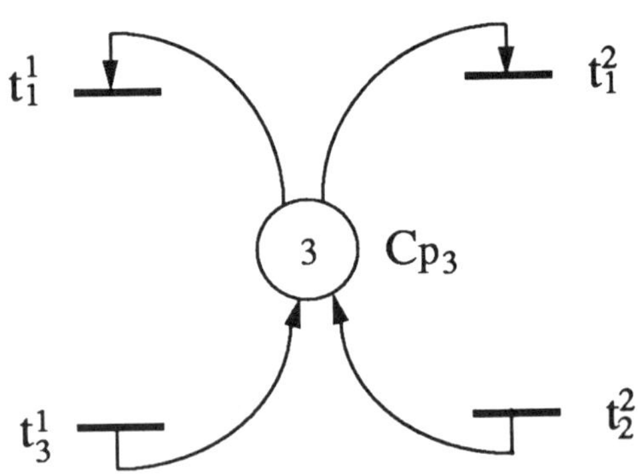

Fig. 5. The Petri net realization of the least restrictive deadlock avoidance policy for the FMS in Example 1 1.

2). The Petri net Realization of DAP ρ_2

Similar to the case of ρ_1, for every deadlock structure $D_i \in \mathcal{D}$, we construct a Petri net $S^i = (P^i, F(D_i) \bigcup B(D_i), I^i, O^i, m_0^i)$ where P^i has only one place, denoted by cp_i,

$$I^i = P^i \times F(D_i),$$
$$O^i = B(D_i) \times P^i,$$
$$m_0^i(cp_i) = L(D_i) - 1,$$

then $S = S^1 \parallel S^2 \parallel \ldots \parallel S^k$ is a Petri net and we can see that the restriction of S on Z is equivalent to that of ρ_2 on Z. Hence S is a Petri net realization of ρ_2, The closed—loop system for Z with Petri net controller S is a live Petri net $S \parallel Z$.

Since the restriction policies can be realized by Petri nets, the system controlled by the restriction policy can be modeled by the live Petri net.

6. Application Illustration

In this section, we illustrate how the deadlock avoidance method proposed in this chapter can be applied to an FMS in which deadlock can occur.

Example 5. We consider an exemplary FMS illustrated in Fig. 6 [4].

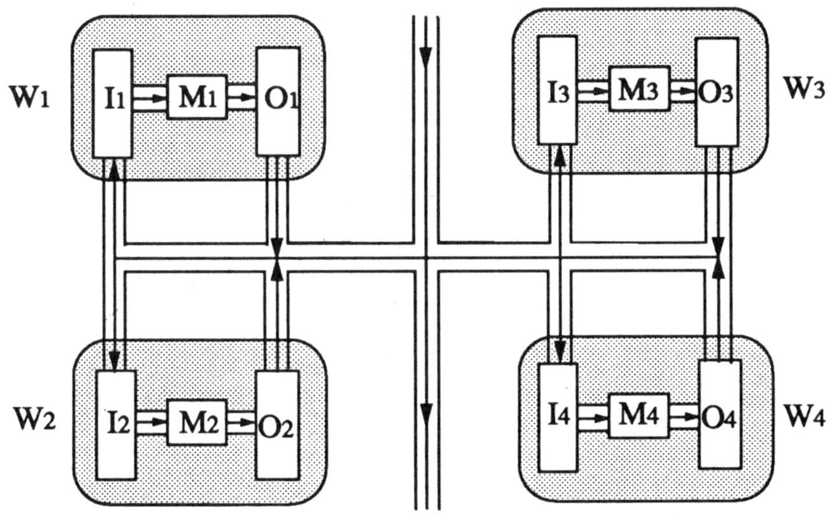

Fig. 6. Flexible manufacturing system with four workstations [4]

There are four work stations W_1, W_2, W_3 and W_4 served by a transport system S and two types of products q_1 and q_2. Each workstation has an input buffer to hold products to be processed and an output buffer to hold products for which the process step has been completed. Lables I_1, I_2, I_3, I_4 and O_1, O_2, O_3, O_4 denote the input and output buffers for workstations W_1, W_2, W_3 and W_4, respectively. Each buffer has a capacity of 5. Suppose that there are a single transport resource in S and a single machine M_i in the workstation W_i, $i=1,2,3,4$, then the set of resources is given by $R=\{I_1,\ldots,I_4,O_1,\ldots,O_4\}$.

Suppose that q_1 and q_2 are processed by sequencing machines M_3, M_1, M_2, M_1 and M_2, M_3, M_4 in order, respectively. Hence the process routings are specified as follows;

$$\sigma(q_1)=(I_3, O_3, I_1, O_1, I_2, O_2, I_1, O_1)$$
$$\sigma(q_2)=(I_2, O_2, I_3, O_3, I_4, O_4)$$

First suppose only one type of products, say q_1, is produced. The PPN model of the system with one type of products q_1 is shown in Fig. 7.

Since the number of each type of resources is 5, hence we need to construct the restriction policy ρ_1 which has the form of the policy in Case I of Section 4 and restricts the number of tokens in ^oD for each deadlock structure D of Z within $\sum_{r\in R(D)} C_r-1$. Deadlock structures of Z are $D_1=\{t_4, t_5, t_6, t_7\}$, $D_2=\{t_5, t_6, t_7, t_8\}$ and $D_3=D_1\cup D_2=\{t_4, t_5, t_6, t_7, t_8\}$. Since $R(D_1)=R(D_2)=R(D_3)=\{I_1, O_1, I_2, O_2\}$ and $D_1\subset D_3$, $D_2\subset D_3$, hence we only consider D_3 in the construction of deadlock avoidance policy, and policy ρ_1 only needs to restrict the number of tokens in the places of $^oD_3$ not greater than $\sum_{r\in R(D_3)} C_r-1=C_{I_1}+C_{O_1}+C_{I_2}+C_{O_2}-1=19$. If this PN is controlled by the DAA restriction policy δ proposed by Banaszak and Krogh [4], then the number of tokens in places of $^0D_3$ is not greater than 10 [4, pp. 731—732]. When the buffers I_1 and O_1 are full, t_3 is not allowed to fire by DAA restriction policy δ, even if the number of tokens in places of $^0D_3$ is less than 10. This shows the strength of our proposed method.

For deadlock structure D_3, $F(D_3)=\{t_3\}$ and $B(D_3)=\{t_8\}$, the Petri net realization of policy ρ_1 is $X=X^3=(P_x, \{t_3, t_8\}, I_x, O_x, m_{x0})$, where $P_x=\{cp\}$, $I_x=\{(cp,t_3)\}$, $O_x=\{(t_8,cp)\}$, and $m_{x0}(cp)=\sum_{r\in R(D_3)} C_r-1=19$ As shown in Fig. 8.

Considering the marking $\overline{m}\in RM(Z,m_0)$

$$\overline{m}(p)=\begin{cases}5 & p\in\{p_3, p_4, p_5, p_6, a_{I_3}, a_{O_3}\}\\ 0 & \text{otherwise.}\end{cases}$$

for which transitions t_4, t_5, t_6, t_7 are in deadlock. where $\{t_4, t_5, t_6, t_7\}=D_1$ is a deadlock structure and $D_1\subset D_3$, therefore the marking $\overline{m}$ is not reachable under the control of X, because X allows at most 19 tokens in

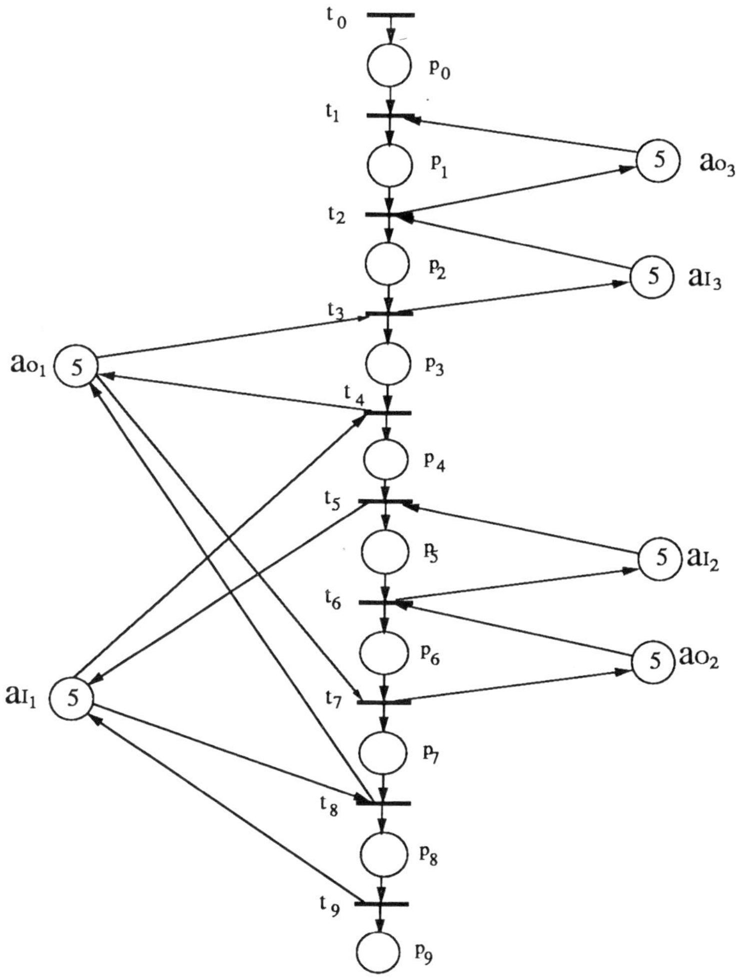

Fig. 7. the PPN model for FMS with one kind of products in Fig. 6.

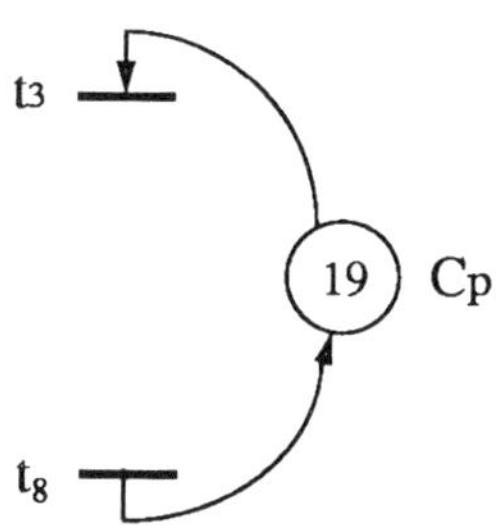

Fig. 8. The Petri net realization of the least restrictive deadlock avoidance policy for the PPN model in Fig. 7.

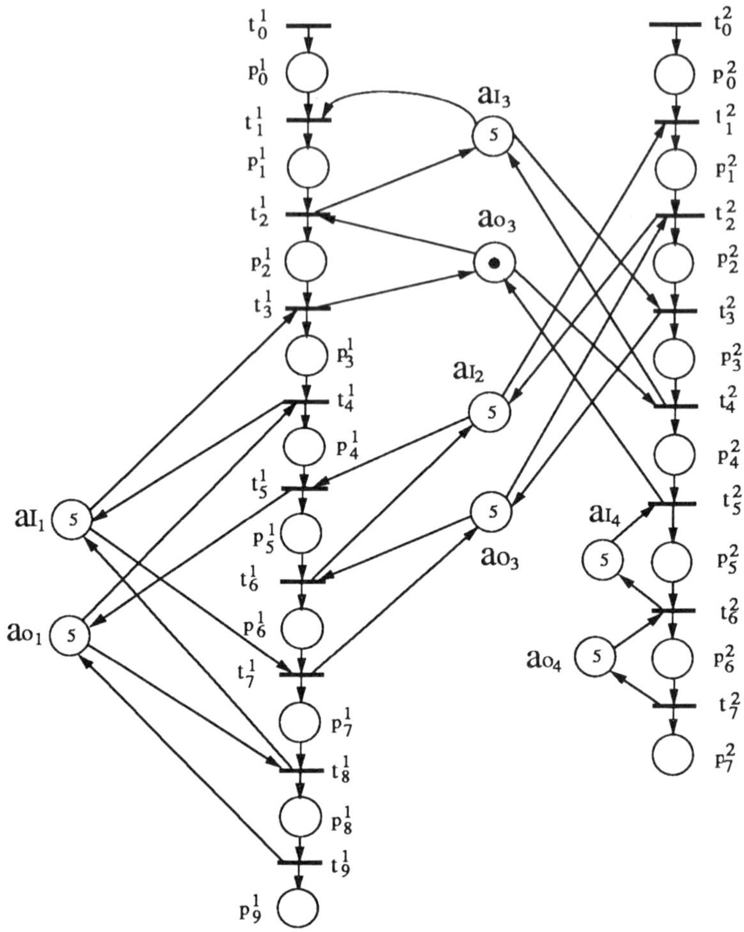

Fig. 9. The complete Petri net model for the FMS in Fig. 6.

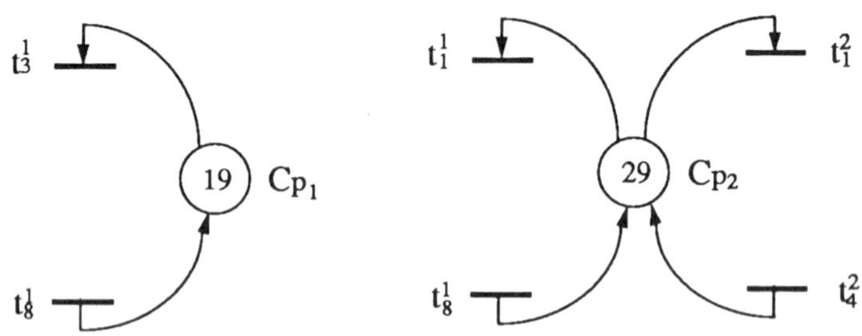

Fig. 10. The Petri net realization of the least restrictive deadlock avoidance policy for the FMS in Fig. 6.

the places of ${}^{\circ}D_3=\{p_3, p_4, p_5, p_6, p_7\}$. This number (i.e. 19) is also the biggest number of tokens which can enter places of ${}^{\circ}D_3$ without leading to deadlock. If the number of tokens in ${}^{\circ}D_3$ is 20, then one of D_1, D_2, D_3 must be in deadlock. For example, under the marking $\overline{m}$ the number of tokens in ${}^{0}D_3$ is 20, D_1 is in deadlock. Thus the restriction policy ρ_1 is the minimally restrictive deadlock avoidance policy.

Now consider the case in which q_1 and q_2 are produced at the same time. The PN model is shown in Fig. 9. Its deadlock structures are specified as follows:

$$D_1=\{t_4^1, t_5^1, t_6^1, t_7^1\},$$
$$D_2=\{t_5^1, t_6^1, t_7^1, t_8^1\},$$
$$D_3=\{t_2^1, t_3^1, t_4^1, t_5^1, t_2^2, t_3^2\},$$
$$D_4=\{t_3^1, t_4^1, t_5^1, t_6^1, t_3^2, t_4^2\},$$

and the various unions of D_1, D_2, D_3 and D_4.

Since $R(D_1)=R(D_2)=R(D_1\bigcup D_2)=\{I_1, O_1, I_2, O_2\}$, $R(D_3)=R(D_4)=R(D_3\bigcup D_4)=\{I_1, O_1, I_2, O_2, I_3, O_3\}$, $D_1\subset D_1\bigcup D_2$, $D_2\subset D_1\bigcup D_2$ and $D_3\subset D_3\bigcup D_4$, $D_4\subset D_3\bigcup D_4$, we consider only $D_5=D_1\bigcup D_2$ and $D_6=D_1\bigcup D_2\bigcup D_3\bigcup D_4$ in the construction of deadlock avoidance policy ρ_1. Thus the policy ρ_1 only guarantees that the numbers of tokens in ${}^{\circ}D_5$ and ${}^{\circ}D_6$ are not greater than $\sum_{r\in R(D_5)} C_r-1=19$ and $\sum_{r\in R(D_6)} C_r-1=29$ respectively. With these two conditions holding true, the closed loop system controlled by ρ_1 is live. Since $F(D_5)=\{t_3^1\}$, $B(D_5)=\{t_8^1\}$, $F(D_6)=\{t_1^1, t_1^2\}$, $B(D_6)=\{t_8^1, t_4^2\}$, the Petri net controller X can easily be specified as shown in Fig. 10 and X is the minimally restrictive deadlock avoidance controller for Z. But in the system controlled by DAA restriction policy δ in [4], if one of the buffers I_1, O_1, I_2, O_2, I_3, O_3 is full, t_1^1 is not allowed to fire. If one of the buffers I_2, O_2, I_3, O_3 is full, t_1^1, t_1^2 are not allowed to fire, in this case only 5 products are being concurrently produced in the whole system. The policy ρ_1 is the least restrictive and so more permissive than those in [2], [4] and [11]. This has a clear implication in improving the machine utilization and system productivity.

7. Conclusion

In this chapter we have formulated the concept of deadlock structure and studied the problem of deadlock avoidance in Petri net models which represent the flexible manufacturing systems with shared resources. We have presented an efficient method for synthesizing deadlock avoidance policies.

When the number of any key kind resources is greater than 1, the necessary and sufficient conditions are derived which guarantee against occurrence of any deadlock, and the minimally restrictive deadlock avoidance policy is obtained. This policy will potentially maximize the usage of resources and productivity. In general case we give a very little restrictive deadlock avoidance policy. For these deadlock avoidance policies we have presented the Petri net realizations. Therefore the models of the closed loop systems are live Petri nets. These nets can be used to design deadlock—free discrete event controllers which are very important in automated manufacturing. An advantage of the restriction policy synthesis procedure is that it is based on the Petri net model of the open—loop system. Thus the new policy can be computed quickly when the types of products or the number of resources are changed.

Although the proposed method focuses a class of manufacturing systems which can be modeled as production Peri nets, it could be extended to the cases where multiple routes are allowed for a type of products. Petri net modeling methods presented in [10] will be useful in building up open—loop system models. The deadlock structures and related necessary and sufficient conditions for a live Petri net will be more complex. The evaluation of the proposed method and other deadlock avoidance methods [4,10,11] in the context of deterministic /stochastic timed operations is an interesting research topic.

8. References

[1] J. L. Peterson, *Petri Net Theory and the Modeling of Systems*, Englewood Cliffs, NJ: prentice hall, 1981.

[2] N. Viswanadham, Y. Narahari and T. L. Johnson, "Deadlock prevention and deadlock avoidance in FMS using Petri net models," *IEEE Trans. Robotics and Automation*, Vol. 6, No. 6. 1990.

[3] Minoura and C. Ding, "A deadlock prevention method for a sequence controller for manufacturing", *International Journal of Robotics and Automation*, Vol. 6, No. 3, 1991.

[4] Z. A. Banaszak and B. H. Krogh, "Deadlock avoidance in flexible manufacturing systems with concurrently competing process flows," *IEEE Trans. Robotics and Automation*, Vol. 6, No,6,1990.

[5] T. Murata, B. Shenker and S. M. shatz, "Detection of ada static deadlock using Petri net invariants", *IEEE Trans. Software Eng.*, Vol. 15, No. 3, 1989.

[6] D. Hauschildt and R. Valk, "Safe states in banker like resource allocation problems", *Lecture Notes Comp. Sci.*, 1985.

[7] E. G. Coffman Jr., M. Elphick and A. Shoshani, " System deadlocks," *Comp. Surveys*, Vol. 3, No. 2, pp. 67—78. June 1971.

[8] B. H. Krogh and L. E. Holloway, "Synthesis of feedback control logic for discrete manufacturing systems," *Automatica*, Vol. 27, No. ,1991.

[9] K. Y. Xing, *The research on control theory and methods for discrete event system*, Ph. D dissertation, Xian Jiaotong University, P. R. China, 1994.

[10] M. C. Zhon and F. DiCesare, *Petri Net Synthesis for Discrete Event control of Manufacturing Systems*, Kluwer Academic, Boston, MA, 1993.

[11] D. Y. Hsich and S. C. Chang, "Dispatching—driven deadlock avoidance controller synthesis for flexible manufacturing systems, " *IEEE Trans. Robotics and Automation*, 10(2), pp 196—209, 1994.

10

Discrete Event Control Design for Manufacturing Systems Via Ladder Logic Diagrams and Petri Nets: A Comparative Study

Kurapati Venkatesh*, MengChu Zhou#, Reggie J. Caudill*
* Department of Mechanical Engineering,
\# Department of Electrical and Computer Engineering,
Center for Manufacturing Systems,
New Jersey Institute of Technology, Newark, NJ 07102, USA
Zhou@ece.njit.edu

Abstract

Design methods for sequence controllers play a prominent role in advancing industrial automation. The increasing complexity and varying needs of modern discrete manufacturing systems have challenged the traditional design methods such as the use of ladder logic diagrams (LLDs) for programmable logic controllers. The methodologies based on research results in computer science have recently received growing attention by academic researchers and industrial engineers in order to design flexible, reusable, and maintainable control software. Particularly, Petri nets (PNs) are emerging as a very important tool to provide an integrated solution for modeling, analysis, simulation, and control of industrial automated systems. However, in order to establish PNs as alternative to LLDs there is a need for benchmark studies to formally compare them. The contribution of this chapter is two fold. First, certain criteria are identified to compare LLDs and PNs in designing sequence controllers subject to the changing control requirements. The comparison is performed through a practical system after introducing 'real-time Petri nets' for discrete-event control purpose. Secondly, some analytical formulas and a methodology are developed to estimate the number of basic elements used in the PN and LLD designs prior to their constructions. The results will be useful for researchers and engineers to design control systems for complex industrial automated systems.

1. Introduction

A discrete manufacturing system consists of such units as machines, robots, automated guided vehicles, programmable logic controllers, and computers which function asynchronously and concurrently to meet the dynamically changing needs of the market. Integrated software development methods aiming for modeling, analysis, control, and simulation of such systems are important. The typical functions of the control software are to sequence the operations, monitor the system functioning, and determine the states of different elements in a discrete manufacturing system with respect to real time. Traditionally, ladder logic diagrams (LLDs) are used to capture the sequence of operations executed by the system's control software. They specify the I/O procedures of the Programmable Logic Controller (PLC) that drive and perform the repetitive operations of the system. These diagrams grow so complex that locating the cause when a problem is detected becomes extremely difficult [2]. Furthermore, their usage is limited only to control the system but not to analyze and evaluate the qualitative and performance characteristics. They often need significant changes as the specification changes resulting in difficult maintaining problem.

Hence, researchers are constantly pursuing to develop integrated tools that eliminate the limitations of LLDs. These tools are aimed not only for control but also system analysis, evaluation, and simulation. Petri nets (PNs) originated from the area of computer science are such a tool as evident from its numerous applications in manufacturing systems [14,24]. Realizing the potential of PNs for control, in France, GRAFCET - a PN like representation tool is proposed as a standard specification of sequence controllers [5]. Its international standard version is called sequential function charts [7]. Hitachi developed a commercial product based on an enhanced PN that interacts with the physical system to control [11]. More details and advantages of PNs for controlling manufacturing systems can be found in [1]. More recent studies of PNs for discrete event control can be seen in [8,17-24]. Different implementation schemes were proposed to apply PNs for sequence control as shown in Table 1.

It is observed that none of the earlier studies has in detail and formally compared LLDs and PNs for the design of sequence controllers. Boucher, *et al.* [1] used LLDs and PNs to control a same manufacturing system and reported the graphical representation by PNs makes the controller more tractable than that of LLDs. However, they have not formally quantified the comparison between PN and LLDs to design sequence controllers. The sequence controller in this chapter means a class of discrete event controllers without choices in executing the operations/activities. The detailed comparison of LLDs and PNs is very important in realizing their advantages and disadvantages and, particularly, in establishing PNs as an emerging design technique for effective sequence control of industrial automated systems. Some of the problems associated to compare PNs and LLDs are: (1) unlike in case of LLDs, there exist several classes of PNs with various implementation schemes for discrete control as shown in Table 1, and (2) identification of the criteria with respect to which the comparison should be performed.

Table 1. Various methods of Petri net based sequence control [19]

Implementation scheme of Petri net controller	Hardware used	Reference
Petri net → High-level language description → Compiler → Control sequence tables and decision programs → control	Intel 8085 Assembler	[3]
Petri net → High-level language description → Interpreter → Control tables → control	A PLC based on ZILOG 80A microprocessor	[16]
Control Petri net → Interpreter → control	16 bit microcomputer	[11]
Petri net → Execution algorithm → control	VAX 11/780 and IBM-PC/XT	[4]
Petri net → Software → control Note: Details of implementation are not given	IBM-PC/XT	[1]
Petri net → Petri net description language → Token player → control	IBM-PC/XT	[22]
Control Petri net → Petri net simulator → control	8086 CPU and DSP (TMS32020)	[15]
Real-time Petri net → Software system based on token player → control	IBM-PC/AT	[19]

The goal of the research is to compare LLDs and PNs when they are used to design discrete event controllers for manufacturing systems. The objectives of this chapter are:

1. To introduce *Real-time PNs (RTPNs)* which closely resemble ordinary PNs as an integrated tool to develop discrete event controllers,
2. To identify the criteria to compare LLDs and RTPNs for design of sequence control,
3. To compare LLDs and RTPNs in designing sequence controllers that respond to specification changes,
4. To formulate mathematical formulas to calculate the number of basic elements to model certain building blocks of logic models using PNs and LLDs, and

5. To present a methodology that synthesizes these analytical formulas for estimating the total number of basic elements required to design sequence controllers using PNs and LLDs.

Section 2 presents the concepts of Real-time PNs with an example illustrating how they can be used for real-time control. In Section 3, comparison criteria for control logic design by PNs and LLDs is presented. The comparison between LLDs and RTPNs is performed through a practical system as a case study in Section 4. Section 5 presents the analytical formulas to estimate the basic elements required to implement certain building blocks of control logic. A methodology that uses these formulas to estimate the number of basic elements required to implement a given control logic specification using PNs and LLDs is presented in Section 6. It is illustrated through two examples and the case study considered in Section 4. Finally conclusions and further research are reported in Section 7.

2. Real-time PNs

An excellent tutorial on PNs and their applications is given in [11] and their applications in manufacturing automation are reported in [6,14,24]. PNs have been augmented and implemented in a variety of ways to achieve real-time control as shown in Table 1. Based on the research in PN control literature, this chapter proposes a class of PNs called Real-time PNs (RTPNs) for sequence controller design. A simple and straight-forward procedure is demonstrated to implement them [17].

RTPNs can be obtained by associating timing, I/O sensory information to the untimed PNs. It is an eight tuple and defined as: RTPN=(P, T, I, O, m, D, X, Y) where:

1. P is a finite set of places;
2. T is a finite set of transitions with $P \cup T \neq \emptyset$ and $P \cap T = \emptyset$;
3. I: $P \times T \rightarrow N$, is an input function that defines the set of directed arcs from P to T where $N = \{0, 1, 2, ...\}$;
4. O: $P \times T \rightarrow N$, is an output function that defines the set of directed arcs from T to P;
5. m: $P \rightarrow N$, is a marking whose i^{th} component represents the number of tokens in the ith place. An initial marking is denoted by m_0;
6. D: $T \rightarrow R^+$, is a firing time function where R^+ is the set of non-negative real numbers;
7. X: $P \rightarrow \{-, 0,1,2, ...K\}$ and $X(p_i) \neq X(p_j)$, $i \neq j$, is an input signal function, where K is the maximum number of input signal channels, and "-" is the dummy attribute indicating no assigned channel to the place.
8. Y: $T \rightarrow L$, is an output signal function, where L is a set of integers.

In a RTPN, the first five tuples represent the untimed PN and the last three are extensions added to it and explained below:

1. Timing vector (D) is intended to associate time delays to transitions modeling the activities in the system,

2. Input signal vector (X) reads the state of the input signals from digital input interface. X associates attributes to every place. X_i=$X(p_i)$ and is an attribute associated with place p_i, representing the input channel number associated with p_i. For example, if p_i models a limit switch, the RTPN reads the status of that switch from the digital input interface through the channel number represented by X_i. The initial marking, $m_0(p_i)$ is considered as the first attribute of p_i and X_i is the second one. The contents of any input channel X_i are either 0 or 1.
3. Output signal vector (Y) is intended to send output signals through digital output interface. Y associates attributes to every transition. Y_i=$Y(t_i)$ and is the attribute associated to transition t_i which represents the number that is to be sent to the digital output interface. For example, t_i may model the activity "send signal to actuate solenoid A", or "execute a procedure to control a robot". Each solenoid is activated by writing a specific number on to the digital output interface. During execution of the program, when a transition fires, RTPN writes the decimal number corresponding to the output channel to digital output interface. The contents of any output channel are either 0 or 1. The usage of this vector is later detailed in the example system.

There are two events for a transition firing, *start_firing* and *end_firing*. Between these the firing is in progress. The removal of tokens from a transition's input place(s) occurs at *start_firing*. The deposition of tokens to a transition's output places(s) occurs at the *end_firing*. While transition firing is in progress, the time to end firing, called the *remaining firing time*, decreases from firing duration to zero at which its firing is completed. The execution rules of a RTPN include enabling and firing rules:

1. A transition $t \in T$ is *enabled* iff $\forall\ p \in P$ and $I(p,t) \neq 0$, $m(p) \geq I(p, t)$ and X(p) has content 1.
2. Enabled in a marking m, t *fires* and results in a new marking m' following the rule:
   $m'(p) = m(p) + O(p,t) - I(p,t), \forall\ p \in P.$

The design procedure for formulating a RTPN based controller is shown in Fig. 1 and is briefed in the following five steps:

1. Model the control sequence using PNs to obtain the PN model of the sequence controller.
2. Assign input channels to inputs of the system such as limit switches, sensors, etc. to formulate an input mapping table.
3. Assign output channels to outputs of the system such as solenoids, switches etc. Also, identify timing information for activities to obtain an output mapping table.
4. Using the input mapping table assign input channel number to each place in PN based controller. The initial state of the system decides the initial marking of RTPN. In the PN model some places do not represent the inputs of the system as they represent the intermediate states of system or logical places to model counters in the sequence. Hence, no channel is assigned to these places, represented by "-".

5. Using the output mapping table and the action(s) that are modeled by a transition, assign a number to each transition in a PN based controller. The operations and the time delays given in the sequence to be controlled decide firing time function of RTPN. In the PN model some transitions may represent concurrent actions. Hence, care should be taken to assign the numbers for such transitions.

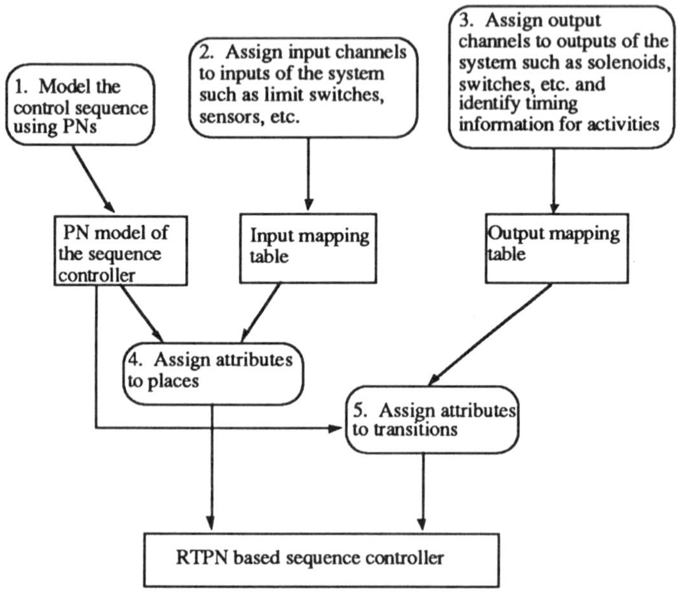

Figure 1. Procedure for formulating a RTPN based controller

By following the above procedure a RTPN based controller can be formulated for a given sequence. There are several ways to use PNs to perform the sequence control. One is based on "token game" [17,22] and the other converts the net into either Programmable Logic Controllers [16] or control code directly [24]. The first scheme is used here (see Fig. 2) and explained here. RTPN based controller can be embedded in and executed by a computer. As the execution of RTPN starts and continues, the system being controlled also starts and continues to perform operations corresponding to the sequence modeled by RTPN. There are digital I/O interfaces that act as a bridge between the RTPN and the system being controlled. For simplicity and to focus the objective of this chapter, breakdowns are assumed not to occur. However, by directly extending the concepts of "Augmented timed PNs" [18], the detailed modeling of breakdown handling is possible with RTPNs.

In order to illustrate the control by RTPNs a simple assembly system shown in Fig. 3 is considered.

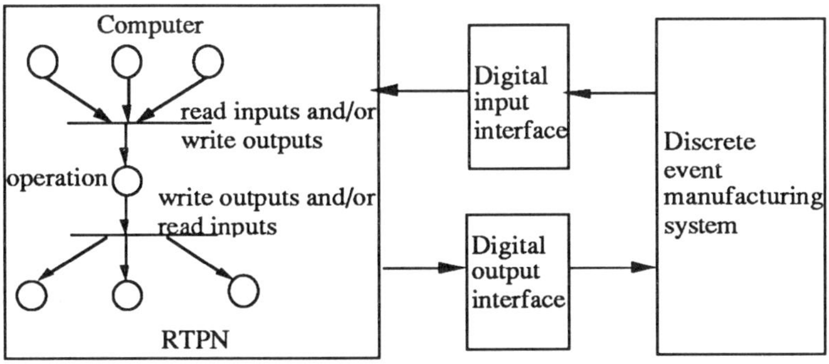

Figure 2. Controlling a system using a RTPN based controller

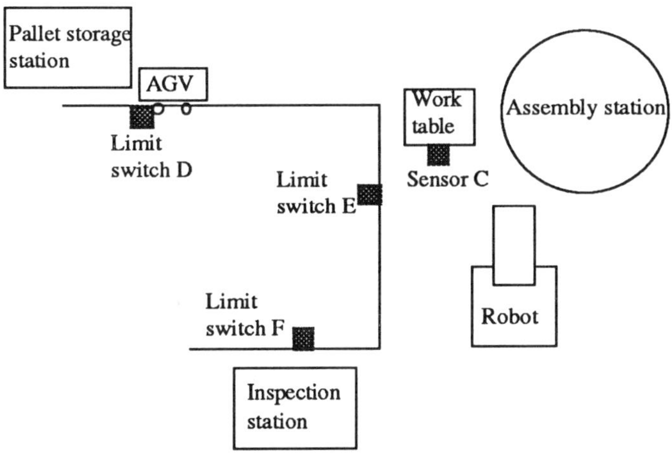

Figure 3. Schematic of an automatic assembly system [21]

It consists of a pallet storage station, assembly station, an inspection station, an AGV, and a robot. Two workpieces are assembled by the robot at the assembly station. After the assembly is completed, the robot unloads the finished product on a work table. The AGV transports empty pallets from pallet storage station to assembly station. Robot is also used to unload the finished product from work table and place it on the pallet which is on the AGV. Finally, the AGV transports the finished product loaded on pallet to inspection station for inspection. The availability of Workpieces 1 and 2 is sensed by two sensors A and B (not shown in figure), respectively. Limit switches D, E, and F sensing the presence of AGV are at pallet storage station, assembly station, and inspection station respectively. Sensor C recognizes the loading of finished product on work table by robot. Basically there are three operations to be executed in this system as follows:

1. Robot's assembly and transfer of the finished product on to work table,
2. AGV's travel from pallet storage station to assembly station, and
3. AGV's travel from assembly station to inspection station.

Assume the time duration for operation 'i' is 'τ_i' time units and is executed when output relay 'i' is high. There is a digital I/O interface to control this system with details as shown in Tables 3 and 4.

Table 3. The input mapping table where X_i is input channel number.

Limit switch/Sensor	A	B	C	D	E	F
X_i	0	1	2	3	4	5

Table 4. The output mapping table where Y_i is number sent to the digital output interface.

Output relay	Output Channel number
1	0
2	1
3	2

Assume this system is controlled by PN with places A, B, C, D, E, and F modeling the corresponding sensors, and transition t_i models operation i. The attributes of places and transitions for this RTPN are listed in Table 5.

Table 5. Attributes of places and transitions in the RTPN

Place	1st attri-bute	2nd attri-bute	Tran-sition	Output relay	Output channel to be activated (x)	1st attri-bute	2nd attri-bute (2^x)
A	1	0	t_1	1	0	τ_1	1
B	1	1	t_2	2	1	τ_2	2
C	0	2	t_3	3	2	τ_3	4
D	1	3					
E	0	4					
F	0	5					

The first (second) attribute of a place is initial marking (assigned input channel number) and for a transition the first attribute is time duration. The second attribute of a transition is the number to be sent to the I/O interface to activate the corresponding output relay. Following the procedure illustrated in Fig. 1, the RTPN model obtained is shown in Fig. 4. Observe that time durations for these transitions need not be known to execute RTPN (or to control the system). However, they are needed when the same RTPN model is used for performance evaluation. For more details of the software implementation of RTPN based controllers, refer to [17].

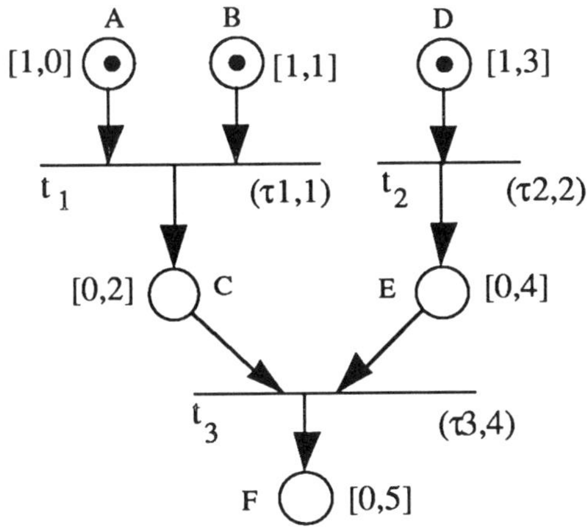

Figure 4. RTPN model for the system

Even though RTPNs and earlier classes of PNs for control share similar principles, some of the differences between them are listed below:

1. Earlier studies use a variety of places are to model timers and counters [11]. This might cause the model difficult to understand. In RTPNs neither new places nor transitions are introduced for modeling them. Timers are modeled by assigning attributes to transitions and counters by places with initial markings and weights on certain arcs. In [15] new sets of places are introduced for modeling I/O signals which increases the number of places in PN model. In RTPNs, I/O signals are modeled as attributes for places and transitions respectively. Hence, due to the use of attributes RTPNs have less nodes and links compared to PNs in [11,15], thereby reducing the graphical complexity.
2. In earlier works, e.g., [1,3,15], the resetting of timers, counters, and emergency stop are not explicitly modeled. Furthermore, often they use additional functions to model and implement timers and counters [15]. Using RTPNs all these can be clearly modeled. The automatic resetting of timers and counters is also embedded in the execution of RTPNs.
3. RTPNs model the system more realistically by natural mapping of the limit switches, start, and stop buttons as places and can be easily extended to model breakdown handling procedures by using the concepts of Augmented Timed Petri nets [18].
4. RTPNs can be implemented by a simple implementation scheme. For example, RTPNs eliminate the usage of high level net description languages used in [3,16,22] since the RTPN model can be directly used for control with the help of a token player. By adopting this implementation scheme, the need to translate the PN model to higher level net description language is avoided. The actual implementation of the token player in RTPNs is transparent to the users and hence their only task to control a system is simplified to model the control logic.

3. Comparison criteria for control logic design by PNs and LLDs

The application of LLDs for sequence control is widely known as they are used by several industries [9,12]. Graphically, RTPNs are same as PNs except that in the former, places are given attributes to model input sensory information and transitions are associated with attributes to model output and timing information. Hence, the comparison presented in this chapter corresponds to RTPNs which are referred to PNs hereon for short. In order to start an operation in a system some conditions have to be fulfilled which are called as pre-conditions. Upon its completion, an operation results in some conditions which are called as post-conditions. A transition in a PN models an operation. The input and output places of a transition model the pre- and post-conditions respectively.

In LLDs as shown in the previous section, an operation is modeled by activating either an output coil corresponding to a relay or a solenoid. In addition to an output coil a relay consists of number of contacts, some normally open (N.O) and some normally closed (N.C). In order to energize the output coil of a relay, the corresponding contacts of that relay should be switched, i.e., an N.O. contact closes, while an N.C. one opens. LLD also uses various input and output elements. A typical output element is a solenoid which is usually used to actuate pneumatic or hydraulic solenoid valves. Push-button and limit switches constitute as inputs elements. The contacts of relay along with the input elements constitute pre-conditions. The solenoids and output coils are referred to as post-conditions.

The logic and other basic building blocks used in sequence control are modeled by PNs and LLDs as shown in Table 2. The first four rows show the basic PN elements to model conditions, status, activity, information and material flow, and resources. Note that LLDs do not have the corresponding explicit representations. Logical AND and logical OR can be easily modeled by both PN and LLD with similar complexity. Other important concepts, e.g., concurrency, time delay, and synchronization are also illustrated in Table 2.

The systematic methods to formulate PN models can be seen in [23,24] and the methods of developing LLDs can be seen in [9,12]. Two of the important factors for comparison of PN and LLD for discrete event control are identified as design complexity and response time.

Design Complexity

Design complexity is defined as the complexity associated in designing the control logic for a given specification. Since it is influenced by many factors, e.g., the experience of designers, size of control program, and number of dynamic steps necessary for coding or changing control program, it is very hard to formally quantify. However, it can be characterized by two factors namely graphical complexity and adaptability for change in specification.

Table 2. Representations by Petri nets and ladder logic diagrams

Logic constructs	Petri nets	Ladder logic diagrams
Condition or the status of a system element	O Place	No explicit representation
An activity	——— Transition	No explicit representation
Flow of information or material.	——→ Directed arc	No explicit representation
Objects such as machines, robots, pallets, etc.	O Token(s) in place(s)	No explicit representation
Logical AND IF A = 1 and B = 1 and C = 1 THEN D = 1	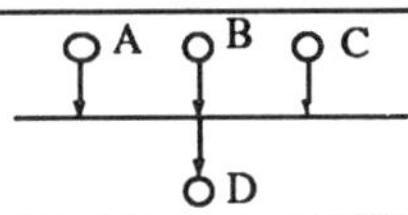	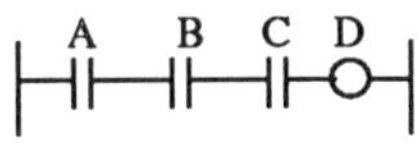
Logical OR IF A = 1 or B = 1 or C = 1 THEN D = 1	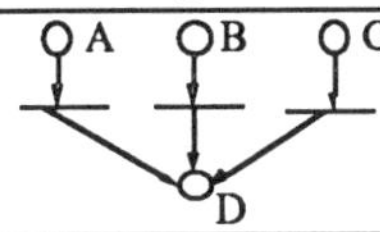	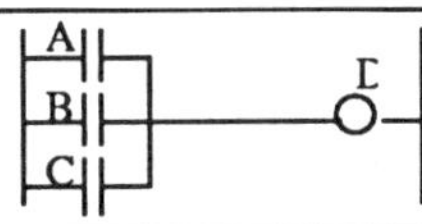
Concurrency IF A = 1 and B = 1 THEN C = 1 and D = 1 and E = 1	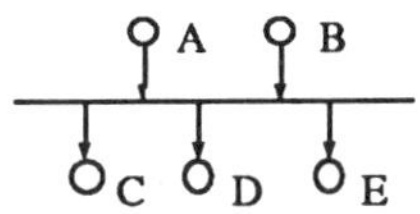	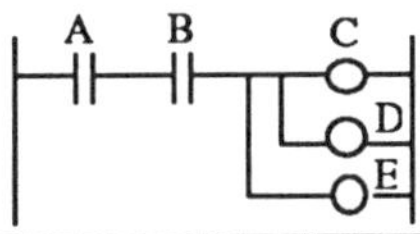
Time delay IF A = 1 THEN delay "τ time units" B = 1	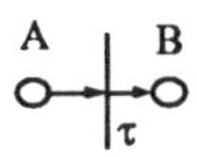	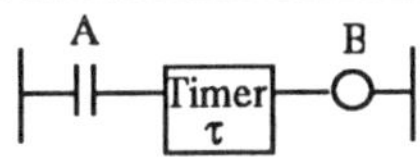
Synchronization IF A = 1 THEN delay "τ1 time units"; D = 1 IF B = 1 and C = 1 THEN delay "τ2 time units"; E = 1 IF D = 1 and E = 1 THEN delay "τ3 time units"; F = 1	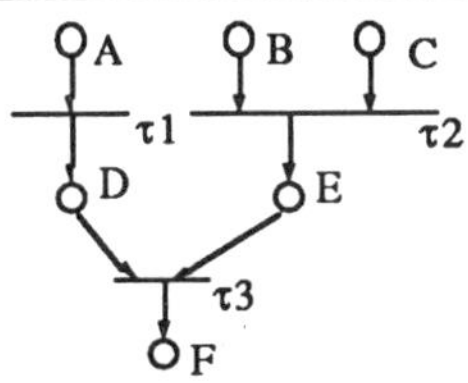	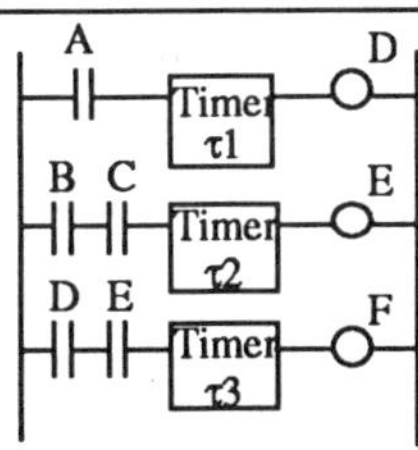

Graphical complexity: It is mainly determined by the number of nodes and links for a given graphical control logic design. Graphical complexity influences the understandability of control logic by people who do not have knowledge of either PN or LLD. Hence, it is an important factor in designing the logic at the initial stages and subsequently debugging the errors during its implementation. The graphical complexity in terms of the net size is a major issue in manufacturing systems [22] and it was reported that the simpler the graphical representation of control logic, the easier to track the controller [1]. Graphical complexity may also influence response time as described later. For example, in the case of LLD, the response time depends on the size of the LLD. Hence, a shorter LLD results in a faster controller [12].

Adaptability for change in specifications: This factor is gaining much importance in the context of agile manufacturing in which control sequences need to be changed often to meet the dynamically changing requirements of the market. The control software should be easily adaptable to changes in specifications in order to improve the software productivity and thus keep minimal development time. One of two designs is said to be more adaptable if it needs fewer changes compared to another in order to fulfill a specification change.

Response Time

Response time is termed as scan time in LLD literature and execution time in PN. Its importance to control real-time systems is clear since it decides how fast the control system responds to an event in the system/process under control.

The important factor that influences graphical complexity and adaptability is the physical appearance (size) of the model, whereas the response time is influenced by not only physical appearance but also the method of implementation. Methods of implementation constitute the software and hardware used to control the system using either PNs or LLDs. Graphical complexity and adaptability cannot be quantified whereas response time can be measured accurately, given a logic design and implementation. However, since there are several ways to implement PNs as shown in Table 1 and LLDs [9, 12] both in terms of hardware and software, it is very difficult for a fair comparison of LLDs and PNs solely on the response time criterion.

The reasons discussed above motivate to find common measures that give an idea about the graphical complexity, adaptability, and response time. One of such measures is the number of nodes and links used in a control logic model. For PNs nodes are places and transitions and links are arcs; whereas in LLDs, nodes are normally opened/closed switches, timers, counters, relays, and push buttons, and links are connections. If more nodes and links are used in a design, it is graphically more complex and thus may need more response time. In the similar manner, a control logic is more adaptable if it needs fewer changes in the number of nodes and links compared to another logic to meet a change in specification. Hence, this study uses the number of nodes and links in LLD and PN as a measure to compare their design complexity and response time. For the sake of convenience nodes and links are called as basic elements. Future work is needed to address other common measures.

4. Comparison through an electro-pneumatic system

4.1. System description

It is found that one effective way to perform the comparison between LLDs and PNs is through an actual industrial automated system. The system considered in this chapter is shown in Fig. 5. It consists of four pneumatic pistons (A, B, C, and D) which are operated by spring-loaded five ports and two-way solenoid valves. Each piston has two normally open limit switches. For example, when the end of piston A contacts limit switch a0 (a1), a0 (a1) is closed indicating that the piston A is at the end of its return stroke (forward stoke). The time that a piston takes for completion of either forward or backward stroke is one second. In manufacturing, typical functions of these pistons can be to load/unload the part from the machine table, to extend/retract a cutting tool spindle, etc. Three push buttons are provided to start the system (switch SW1), to stop the system normally (switch SW2) and to stop the system immediately in emergency (switch SW3, ES).

The system has 11 inputs corresponding to 8 limit switches (two for each piston) and 3 push buttons. The system has 6 outputs corresponding to 4 solenoid valves and two lights that indicate the status of the system. "Interlock logic" also referred to as "double command" is an error normally encountered in designing LLDs due to which both sides of the solenoid are activated simultaneously. Unfortunately, the system under study does not exhibit such problems because no double activated solenoids are used. The comparison between LLD and PN for complex interlock logic is important and should be addressed in the future. In this study, LLDs are implemented in a Modicon PLC and PNs are implemented through IBM PC and digital I/O interface as shown

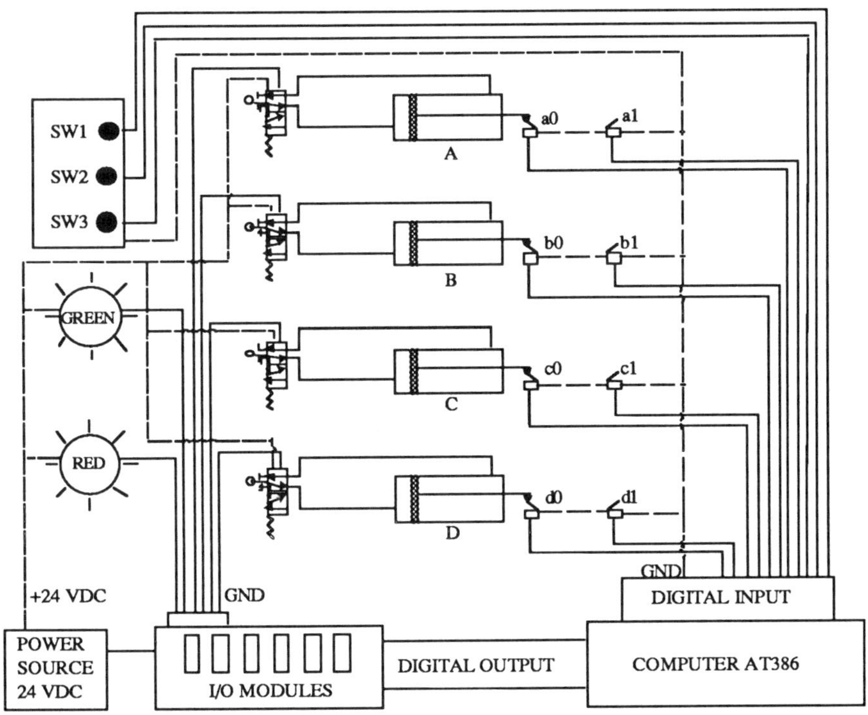

Figure 5. Schematic of an electro-pneumatic system considered [19]

in Fig. 5 using the principle shown in Fig. 2. The procedure given in Fig. 1 is followed to design the control logic by PNs. Tables 6 and 7 show the I/O mappings of the system to PN respectively. For more details on this system and its implementation with PNs refer to [17]. The following focuses on the design complexity when the control specification varies.

Table 6. The input mapping table where X_i is input channel number [19]

Switch	a1	b1	c1	a0	b0	c0	SW1	ES
X_i	0	1	2	4	5	6	8	10

Table 7. The output mapping table [19]

Solenoid or Light	Output Channel number	Y_i NA	ND
A	0	1	-1
B	1	2	-2
C	2	4	-4
D	3	8	-8
Light 1	4	32	-32
Light 2	5	64	-64

Note: 1. Y_i is number sent to the digital output interface.
2. NA (ND) means the number that is to be written on digital output interface to activate (deactivate) the solenoid/light

4.2. Sequence Controller Design

Sequence 1: START, A+, B+, {C+, A-}, {B-, C-}

Consider that the system has to be controlled to execute the above sequence where A+ represents that the piston has to do forward stroke and A- return one. {C+, A-} represents two concurrent actions taking place simultaneously: Piston C to do forward stoke and Piston A to do return one. Fig. 6 (a) shows the LLD and Fig. 6 (b) shows the PN corresponding to this sequence. Note that in the PN, a place has attributes $[n_1, n_2]$ where n_1 is the first attribute representing initial number of tokens and n_2 is the second one mapping input channel number. Similarly, a transition has attributes (n_1', n_2') where n_1' is the firing duration and, n_2' is the number to be written on the digital output interface. When concurrent actions such as {C+, A-} are to be modeled, care should be taken to associate the second attribute to transitions as shown in Table 8.

As discussed earlier, basic elements in a LLD or PN are nodes and links. In LLD, nodes are push buttons, normally opened switches, normally closed

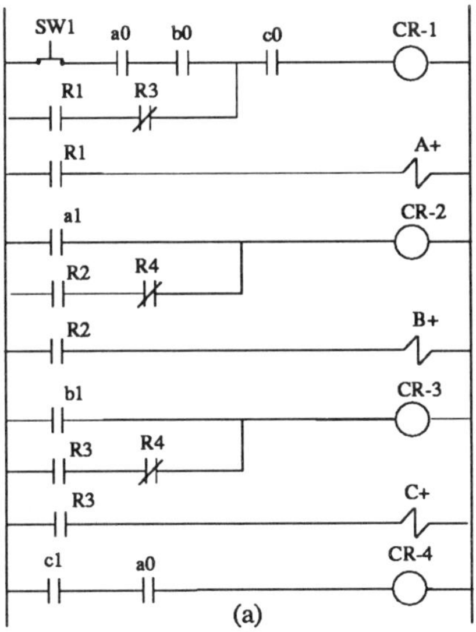

(a)

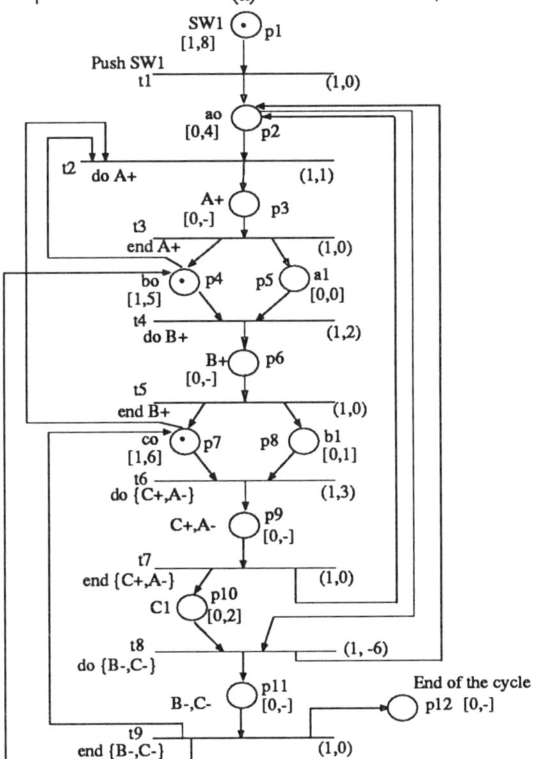

(b)

Figure 6. (a) LLD and (b) PN for Sequence 1: ST, A+,B+,{C+,A-}, {B-,C-} [19]

Table 8. Attributes of transitions modeling actions [19]

Action	Output channel to be activated	Output channel to be deactivated	Transition modeling action	Second attribute of transition
do A+	0	-	t_2	$2^0 = 1$
do B+	1	-	t_4	$2^1 = 2$
do {C+, A-}	2	0	t_6	$2^2 - 2^0 = 3$
do {B-, C-}	-	1, 2	t_8	$-2^1 - 2^2 = -6$

switches, output relay coils, timers, and counters and links are connections which connect these nodes. In PN, nodes are places and transitions and links are arcs connecting them. The LLD shown in Fig. 6 (a) has 58 basic elements (24 nodes and 34 links), whereas the PN shown in Fig. 6 (b) has 50 basic elements (21 nodes and 29 links). At this stage, even though PN uses less number of basic elements, it looks more complex due to the fact that all loops have to be closed to represent repetitive processes. This complicates the graphical appearance of PN compared to LLD.

4.3. Control for other sequences

In order to compare the LLDs and PNs various sequences with increasing complexity are considered. These sequences will involve emergency stop, counters for counting the number of repetitive operations, and timers for providing delays between certain operations. In order to highlight the changes in a model (PN or LLD) from one sequence to next sequence, given a model for a sequence, the additional basic elements needed to model the next sequence are shown by bold lines.

Sequence 2: START, 5 [A+, B+, {C+, A-},{B+, C-}] (with emergency stop and counter)

Now, consider that the specification is changed such that the new control sequence is indicated as above. In this sequence, there is a need to provide emergency stop and a counter. Both LLD and PN are implemented such that when emergency stop switch, ES is pressed, the whole system including the active elements are immediately stopped. In other words, when switch ES is pressed all the solenoids are immediately deactivated. Fig. 7 (a) shows the LLD and Fig. 7 (b) the PN corresponding to this sequence. In order to incorporate emergency stop, PN uses a place p_{13} with an inhibitory arc as an input place for t_2-t_9 whereas LLD uses an output coil CR-5. When ES is pressed, CR-5 would become active and deactivates the coils CR-1, CR-2, and CR-3 which drive pistons A, B, and C respectively. The LLD shown in Fig. 7 (a) has 87 basic elements (37 nodes and 50 links), whereas the PN shown in Fig. 7 (b) has 64

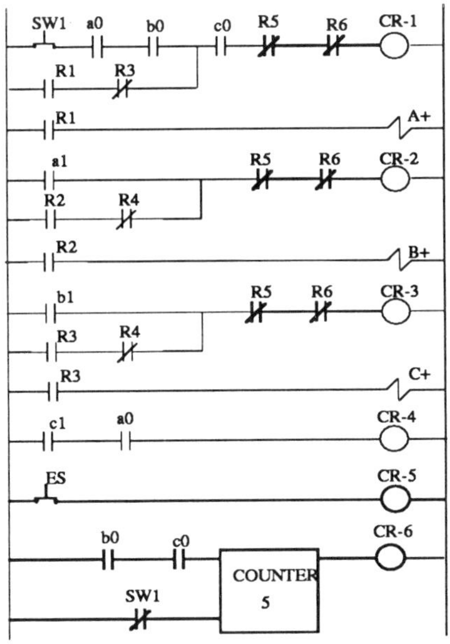

(a)

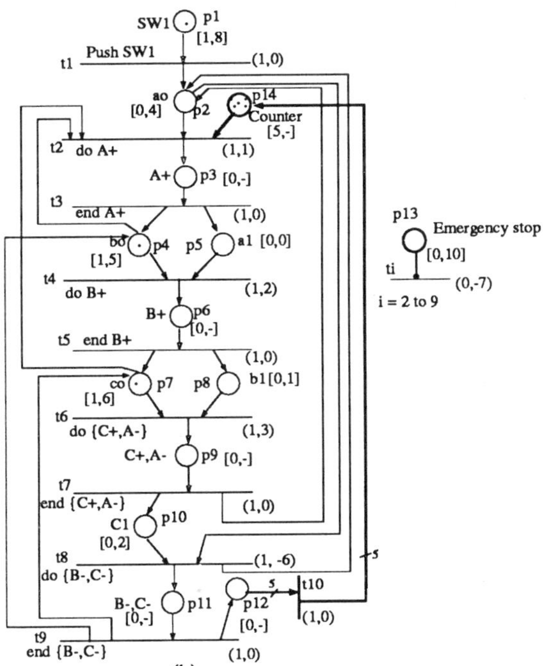

(b)

Figure 7. (a) LLD and (b) PN for Sequence 2 [19]

basic elements (24 nodes and 40 links). Notice that there is no significant change in the physical appearance of LLD or PN compared to Sequence 1.

Observe that LLD needs more additional basic elements compared to the PN. This is because the PN needs only one place p_{14} with an arc as an input to t_2 to implement the counter. The counter resetting is modeled by t_{10}. In contrast to this, LLD additionally needs two normally opened switches b0, c0 as inputs for counter, four normally closed switches (SW1 as a reset signal for counter, and switch R6 as an input for CR-1, CR-2, and CR-3), a counter, and 16 more links to implement this sequence.

Sequence 3: START, 5 [A+, B+, {C+, A-}, 5 sec, {B+, C-}] (with emergency stop, counter, and timer)

The sequence is changed such that there is a need to incorporate a timer in the control logic to provide 5 seconds delay in between {C+, A-} and {B+, C-}. Fig. 8 (a) shows the LLD and Fig. 8 (b) the PN accordingly. The LLD shown in Fig. 8 (a) has 94 basic elements (40 nodes and 54 links), whereas the PN shown in Fig. 8 (b) is same as the one shown in Fig. 5 (b) with 64 basic elements. It is observed that LLD needs more additional basic elements compared to the PN. This is because the same PN used in the earlier sequence is used without changing the physical appearance. In the PN shown in Fig. 7 (b), only the first attribute of t_8 is changed to obtain the PN shown in Fig. 8 (b) to incorporate time delay of 5 seconds in the sequence. On the other hand, LLD needs additionally two normally closed switches, a timer, and four links to implement this sequence.

Sequence 4: START, 3[A+, B+, {C+, A-}, 5 sec, {B+, C-}], 10 sec, 2[A+, B+, {C+, A-}, 5 sec, {B+, C-}] (with emergency stop, counters, and timers)

This new sequence represents a complex one in which Sequence 3 is divided into two segments (one with three cycles and another with two cycles) with 10 seconds time delay between them. The LLD shown in Fig. 9 has 130 basic elements (54 nodes and 76 links), whereas the PN in Fig. 10 has 69 basic elements (26 nodes and 43 links). In this case also note that LLD needs more additional basic elements compared to PN. This is because the PN needs only one additional transition t_{11} with an input arc from a new place p_{15} and an output arc to p_{14} to reset the counter.

4.4. Discussions

As mentioned in Section 2 the number of basic elements is a common measure that gives an idea about graphical complexity, adaptability, and response time. It is observed that as the specification changes, PN requires a fewer changes compared to LLD.

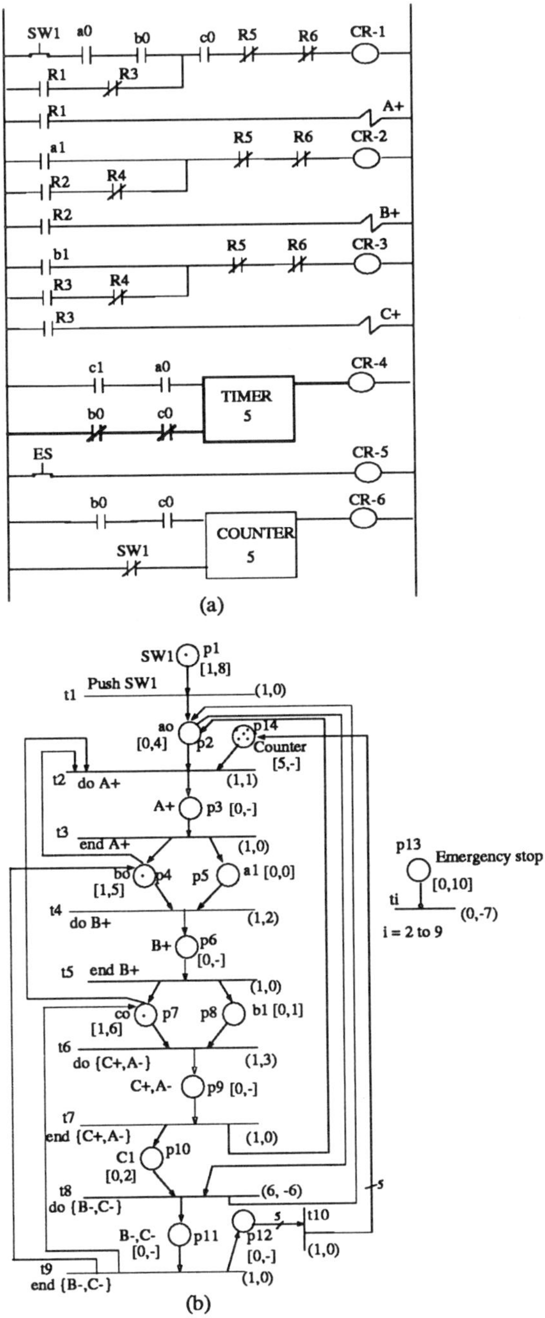

Figure 8. (a) LLD and (b) PN for Sequence 3 [19]

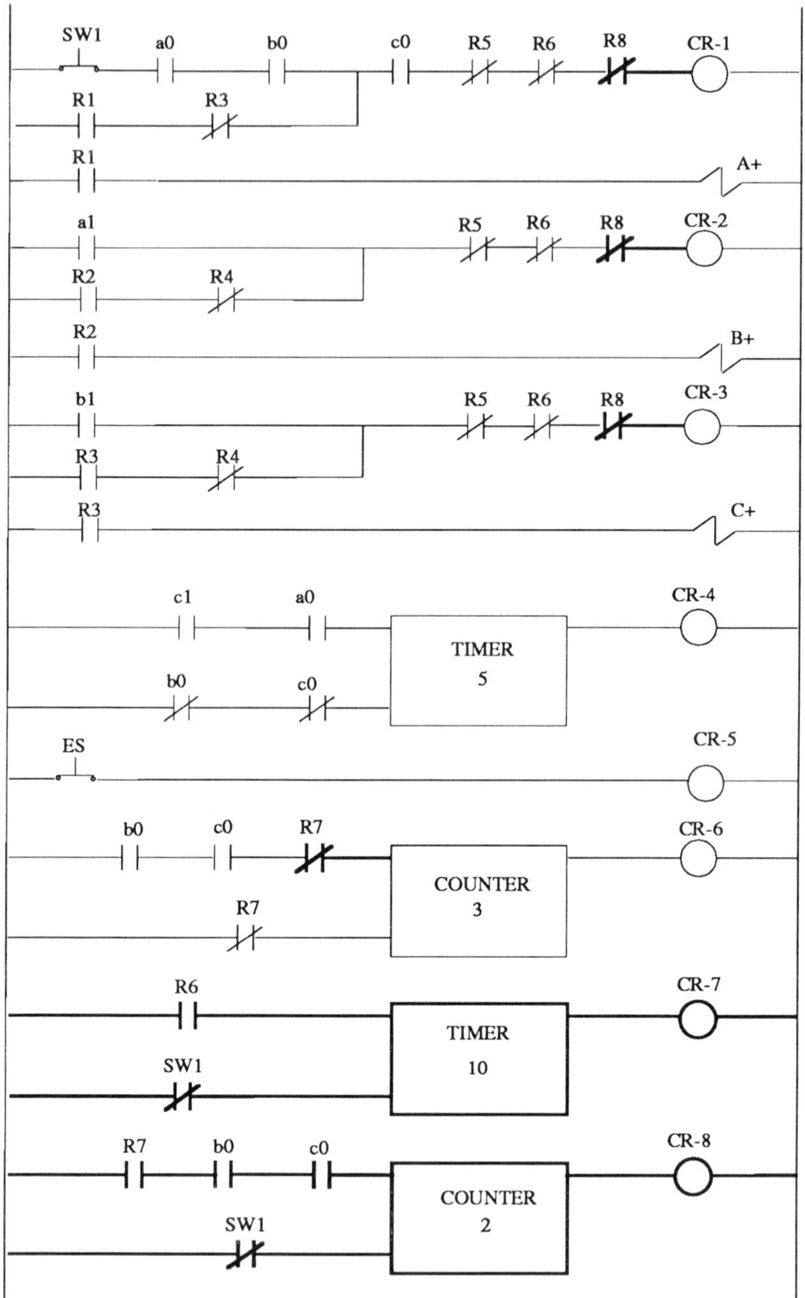

Figure 9. LLD for Sequence 4 [19]

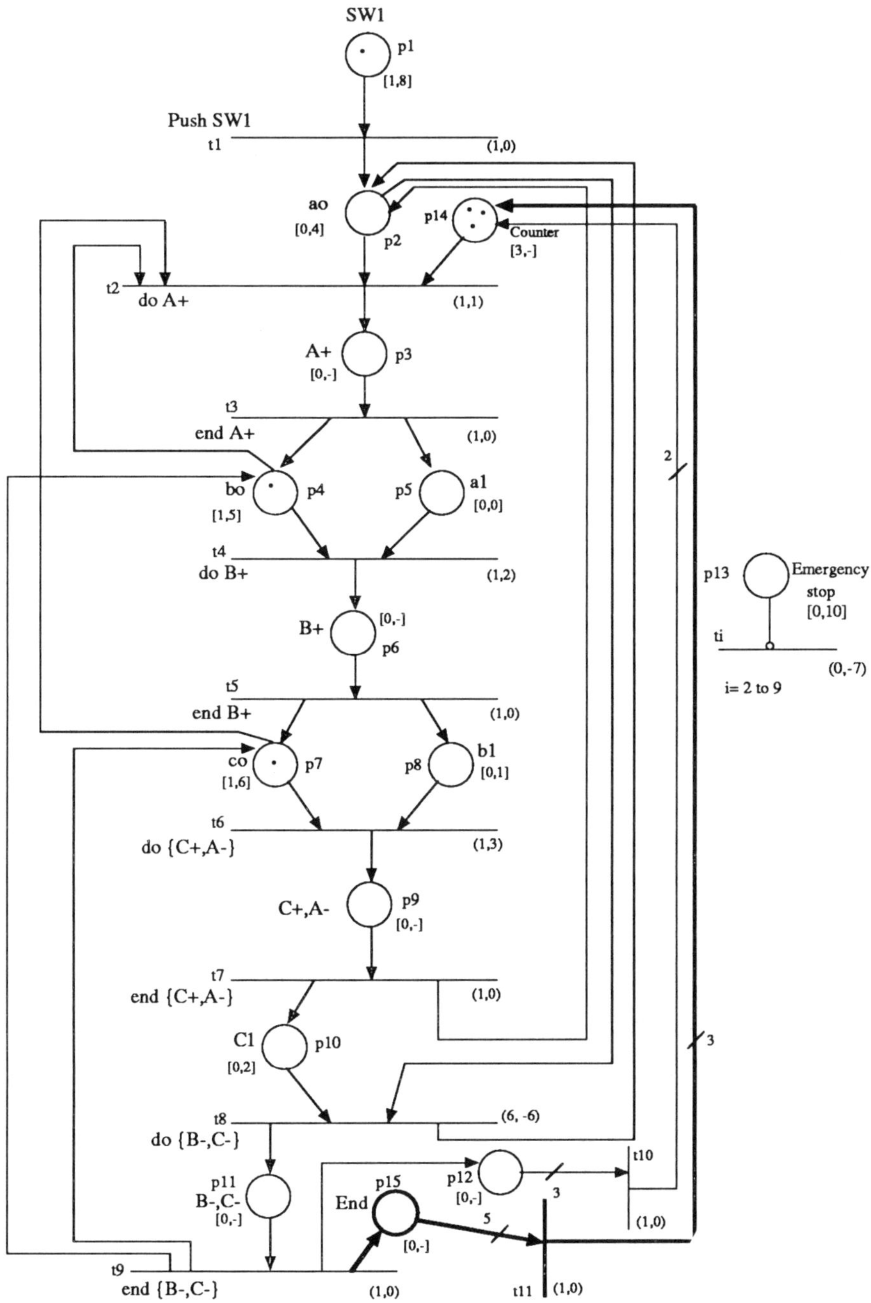

Figure 10. PN for Sequence 4 [19]

Table 9 summarizes how the number of basic elements increases in LLD and PN as the complexity of sequence control specifications increases.

Table 9. Comparison of the basic elements in LLD and PNs

Sequence #	Basic elements in LLD			Basic elements in PN		
	Nodes	Links	**Total**	Nodes	Links	**Total**
1	24	34	**58**	21	29	**50**
2	37	50	**87**	24	40	**64**
3	40	54	**94**	24	40	**64**
4	54	76	**130**	26	43	**69**

It can be inferred that the PN shown in Fig. 6 (b) used to execute the first sequence is slightly modified to get the PN shown in Fig. 10 corresponding to the last sequence. However, the LLD shown in Fig. 6 (a) is significantly modified to get the LLD shown in Fig. 9. The modifications in terms of basic elements can be quantified using Table 9. Also, observe that the physical appearance of PN is preserved (with slight modifications) starting from the first sequence to the last one. This is not true in the case of LLDs as in the previous figures. Furthermore, this case study reveals that PNs and LLDs do not differ much when the control sequence is relatively simpler as seen in the first sequence. In fact, PN model may appear more complex than LLD at the first sight as shown for the first sequence. However, when this sequence is gradually modified to result in a complex one, PNs are more easily modifiable and hence maintainable than LLDs. Ease in modifiability and maintainability yields several advantages such as improvement in readability, understandability, and reliability as concluded in [11]. In LLDs, nodes appear multiple times which may lead to difficulty in understanding the logic and cause errors in developing the logic. LLDs needs more basic elements to model timers and counters compared to PNs. In addition to these findings, the following points are experienced during the design and implementation of sequence controllers using LLDs and PNs:

1. Using PNs the control logic can be qualitatively analyzed to check properties such as absence of deadlocks and presence of reinitilizability in the system. Using LLDs qualitative analysis is not possible until it is simulated or implemented.
2. During implementation of control Sequences 3 and 4 it is found that debugging of the control logic with LLDs is difficult compared to PN. This is because PNs help to dynamically track the system with the help of the states of places and transitions [17].
3. Using PNs, the initial state of the system can be directly represented by its initial marking.

5. Analytical formulas to evaluate the complexity of PNs and LLDs

In general, the fewer the basic elements in a controller, the better the model used to implement the controller [1,12-13,19]. The number of basic elements decides the length of the control model. A shorter model uses less number of basic elements and is usually easier to understand, check, diagnose, and maintain. It also takes less time to enter the controller/computer [13,19]. Moreover, when low cost controllers with short memory are used it is possible that they may run out of memory if the control model is too large [13]. Counting of basic elements in a PN or LLD becomes cumbersome when their control specification becomes complex. Also, before physically modeling the given control logic with either LLD or PN, there is a need for a method that can help control engineers select between PN and LLD by estimating the basic elements used in these two models. Motivated by these reasons, this section presents some analytical formulas to estimate the basic elements used in a PN or LLD.

Most of the control specifications can be modeled by using the logic constructs such as logical AND (NAND), logical OR (NOR), sequential model, and timed sequential model. It is obvious that AND/OR basically requires the same number of basic elements as NAND/NOR. Hence, the analytical formulas presented for AND/OR can also be applied to NAND/NOR cases. The formulation of these analytical formulas consists of their derivation and verification. This is done by physically modeling each logic construct using PN and LLD as described below.

For the sake of convenience, the number of basic elements used in PN and LLD are represented by α and β, respectively. α is the summation of α_n and α_l, where α_n and α_l represent the number of nodes and links used in PN respectively. Similarly, β is the summation of β_n and β_l. We also define a function Δ which is defined as $\Delta = \alpha - \beta$ which helps to decide whether PN is better or LLD or better in terms of design complexity. For example, if Δ is negative, PN is preferred as it uses a smaller number of basic elements. On the other hand LLD is preferred when Δ is positive. In all the formulas presented in this chapter, m stands for number of pre-conditions, and n for number of post-conditions.

5.1. Logical AND

Figure 11 shows the PNs and LLDs for deriving the formulas to obtain α and β for logical AND. They are given as follows:

For PN: $\alpha = 2(m + n) + 1$, where

$\alpha_n = m + n + 1$, and

$\alpha_l = m + n$

For LLD: $\beta = 2m + 3n$ where

$$\beta_n = m + n, \text{ and}$$
$$\beta_l = m + 2n$$

$\Delta = 1 - n$. This indicates when $n = 1$, both PN and LLD yield the same number of basic elements and if $n > 1$, PN is preferred.

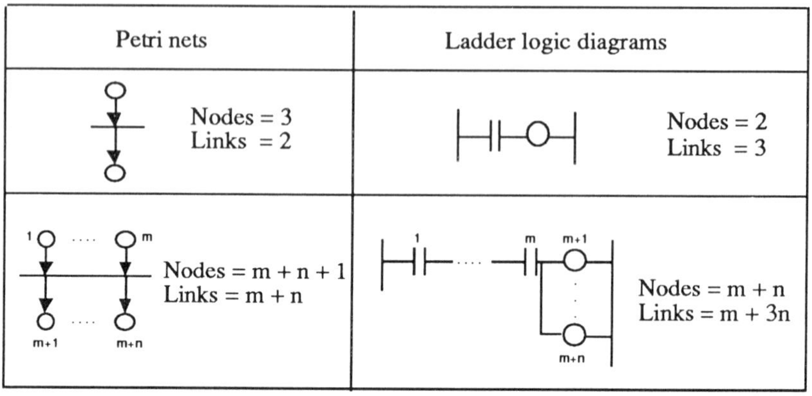

Figure 11. PNs and LLDs modeling logical AND [21]

5.2. Logical OR

Figure 12 shows the PNs and LLDs for deriving the formulas to obtain α and β for logical OR.

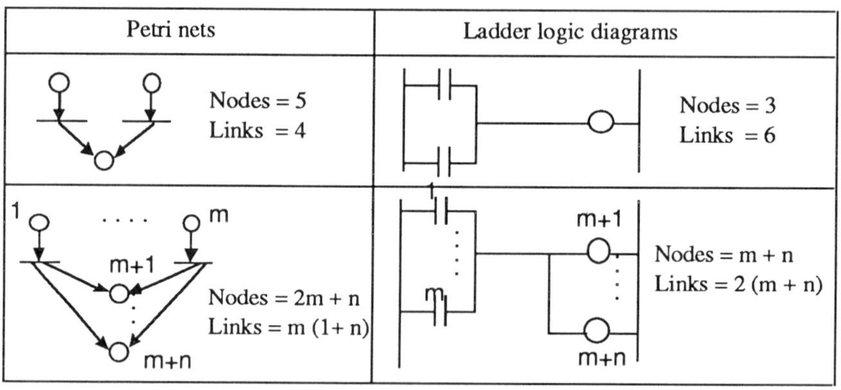

Figure 12. PNs and LLDs modeling logical OR [21]

They are given as follows:

for PN: $\alpha = 3m + n + mn$, where

$$\alpha_n = 2m + n, \text{ and}$$
$$\alpha_l = m\,(1 + n)$$

For LLD: $\beta = 3\ (m + n)$, where

$$\beta_n = m + n, \text{ and}$$
$$\beta_l = 2\ (m + n)$$

$\Delta = -n\ (2-m)$. LLD is better than PN for $m > 2$.

5.3. Sequential modeling

A condition is called a sequential condition when it acts as a pre-condition for its output operation (say operation $i+1$ and post conditions for its input operation (say operation i). The assumption here is that all operations except the first and last operations in the sequence have only one pre-condition and post-condition. For sequences that violate this assumption, number of basic elements can be found by decomposing the sequence with logical AND where some portions of the sequences have $n = 1$ and others have $n > 1$. Figure 13 shows the PNs and LLDs to derive the formulas to obtain α and β for sequential models.

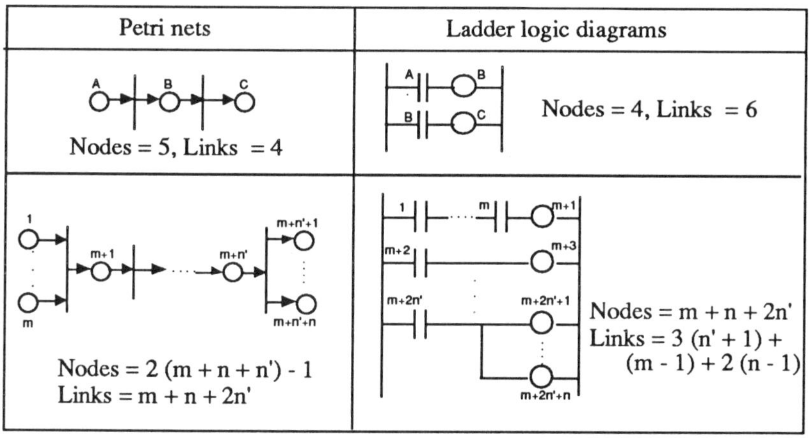

Figure 13. PNs and LLDs for sequential modeling [21]

For PN: $\alpha = 3(m + n) + 4n' - 1$, where

$$\alpha_n = 2\ (m + n + n') - 1, \text{ and}$$
$$\alpha_l = m + n + 2n'.$$

Here, n' represents the number of sequential conditions.

For LLD: $\beta = 2m + 3n + 5n'$, where

$$\beta_n = m + n + 2n', \text{ and}$$
$$\beta_l = 3\ (n' + 1) + (m-1) + 2\ (n-1)$$

$\Delta = f(m,n,n') = m - n' - 1.$

5.4. Timed logical AND

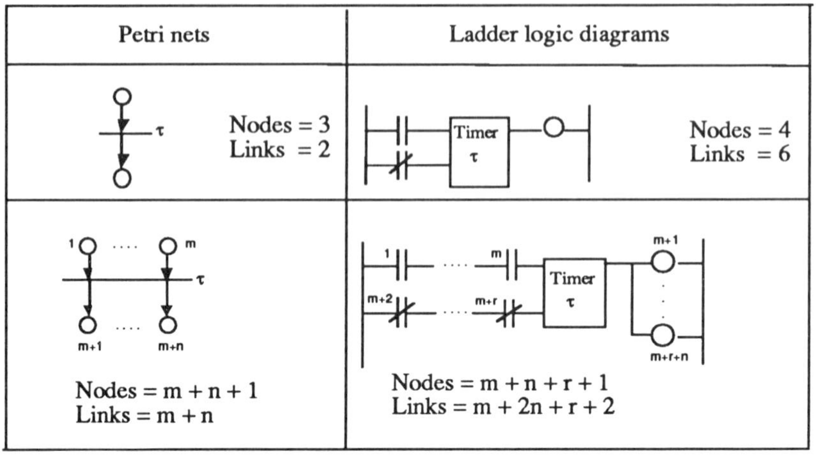

Figure 14. PNs and LLDs for timed logical AND [21]

In this model delays are associated with operations of the logical AND considered before. Figure 14 shows the PNs and LLDs to derive the analytical formulas to obtain α and β for timed logical AND. They are given as follows:

For PN: $\alpha = 2\ (m + n) + 1$, where

$\alpha_n = m + n + 1$, and

$\alpha_l = m + n$

For LLD: $\beta = 2m + 3m + 2r + 3$, where

$\beta_n = m + n + r + 1$, and

$\beta_l = m + 2n + r + 2$

$\Delta = -n - 2r - 2$. This indicates that PN is always preferred to model timed logical AND.

5.5. Timed Logical OR

In this model it is assumed that for each timer there is only one reset signal. Figure 15 shows the PNs and LLDs to derive the analytical formulas to obtain α and β for timed logical OR. α and β are are given as follows:

For PN: $\alpha = 3d + n + dn$, where

$\alpha_n = 2d + n$, and

$\alpha_l = d\ (1 + n)$.

In this chapter d represents the number of delays in the sequence.

For LLD: $\beta = 8d + 3n$, where

$\beta_n = 3d + n$, and

$$\beta_l = 5d + 2n$$

$$\Delta = f(d,n) = -5d - 2n + dn.$$

This indicates that for d and n values which satisfy 5d + 2n > dn, PN is preferred. For d and n values that do not satisfy the above condition, LLD is preferred. For example, for the pairs, $d = 3, n = 4$; and $d = 4, n = 3$, PN is preferred. On the other hand for $d = 7, n = 8$, LLD is preferred. Similar analysis can be performed for any given d and n values using Δ function.

Petri nets	Ladder logic diagrams
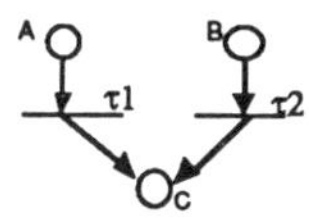 Nodes = 5 Links = 4	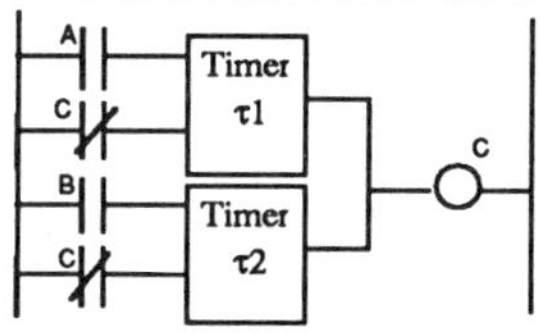 Nodes = 7 Links = 12
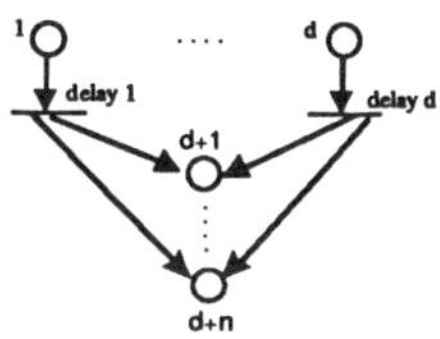 Nodes = 2d + n Links = d(1+ n)	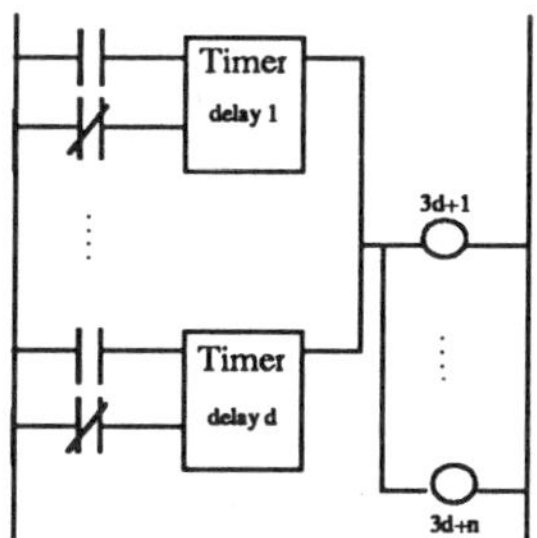 Nodes = 3d + n Links = 5d+ 2n

Figure 15. PNs and LLDs for timed logical OR [21]

5.6. Timed Sequential Modeling

In this model it is assumed that for each timer there is only one reset signal. Figure 16 shows the PNs and LLDs to derive the analytical formulas to obtain α and β for this case. They are given as follows:

For PN: $\alpha = 4d + 2(m + n) - 3$, where

$$\alpha_n = 2d + m + n - 1, \text{ and}$$

$$\alpha_l = 2d + m + n - 2.$$

For LLD: $\beta = 10d + 2m + 3n - 5$, where

$$\beta_n = 4d + m + n - 2 \text{ and}$$

$$\beta_l = 6d + m + 2n - 3$$

$$\Delta = f(d,m,n) = -6d - n + 2$$

Δ indicates that PN is always better than LLD.

Petri nets	Ladder logic diagrams
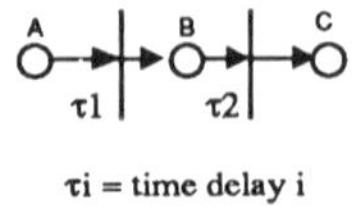 Nodes = 5, Links = 4	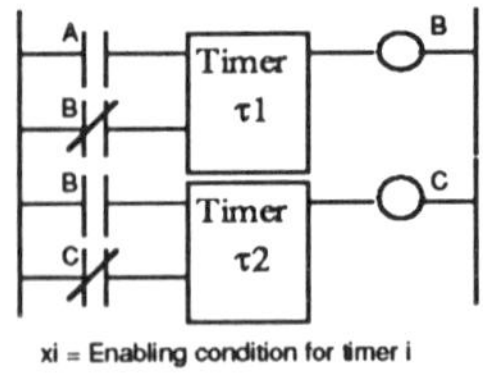 Nodes = 8 Links = 12
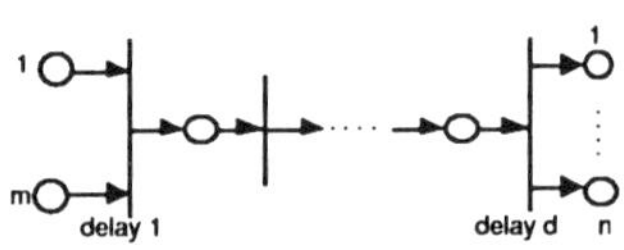 Nodes = 2d + m + n - 1 Links = 2d+ m + n - 2	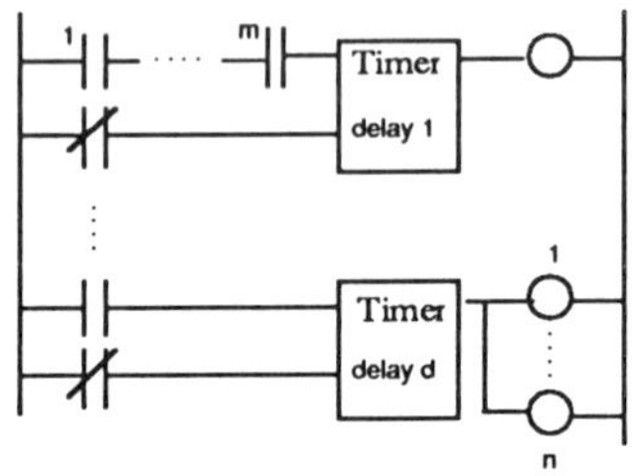 Nodes = 4d + m + n - 2 Links = 6d+ m + 2n -3

Figure 16. PNs and LLDs for timed sequential modeling [21]

5.7. Other formulas for estimating basic elements in PN or LLD

The models presented till now are common for both PNs and LLDs. However, there are certain models that are specific to PN or LLD. For example, the implementation of timer and counter is similar in LLDs. In PNs, timer is implemented by associating delays to certain transitions and hence the implementation of timer and counter are not the same. Hence, there is a need for separate formulas to estimate the number of basic elements to model a counter in case of a PN.

Also, to control certain systems, relays have to be used in LLDs. This is explained here. A discrete event system can be controlled by controlling two types of elements: those that require sustained actuating signals, and those that need only a momentary or pulsed actuating signal [13]. The first type of elements can be exemplified by an on-off solenoid valve with return spring and shown in Fig. 17 (a). As long as the solenoid is actuated, the valve remains open (or closed, as the case may be). As soon as the solenoid is released, the return spring returns the valve to its original position. Hence, the solenoid needs a sustained actuating signal to keep the valve open. This requires the use of relays in LLDs. The second type of elements can be exemplified by a solenoid valve with two opposing solenoids but without a return spring and is shown in Fig. 17 (b). Here, a momentary solenoid signal is sufficient to shift the valve into its other position, and the valve will remain there until the opposing solenoid is actuated. It is strictly not allowed to actuate both solenoids at the the same time. Actuating both solenoids simultaneously results in an undefined state and termed as 'interlock' as the two solenoids fight each other causing heat up and burn out.

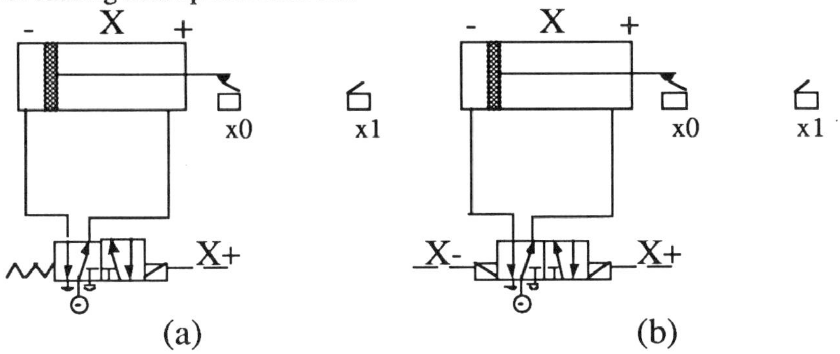

Figure 17. Typical cylinder-actuating circuits [13]
(a) requiring sustained solenoid signals
(b) not requiring sustained solenoid signals

Using a single PN model, both types of systems in Fig. 17 can be controlled. This can be accomplished by changing the corresponding attributes for transitions as reported in PN based control using Real-time PNs. However, when using ladder logic diagrams, two separate design procedures have to be followed to design two separate diagrams as shown in [13]. The formulas presented earlier are relatively simple compared to the ones described below. The formulas that use earlier formulas and needs certain description for estimating basic elements in PN or LLD are presented below.

Emergency stop modeling

Emergency stop can be implemented in two different ways. In the first, when the push button corresponding to emergency stop is pushed, all the operations that are in progress and those that are ready to start are immediately stopped. In other words, the pistons executing the operations are stopped at the position where they are. In the second implementation, when the push button is pushed, the system is brought to initial condition. That is, all the pistons that are executing operations are retracted. The second method of implementing emergency stop is very common and hence it is implemented and studied in this work. Denote the number of basic elements needed to model emergency stop (ES) modeling by PN and LLD as α_{ES} and β_{ES} respectively. In a PN, emergency stop is modeled by a place and inhibitory arcs from this place to transitions that model execution of actions. Fig. 18 shows the implementation of ES using PN and LLD. Given any sequence, the additional basic elements needed are shown with bold lines. α_{ES} is given as $\alpha_{ES} = 1 + k\,\Gamma$ where $k = 1$ if each action is modeled by one transition, and $k = 2$ if each action is modeled by two transitions (one to represent start and second to represent end) and Γ stands for number of actions in a given specification. β_{ES} is given as $\beta_{ES} = 5 + 2k'\Gamma'$ where $k' = 1$ if each action is executed by sustained actuating signal, and $k' = 2$ if each action is executed by a pulsed actuating signal and Γ' stands for number of pistons in a given specification.

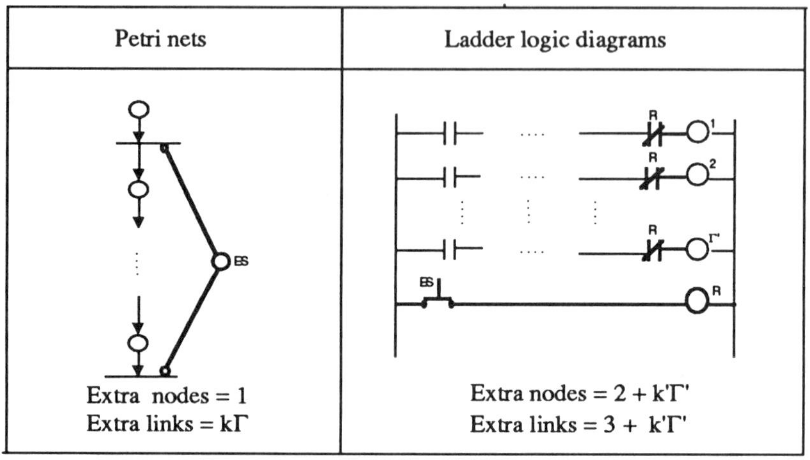

Figure 18. PNs and LLDs modeling emergency stop

Basic elements needed to model a counter

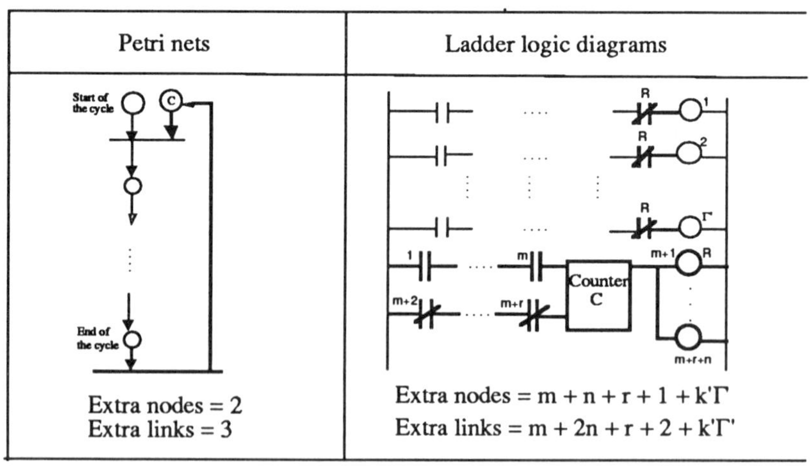

Figure 19. PNs and LLDs modeling counter

Irrespective of the given sequence, the implementation of a counter is unique as shown in Fig. 19. In Fig. 19, given any sequence, the additional basic elements to implement a counter are shown with bold lines. As it can be observed from this PN model three new arcs and and two new nodes are needed to incorporate a counter in a sequence. Hence, the number of basic elements needed to model a counter using PNs is five. In LLDs, even though the implementation of timer and counter are similar, the output of counter may have to be used as a precondition to execute certain actions. This is usually done by adding the output of counter as a precondition for certain actions with normally closed contact. This results to include an extra node and a link for each action. Hence, The number of basic elements to implement counter in LLD is given as $\beta = (2m + 3n + 2r + 2k'\Gamma + 3)$ where r = *number of reset signals*, $k' = 1$, if each action is executed by sustained actuating signal and $k' = 2$, if each action is executed by a pulsed actuating signal and Γ stands for number of pistons in a given specification.

Basic elements needed to model a relay

When LLDs are used to control a system with sustained actuating signals, relays are needed to implement the sustained actuating signals. Fig. 20 shows the implementation of a relay. It can be observed that the number of basic elements needed to model a relay as $2m + 2r + 6$ where r = *number of reset signals.*

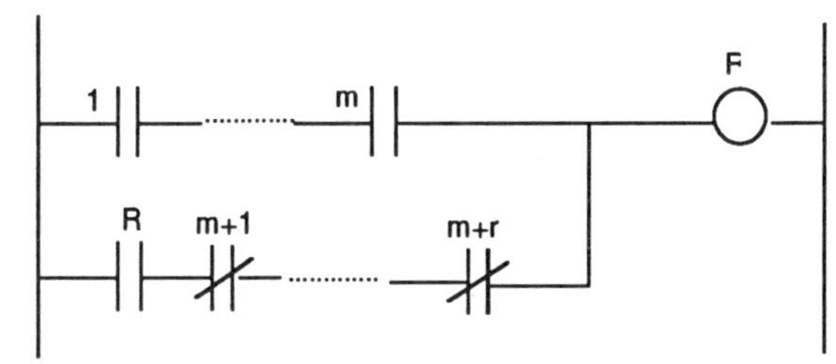

Figure 20. LLD modeling a relay

6. Methodology to use the analytical formulas

Most of the control logics can be represented using the basic building blocks of logic constructs described in the earlier section. Hence, in order to find the basic elements to model a given control logic using the models developed, first it has to be decomposed into the logic constructs or segments. Then for each segment, the corresponding analytical formula is applied to find the number of basic elements.

The basic elements corresponding to all such segments are then added to obtain the total number of basic elements *(β)* needed to model the given control logic using a LLD.

$$\beta = \sum_{i=1}^{k} \beta_i$$

where β = Total number of basic elements to model the control logic using PN, β_i = Number of basic elements in segment i, and k = Total number of segments.

The above procedure is slightly different in the case of PN due to its physical model. This is because when modeling a control logic, intersegment conditions and connections exist in a PN model. They are explained below.

For some cases in the given control logic, the output condition for segment i may be one of the input conditions for segment j. These conditions are called as intersegment conditions. In LLD, such conditions are separately modeled in both segments i and j. Even though such conditions appear in more than one segment, the place modeling such condition physically appears only once in the PN. Therefore, in case of PN to account for the repetitive count of these intersegment conditions, the total number of such conditions has to be subtracted from the summation of basic elements in all segments.

For some control specifications it may possible that an output condition of segment i causes an action which produces an input condition(s) for segment j. That means intersegment connections exist between segments i and j. Basically, intersegment connections model the power/control flow from one segment to another. In LLD these connections need not be explicitly modeled since the vertical line at the left hand side of LLD always models the existence of electric power. Depending upon the logic of each rung, the current flows from this power line and energizes the output of rung connected to the right hand side line in LLD. However, in case of PN the power flow (called as control flow in PN terminology) across segments is modeled by the movement of tokens in PN. Since, transition firings cause the flow of tokens through arcs, the intersegment connections have to be explicitly modeled. The total number of basic elements to model intersegment connections is estimated as shown below.

Let θ_{ij} = The number of basic elements used to model intersegment connections between segments i and j

$$\theta_{ij} = \theta_{ijn} + \theta_{ijl}$$

θ_{ijn} = Nodes (typically transitions) used to model intersegment connections between segments i and j

θ_{ijl} = Links used to model intersegment connections between segments i and j, then

$\theta_{ijn} = q$, where q represents the total number of output conditions in segment i producing inputs for segment j.

Sometimes, it may happen that an action corresponding to the output condition in segment *i* produces several inputs for segment *j*. Considering this, q_{ijl} is given as follows:

$$\theta_{ijl} = q + \sum_{r=1}^{q} w_r$$

where w_r represents the number of inputs for segment *j* produced by output condition r in the segment *i*. The first term in θ_{ijl} corresponds to the output arcs from the places modeling output conditions in segment *i* and the second term corresponds to the input arcs to the places modeling input conditions in segment *j*. Hence, θ_{ij} is given as:

$$\theta_{ij} = 2q + \sum_{r=1}^{q} w_r$$

The total number of basic elements modeling intersegment connections in the whole control logic is then given as follows:

$$\theta = \sum_{i \neq j}^{k} \theta_{ij}$$

Considering intersegment connections, in case of PN, the total number of basic elements used to model the total number of intersegment conditions have to be added for the summation of basic elements in all segments. Hence, considering both intersegment conditions and connections, the total number of basic elements needed to model the control logic using PN is termed as α and given as follows:

$$\alpha = \sum_{i=1}^{k} \alpha_i - \phi + \theta$$

where, α_i = Number of basic elements in segment *i*,
k = Total number of segments,
ϕ = Total number of intersegment conditions,
θ = Total number of basic elements needed to model intersegment connections.

Once, α and β are known, we can select PN or LLD based on $\Delta = \alpha - \beta$. If Δ is positive LLD is preferred, otherwise, PN is preferred from the design complexity view point.

Application example: an electro-pneumatic system with sustained signals

The electro-pneumatic system considered in Section 4 is considered here. The number of basic elements in PNs and LLDs is estimated instead of manual counting in Section 4.

Estimation of the number of basic elements in PNs

Table 10 presents the calculations to estimate the number of basic elements in the PN modeling Sequence 1.

Table 10. Required basic elements to model the Sequence 1 using PN.

(a) Basic elements in each segment without considering and ϕ

Segment	Description of logic construct	Model to be applied	α_i
1	ST- --> a0, b0, c0	Logical AND m = 1, n = 3	2m+2n+1 = 9
2	a0 --> A+	Logical AND m = 1, n = 1	2m+2n+1 = 5
3	a1.b0 --> B+	Logical AND m = 2, n = 1	2m+2n+1 = 7
4	b1.c0 --> {C+, A-}	Logical AND m = 2, n = 1	2m+2n+1 = 7
5	c1.a0 --> {B-, C-}	Logical AND m = 2, n = 1	2m+2n+1 = 7
6	{B-, C}- --> End	Logical AND m = 1, n = 1	2m+2n+1 = 5

(b) Basic elements representing intersegment conditions

Segments j and k	Intersegment condition	ϕ_{jk}
1 and 2	a0	1
1 and 3	b0	1
1 and 4	c0	1
1 and 5	a0	1
5 and 6	{B-, C-}	1

(c) Basic elements representing intersegment connections

Segments j and k	Intersegment connection	Values of variables	θ_{jk}
2 and 3	A+ --> a1	$q = 1, w_1 = 1$	$2q + w_1 = 3$
3 and 4	B+ --> b1	$q = 1, w_1 = 1$	$2q + w_1 = 3$
4 and 5	C+, A- --> c1, a0	$q = 1, w_1 = 2$	$2q + w_1 + w_2 = 4$
5 and 3	B- --> b0	$q = 1, w_1 = 1$	$2q + w_1 = 3$
5 and 4	C- --> c0	$q = 1, w_1 = 1$	$2q + w_1 = 3$

In calculation, observe that x0 is a precondition to execute action X+ where x and X stands for corresponding limit switch and piston identity. For simplicity, denote the number of basic elements in PN corresponding to

Sequence i as α_{si}. Now, following the formulas presented earlier, α_{si} can be estimated as follows:
α_{s1} = (9+5+7+7+7+5) - (1+1+1+1+1) + (3+3+4+3+3) = 40 - 5 + 16 = 51.

Observe that a0 is presented as a precondition in both Segments 2 and 5. However, in PN each limit switch appears physically once (with several input and output arcs). Hence, to compensate this the previous value of α_{s1} has to be subtracted by one. Then, the final value of is 50 which exactly matches with the value obtained in Section 4.

Sequence 2 is obtained by adding a counter and emergency stop to Sequence 1. Hence, α_{s2} can be obtained by adding the basic elements needed to model a counter (α_C) and emergency stop (α_{ES}) to α_{s1}. Then, $\alpha_{s2} = \alpha_{s1} + \alpha_C + \alpha_{ES}$. By following the models presented earlier, $\alpha_C = 5$. The number of basic elements to model ES can be calculated using the earlier developed formula. Since there are four actions (recall that {C+,A-} and {B-,C-} are treated as two concurrent actions) present in the sequence and each action is modeled by two transitions (one to start and second to end), $\alpha_{ES} = 1 + k\Gamma = 1 + 2.4 = 9$. Therefore, α_{s2} = 58+5+9 = 64 as obtained earlier in Section 4.

Sequence 3 is obtained by adding a delay before {B-,C-} of Sequence 2. However, in PNs this delay is implemented by changing the second attribute of transition modeling 'do {C+,A-}'. Hence, there is no need for extra basic elements. In other words, $\alpha_{s3} = \alpha_{s2} = 64$ as given in Section 4. Sequence 4 is obtained by adding a counter and timer to Sequence 3. Again, we know that $\alpha_C = 5$, and there is no need for extra basic elements to model timer. Hence, $\alpha_{s4} = \alpha_{s3} + \alpha_C = 64 + 5 = 69$ as obtained in Section 4.

Estimation of the number of basic elements in LLDs

When sustained actuating signals are used, the following simple procedure is developed to estimate β.

Step 1: Assign a relay for each segment in the given sequence except the segment containing 'Start' command,

Step 2: Identify set and reset signals for each sustained signal implemented by relay. Note that, setting (resetting) of a relay executes forward (return) stroke of a piston,

Step 3: Treat set signals for relay as preconditions, and apply the formulas to calculate the number of basic elements.

For the case study considered in Section 4, the results obtained after each step by following the above procedure are as follows:

After Step 1: Relay 1 (R1), R2, R3, and R4 are assigned to A+,B+,{C+,A-}, and {B-,C-} respectively.

After Step 2: Set signals for the sustained signal by R1 are ST, a0, b0, c0, R1. Since A- is in Segment 3, R3 is a reset signal for R1. Since R2 has to wait till R1 completes all of its actions, set signals corresponding to R2 are a1 and R2 and reset signal is R4 because B- is in Segment 4. Similarly set signals for R3 are b1 and R3 and reset signal is R4 because C- is in Segment 4. Set signals for R4 are c1 and a0.

After Step 3: Table 11 shows the calculations to estimate the number of basic elements in LLD.

Table 11. Required basic elements to model the Sequence 1 using LLD.

Seg-ment	Description of logic construct	Model to be applied	β
1(a)	ST.a0.b0.c0.R3 --> R1	Relay m = 4, r = 1	2m+2r+6 = 16
1(b)	R1 --> A+	Logical AND m = 1, n = 1	2m+3n = 5
2(a)	a1.R4 --> R2	Relay m = 1, r = 1	2m+2r+6 = 10
2(b)	R2 --> B+	Logical AND m = 1, n = 1	2m+3n = 5
3(a)	b1.R4 --> R3	Relay m = 1, r = 1	2m+2r+6 = 10
3(b)	R3 --> C+	Logical AND m = 1, n = 1	2m+3n = 5
4	c1.a0 --> R4	Logical AND m = 2, n = 1	2m+3n = 7

From the table, β_{S1} = 16+5+10+5+10+5+7 = 58. In Sequence 2 counter and emergency stop are added to Sequence 1. Denote β_C and β_{ES} the number of basic elements needed to model a counter and emergency stop, respectively. $\alpha_{S2} = \alpha_{S1} + \alpha_C + \alpha_{ES}$. α_C and α_{ES} are calculated by using the derived formulas as follows:

m is 2 because b0,c0 resulting from executing the last action in the given sequence are to be taken as set signals for the counter. n is 1 because there is only one output from the counter. r is 1 because start switch ST can be taken as a reset signal for the counter. k' is 1 because in the system under study each action that drives the piston forward is executed by sustained actuating signal and Γ' is 3 because three pistons are present. Hence, $\beta_C = 2m + 3n + 2r + 2k'\Gamma' + 3$ = (2.2 + 3.1 +2.1+ 2.1.3 + 3) = 18. $\beta_{ES} = 5 + 2k'\Gamma'$ = 5 + (2)(1)(3) = 11. Hence, α_{S2} = 58 + 18 + 11 = 87 which is exactly the same as obtained in Section 4.

Sequence 3 is obtained by including a delay after {C+, A-} in Sequence 2. Hence, the timer is to be set with c1 and a0 as input signals and b0 and c0 as reset signals. There is only one output from timer. Hence, m = 2, n =1, and r = 2. The number of basic elements needed to implement the timer, $\beta_T = 2m + 3n + 2r + 3$ = 2.2 + 3.1. + 2.2. + 3 = 14. Segment 4 in Table 11 already contains the elements and links corresponding to c1, a0, and R4. Hence, to compensate these, $\beta_{s3} = \beta_{s2} + \beta_T - \beta_4$ where $\beta_4 = 7$ denotes the number of basic elements in Segment 4. Therefore, $\beta_{s3} = 87 + 14 - 7 = 94$ as obtained in Section 4.

Since Sequence 4 is obtained by adding a counter and delay to Sequence 3, it can be implemented by adding a counter and timer to the LLD of Sequence 3. Even though the number and identity of set and reset signals for the new counter and timer are different from those used in the earlier sequences, as an approximation we can still use the number of basic elements that were calculated earlier here. In other words, $\beta_{s4} = \beta_C + \beta_T$. Since, β_C and β_T are 18 and 14 respectively, $\beta_{s4} = 94 + 18 + 14 = 126$ which is very close to 130 as obtained in Section 4.

Even though we could accurately estimate the number of basic elements in the last case by accurately identifying the set and reset signals for timer and counter, it may take significant effort to analyze the logic. However, the motivation behind proposing the formulas presented here is to not to exactly calculate the number of basic elements to match that result from manually counting an LLD or PN, but to give an idea of how many basic elements are needed to model the control logic. Using the approach shown here these formulas can be incrementally applied even for complex sequences consisting of timers and counters. In other words, they are first applied to a simple sequence to estimate the number of basic elements. Then the number of basic elements needed for counters and timers is estimated and added to the previous value.

Hence, by using the formulas and following the proposed methodology to estimate the basic elements, control engineers can select between PN and LLD even before they start to build the control model in detail.

7. Conclusions and further research

Development of flexible, reusable, and maintainable control software is important to implement advanced industrial automated systems. Traditional methods of using ladder logic diagrams (LLDs) to design sequence controllers are being challenged by the needs in flexible and agile manufacturing systems. On the other hand, Petri nets (PNs) are an emerging tool that needs to be established for the control of discrete manufacturing systems. A class of PNs called real-time PNs that resemble ordinary PNs are introduced to design sequence controllers. The procedure for controlling a system using real-time PNs is straightforward, simple, and can be applied to control any discrete event system that has digital input/output interfaces and a computer. Also, this chapter identified design complexity and response time as the criteria to compare LLDs and PNs. Design complexity is defined and characterized by two factors namely graphical complexity and adaptability to meet changes in control specifications. By designing and implementing the control of an industrial automated system subject to changing control requirements, LLDs and PNs are compared in terms of a common measure namely, the numbers of basic elements, which signify both design complexity and response time.

Motivated by the fact that any sequence controller can be designed by synthesizing the building blocks of logic models, this chapter proposed analytical formulas to estimate the number of basic elements to model the most commonly used building blocks of logic modeling by both PN and LLD. Furthermore, a methodology that uses the developed analytical formulas to estimate the total number of basic elements to model a control logic even before physically modeling it using PNs and LLDs is presented. The concepts developed in this chapter are demonstrated by considering several examples of sequence controllers. The examples considered here demonstrate the potential for practical application of the research results.

The methodology presented provides an accurate quantitative comparison of PN and LLD in terms of basic elements. By precluding the need for physically building the controllers by either PN or LLD, this methodology serves as an effective aid for a control engineer to select between PN and LLD even before starting to write the control program. The methodology developed is simple and straightforward to apply. However, in case of complex control specifications, decomposing the control specification in terms of the logic constructs may be difficult. This problem can be solved using the traditional methods such as Karnaugh maps, Huffman method, etc. Other factors which have impact on the selection of LLDs and PNs should also be explored in the future work. Real-time PNs can be extended by adding more attributes to places and transitions in order to control complex hierarchical manufacturing systems that use advanced communication protocols and several computers for control. The significance of this present work is to help industry recognize the prominence of PNs as an emerging technology and encourage for more applications.

Acknowledgements

This research is supported partially by Center for Manufacturing Systems at New Jersey Institute of Technology, Newark, NJ and partially by the National Science Foundation under Grant No. DMI-9410386. The Robotics Center, Florida Atlantic University, Boca Raton, FL is acknowledged for the facility used in this work.

References

[1] Boucher, T.O., Jafari, M.A., and Meredith, G.A., "Petri Net Control of an Automated Manufacturing Cell," *Computers in Industrial Eng.*, 17 (1-4), 459-463, 1989.

[2] Chaar, J.K., Voltz, R.A., and Davidson, E.S., "An Integrated Approach to Developing Manufacturing Control Software", *Proceedings of the 1991 IEEE Int. Conf. on Robotics and Automation*, Sacramento, CA, 1991.

[3] Chocron, D., and Cerny, E., "A Petri net Based Industrial Sequencer," *Proc. of IEEE Int. Conf. and Exhibition on Industrial Control and Instrumentation*, 18-22, March, 1980.

[4] Crockett, D., Desrochers, A.A., DiCesare, F., and Ward, T., "Implementation of a Petri Net Controller for a Machining Workstation,"

Proceedings of the IEEE Conf. on Robotics and Automation, 1861-1867, 1987.

[5] David, R., and Alla, H., *Petri Nets and Grafcet*, Prentice Hall, New York, 1992.

[6] DiCesare F., and Desrochers, A.A., "Modeling, Control, and Performance Analysis of Automated Manufacturing Systems using Petri Nets," *Control and Dynamic Systems*, Vol. 47, C.T. Leondes (Ed.), Academic Press, 121-172, 1991.

[7] Falcione, A., and Krogh, B.H., "Design Recovery for Relay Ladder Logic," *IEEE Control Systems Magazine*, 90-98, April, 1993.

[8] Ferrarini, L., "An Incremental Approach to Logic Controller Design with Petri Nets," *IEEE Trans. on Systems, Man, and Cybernetics*, 22 (3), 461-473, 1992.

[9] Michel, G., *Programmable Logic Controllers: Architectures and Application*, John Wiley and Sons, England, 1990.

[10] Murata, T., "Petri Nets: Properties, Applications and Analysis," *Proc. of IEEE* , 77 (4), 541-580, 1989.

[11] Murata, T., Komoda, N., Matsumoto, K., and Haruna, K., "A Petri Net-based Controller for Flexible and Maintainable Sequence Control and its Applications in Factory Automation," *IEEE Trans. on Industrial Electronics*, 33 (1), 1-8, 1986.

[12] Pessen, D.W., *Industrial Automation: Circuit Design and Components*, New York: Wiley, 1989.

[13] Pessen, D.W., "Ladder-Diagram Design for Programmable Controllers," *Automatica*, 25 (3), 407-412, 1989.

[14] Silva, M., and Valette, R., "Petri Nets and Flexible Manufacturing," *Advances in Petri nets* , Lecture Notes in Computer Science, Springer-Verlag, Heidelberg, 374-417, 1990.

[15] Stefano, Di. A., and Mirabella, O., "A Fast Sequence Control Device Based on Enhanced Petri nets," *Microprocessors and Microsystems*, 15 (4), 179-186, 1991.

[16] Valette, R., Courvoisier, M,, Bigou, JM., and Albukerque, J., "A Petri net Based Programmable Logic Controller," *Computer Applications in Production and Engineering*, E.A. Warman (editor), North-Holland Publishing Company, IFIP, 103-115, 1983.

[17] Venkatesh, K., Ilyas, M., "Real-time Petri Nets for Modeling, Controlling, and Simulation of Local Area Networks in Flexible Manufacturing Systems," (in press) *Computers and Industrial Engineering*, Vol. 28, No. 1, 1995.

[18] Venkatesh, K., Kaighobadi, M., Zhou, M. C., and Caudill, R., "Augmented Timed Petri Nets for Modeling of Robotic Systems with Breakdowns," *Journal of Manufacturing Systems*, Vol. 13, No. 4, pp. 289-301, 1994.

[19] Venkatesh, K., Zhou, M.C., and Caudill, R., "Comparing Ladder Logic Diagrams and Petri Nets for Sequence Controller Design through a Discrete Manufacturing System," to appear in *IEEE Trans. on Industrial Electronics*, 1994.

[20] Venkatesh, K., Zhou, M.C., Caudill, R., "A Control Software Design Methodology for CIM Systems," *Proc. of Rutgers Conf. on CIM in the Process Industries*, East Brunswick, NJ, April 25-26, 565-579, 1994.

[21] Venkatesh, K., Zhou, M.C., Caudill, R., "Evaluating the Complexity of Petri Nets and Ladder Logic Diagrams to Design Sequence Controllers in Flexible Automation," to appear in *Proc. of Seiken/IEEE Symp. on Emerging Technologies & Factory Automation*, Tokyo, Japan, Nov. 6-10, 1994.

[22] Zhou, M. C., F. DiCesare, and D. Rudolph, "Design and Implementation of a Petri Net Based Supervisor for a Flexible Manufacturing System," *IFAC Journal Automatica*, 28 (6), 1199-1208, 1992a.

[23] Zhou, M. C., DiCesare, F., and Desrochers, "A Hybrid Methodology for Synthesis of Petri Net Models for Manufacturing Systems," *IEEE Trans. Robotics and Automation*, 8 (3), 350-361, 1992b.

[24] Zhou, M. C., and DiCesare, F., *Petri Net Synthesis for Discrete Event Control of Manufacturing Systems*, Kluwer Academic Publishers, Boston, MA, 1993.

11

From State Transition Models to DFD Extended Methods for Specifying Reactive Systems

Abd-El-Kader SAHRAOUI

Laboratoire d'Analyse et d'Architecture des Systèmes du C.N.R.S
7, Avenue Colonel Roche 31077 Toulouse, France.
and GR "ICARE", IUT-B 1, Place Georges Brassens Toulouse-Blagnac
e-mail : kader@laas.fr

Abstract

The objective of this chapter is to give an insight of the new approaches recently applied to specify reactive automatic systems. State transition based systems such as Petri nets and statecharts are not self-sufficient when dealing with hybrid systems; complementary tools and associated methods are required. The DFD extended and object oriented methods are used by integrating Petri nets or statecharts for specifying behavioural aspects. It is made throughout the whole chapter, an abstraction of implementation issues, hardware or software.

I. Introduction

The term reactive system is based on the notion of permanent interactions between the system and the environment, especially the interactions happening during the execution of the system, which differentiate reactive systems from transformational ones.

The following definition is given in [Pnueli, 1987]

> "*.........reactive systems are systems whose role is to maintain some interaction with their environment, rather than to compute some final result* ...."

In general the reactive system can be considered a set of transformational and reactive operations. The reactive operations maintain the interaction with the environment during the execution and can exchange the information among other operations. The transformational operation is characterized by initial state to terminal state, e.g. optimisation program and all types of numerical computations.

Some reactive systems are becoming more and more complex to develop. Let recall their main features:

- data modelling,
- data transformation (functional aspect),
- event and signals,
- behaviour (event or control transformation),
- time.

When controlling a system, there is a reactive part interacting with the transformational part, both embedded in the whole system.

For each feature there are concepts, and associated tool support:

- Abstraction for data and data transformation
- behaviour models and associated expressive power
- timing constraint expressiveness and validation

There is growing recognition that the conduct of the initial phases of system development, in which the actual requirements and feasibility of a proposed system are studied, has a significant impact upon the success of the whole project. It is widely accepted that errors made in requirements are often not discovered until later stages, when they are generally very expensive to correct. At the early stages it is much easier to correct errors in understanding of the requirements; unfortunately knowledge about the system is also relatively poor.

Problems with commonly used requirements definition techniques are not specific to any part of the process, but relate to the non-trivial difficulties of eliciting, analysing, representing and validating the requirements for the often highly complex systems under development. This is compounded in the present case by the need to describe and reason about requirements involving time. Formal methods are seen as being one way of overcoming these difficulties owing to their characteristics of precise description and reliable verification.

The second part of this chapter discusses requirements; the third part on Petri nets related models such as Grafcet and statecharts; the fouth part structured development methods; the fifth part a case study combining Petri nets and statecharts; the last part actual research issues on Petri nets related models.

II. The Nature of Requirements in Reactive Systems

2.1 Requirements in specification

A statement of requirements is generally a complex document fulfilling a wide range of roles and containing a great diversity of information. The core of the requirements statement is the system model. This contains an abstraction (model) of the real world which is used to express the properties of the system and its environment. Such a model must therefore capture *all relevant properties* if the requirements specification is to be adequate.

Specific difficulties associated with requirements for reactive computer integrated manufacturing (CIM) systems include:

- the need for analysis of real-time aspects, e.g. performance and schedule characteristics associated with functional or non-functional properties;
- the need for analysis of asynchronous couplings;
- the consideration of event durations, e.g. the consideration of state changes *during* the execution of an event;
- the need for analysis of concurrency aspects, e.g. overlapping system events, communication of system components by shared state;

- non-determinism owing to lack of information about the system;
- the existence of mathematical theories, e.g. control theory, relating to the problem and needed in the formal analysis;
- the need for different "views" of a system such as long term behavioural view of some parts and discrete event based view of others.

Confusion also often arises from the fact that requirements exist at several levels of abstraction. A requirement can be:

- an *objective* of the user;
- a *feature* of a system which enables them to achieve that objective;
- an *implementation* which is a description of a system displaying that feature.

Thus a statement of requirements must say "what needs to be said" at appropriate levels of detail.

The requirements definition process usually consists of three stages:

- elicitation, in which the domain expert indicates the intended function of a system;
- specification, in which those requirements are written down in a precise and detailed way;
- validation, in which those written requirements are checked by the expert to see whether they actually capture the desired behaviour or functionality.

Analysis is ongoing during the requirements definition and can be thought of as the creative activity joining the different stages. The most intense analysis is involved when different levels of abstraction are being traversed.

It is common for these elements to be repeated until the specifier is satisfied with the written requirements correct with respect to the application. When the system features are being identified, these will usually require further validation; once determined, they are compiled into a definitive statement of the system requirements.

In practice, however, it is inevitable that information is obtained concerning the features of the system (the empty box in the idealised diagram). The need to take account of those features (whether by ignoring them or by incorporating them into subsequent analysis and design) has a significant impact upon the approach taken in real situations.

There are two broad approaches to dealing with this complex process: 1) the use of formal development with correctness preserving arguments connecting the different levels of abstraction; 2) structured development with the use of semi-formal representations (such as diagrams). This proposal suggests that there are distinct advantages to be gained from the former approach, particularly if the attractive features of the latter are incorporated : both approaches are illstrated in part III and IV. For its recent development, statecharts model is better illustrated than other models such as Petri nets and Grafcet and discussed in other parts.

2.2 Current approaches

There has been much work in recent years by the theoretical cómputer science community on developing formal specification languages. These have been principally applied to the later stages of system development but are also applicable at the requirements stages. Examples include:

The above modelling approaches fall broadly into the following categories:

- state-based or model-based approaches;

- algebraic and functional approaches;
- process algebras;
- temporal logic.

Each of these is discussed in turn below.

State-based approaches are typified by Z, VDM and GYPSY for data transformation; statcharts, Petri nets for event transformation. A specification typically characterises:

- state of a system;
- the effect of particular events upon that state.

These techniques have been most successful to date in the industrial context, principally as specification languages — very few users of these languages use proof techniques. Great emphasis is placed upon "refinement calculi" whereby Z specifications are progressively augmented until pseudo-code is produced.

There have been many attempts to enhance these methods through the use of:

- modularisation schemes (usually limited to introducing conventions on variable scoping);
- adding time (in a variety of ways, but never entirely successfully);
- other extensions such as the addition of object oriented concepts.

The problem with these methods in general has been a failure to recognise the need for abstraction and escape from the simple system view provided by the state-based approaches.

Algebraic and functional approaches are based on the use of datatypes with associated functions which act on the datatypes. The most obvious advantage of the approaches is that they can be animated or executed automatically.

Their main drawback is that, on the whole, they have an even more limited set of views of the world than state-based approach.

The study of process algebras has been the main focus for work in understanding the following issues:

- concurrency and synchronisation;
- causality and time;
- The problem with these approaches is their failure to capture state adequately in the models. As a result, some issues cannot be addressed, e.g, relations between events and state changes, or time and state changes;
- Temporal logics are seen by many as the way forward for reasoning over behaviour but have met with limited use in a practical setting owing to their obscurity and limited expressive power.

This will bring us to our main topics : Petri nets related models and methods to design automatic controllers.

III. State transition models

Petri nets are compared to two major models used in large scale in industry.

3.1 Grafcet and Petri nets

3.1.1 Grafcet [IEC-SC65A, 1990]: Grafcet was developped in France, in 1977, by a group of academicians and industrial engineers. Grafcet is the industrialised

version of binary ordinary interpreted Petri nets where the syntactic entities differs slightly, but there are semantic differences.
It is based on steps and transitions. By connecting between steps and transitions, they are equivalent to places and transitions in Petri nets. Steps are represented by squares (initial steps by double squares) and transitions by bars.

To each step, it is possible to associate actions which are executed when their associated step is active. The semantics of the actions depends mainly on implementation issues (programming language and hardware support as PLC's) [5,13]. Figure 1 gives an illustration of common Grafcet constructs.

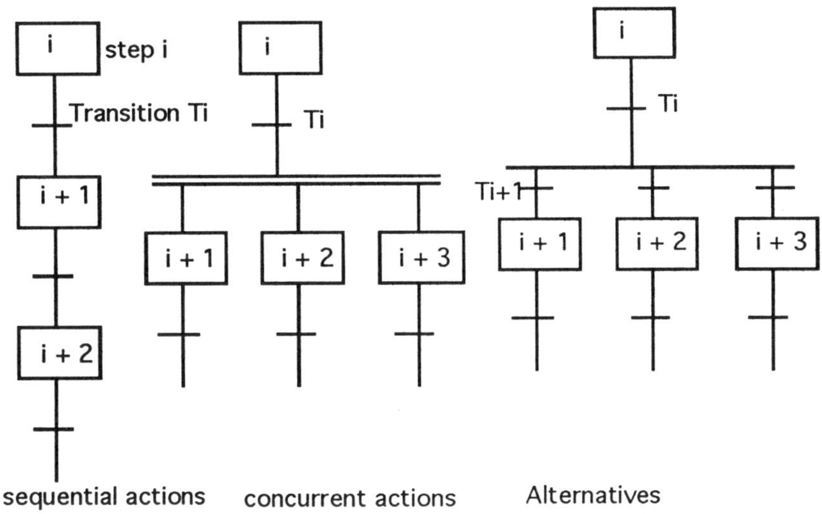

(a) (b) (c)

Figure 1 Grafcet control structure

This shows the common nature between Petri nets and Grafcet except some minor semantic differences.

3.1.2 Semantic differences with Petri nets

The main differences with Petri nets are

- Transition firing is synchronous in Grafcet (asynchronous in Petri nets)
- An active step can be used as a condition asssociated with a transition; this not allowed in Petri nets
- if two transitions having a common precedent step are enabled; then both are fired; this is not the case in Petri nets; since there is a single token in the precedent place.
- if an output step of a transition is active, this transition cannot be fired; this is allowed in Petri nets (numerical Petri nets as opposed to Condition event Petri nets).

3.2 Statecharts and Petri nets

3.2.1 An introduction to Statecharts

Statecharts is a specification and design formalism developed by Harel [7,8,9,20]. It is specific for reactive complex systems. The dynamic change of such systems is subject to event occurrences coming from the external environment and they are under stringe timing constraints. The major problem lies in the difficulty to describe this type of behaviour in an elegant and formal way. Statecharts is an extension of state machines and higraphs. Statecharts is best characterised by the following informal equation :

Statecharts = State diagrams + Hierarchy + Concurrency + Communication

A general description of the main characteristics of Statecharts is given in the sequel.
We present the various concepts of the language through the mechanisms offered to express them.

- Hierarchy

Fig. 2(a) presents the specification of a system using the state machine approach. However this "flat" type specification does not offer clear understanding as details and number of states grows up. An alternative approach with Statecharts is presented in Fig. 2(b).

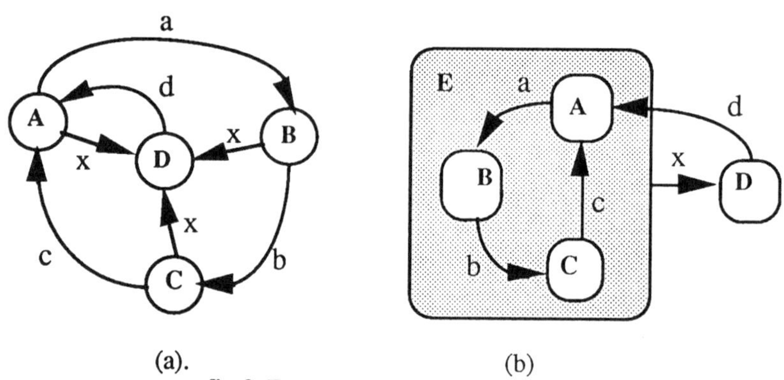

fig 2. From automata to statecharts.

The hierarchy concept introduces the structured design by abstraction and refinement. This concept is very interesting because it allows one to deal with large and complex systems. In Fig. 2b, the global state E is an abstraction of the dynamic of the sub-states encapsulated in it. When the system is in state E and when the event x is triggered, the system goes to state D whatever its active sub-state may be.
Figs. 3(a) to Fig. 3(b) illustrate a bottom-up step (abstraction), e.g. fig 3(b) is an abtraction of figure 3(a) ; Fig. 3(a) to Fig. 3(c) illustrate a top-down step (refinement).

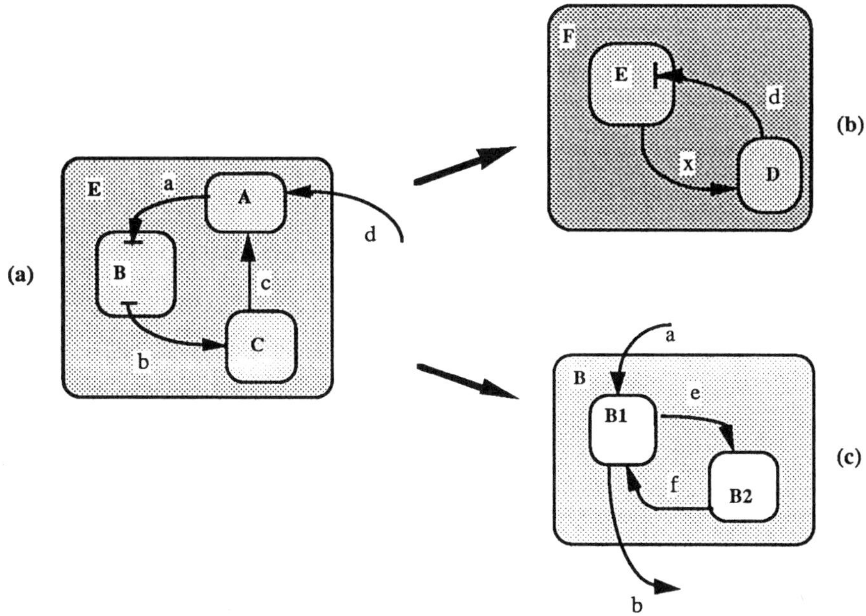

Fig. 3. State abstraction in statecharts

- Concurrency

This construct is denoted by "orthogonality" in the statecharts syntax. In Fig. 4, states X, Y and Z are activated and deactivated simultaneously. Z behaves independently of X and Y.

Synchronisation is ensured by the "in" expression ; as an example, X0 goes to X3 if concurrent state Y is in Y0. On deactivation, event l makes the system leave state S (hence X, Y and Z) ; however, event k makes the system leave S only if X3 and Y1 are active.

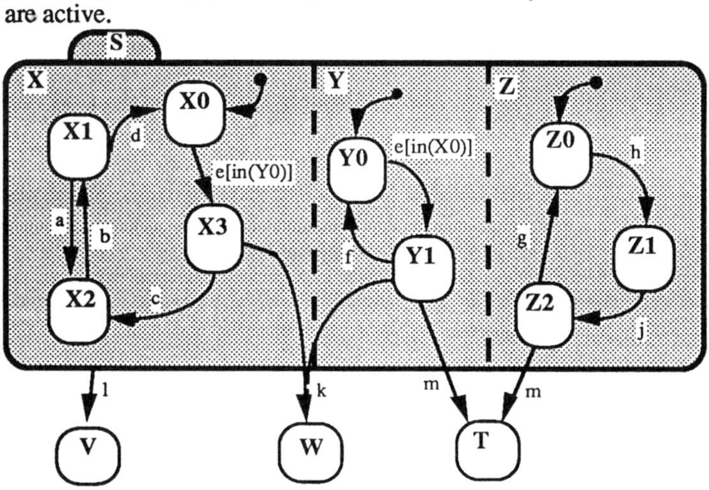

Fig. 4. Concurrency in statecharts

- Communication

When a message is produced, it is broadcast. The event/action model is synchronous, i.e. the event and action happen at the same time; this assumption is the same applied for the synchronous languages as ESTEREL.

The general form of the label associated with transition is : e [c] / a

where : - e = event triggering the transition ;

- c = condition allowing to trigger the transition, it is a Boolean expression ;

- a = resulting action (broadcasted through the system).

Action a can be either an event for another transition or sent to the external environment to trigger an actuator.

This part is well illustrated by the example shown in Fig. 5. The system consists of four concurrent subsystems X, Y, Z and W. The initial state is {C, E, G, J}. Event e triggers the transition C-D, action f which is generated is an event for the orthogonal states putting the system in a stable state {D, F, H, K} ; the system goes to {B, E, I, J} following the occurence of event j and to {C, E, G, J} following the occurence of event g.

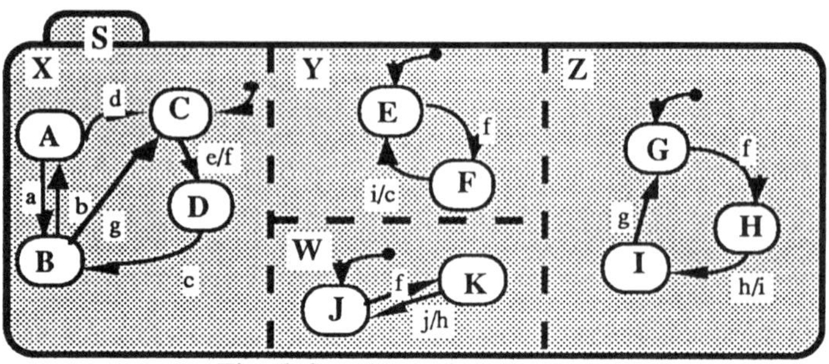

Fig. 5. Synchrony in statecharts

The Statecharts formalism has a flexible approach concerning the entrance into a state. Nevertheless, one has to know the default entry which constitutes the construct used to initialise a state (equivalent to double circled state in automata specification or initial marking in Petri nets).

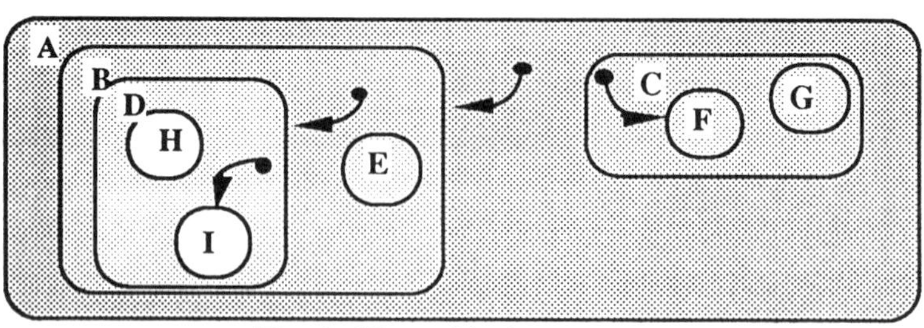

Fig. 6. Hierarchical entry states

In Fig. 6, the system goes to state I through state A, B, D which themselves are default initial states at different levels of hierarchy.

3.2.2 - Statecharts and AMS specification

- Introduction : Five common constructs are fundamental in control design :
  - sequence,
  - alternative,
  - loop,
  - concurrency (and synchronisation),
  - resource sharing.

We will base on an example to illustrate :
- how Statecharts can express these constructs,
- how these constructs can be implemented and combined in case of specification of a manufacturing system. We choose a rather complex example in order to apply all the constructs and show how difficult cases can be solved.

Specifying the control of flexible manufacturing cell

Our running example will be a flexible manufacturing cell (FMC). As shown in Fig. 7, it consists of 9 stations and 4 pallets.
The stations are :

- station A = Resource MA (a tool centre) + post A ;
- station B = Resource MB_I and MB_{II} (two special machines) + post B ;
- station M = the Manager of the cell ;
- station O = Resource RO (an Operator) + post O + set of parts α + set of parts β ;

- stations S1, S2, S3, and S4 = posts S1, S2, S3, and S4 ;
- station T = Resource MT (an automated transfer carriage) + post T + circulation net.

The four pallets P1, P2, P3 and P4 are identical ; each one is an universal support on which the operator can fix a part. Pallets never leave the cell. The places that each pallet Pi can occupy at any moment is one of the following posts :

- post A : the machine MA executes an operation OA on the part fixed on the pallet ;
- post B : machines MB_I and MB_{II} execute simultaneously operations OB_I and OB_{II} on the part fixed on the pallet ;
- post O : the operator RO takes the finished part off the pallet (unloading : operation OOu), and put a new part on the pallet (loading : operation OOl) ;
- post Si : the pallet is kept in a waiting state ; Si is reserved for Pi ;
- post T : it is a moving post ; the carriage is used to transfer the pallet from a static post to an another.

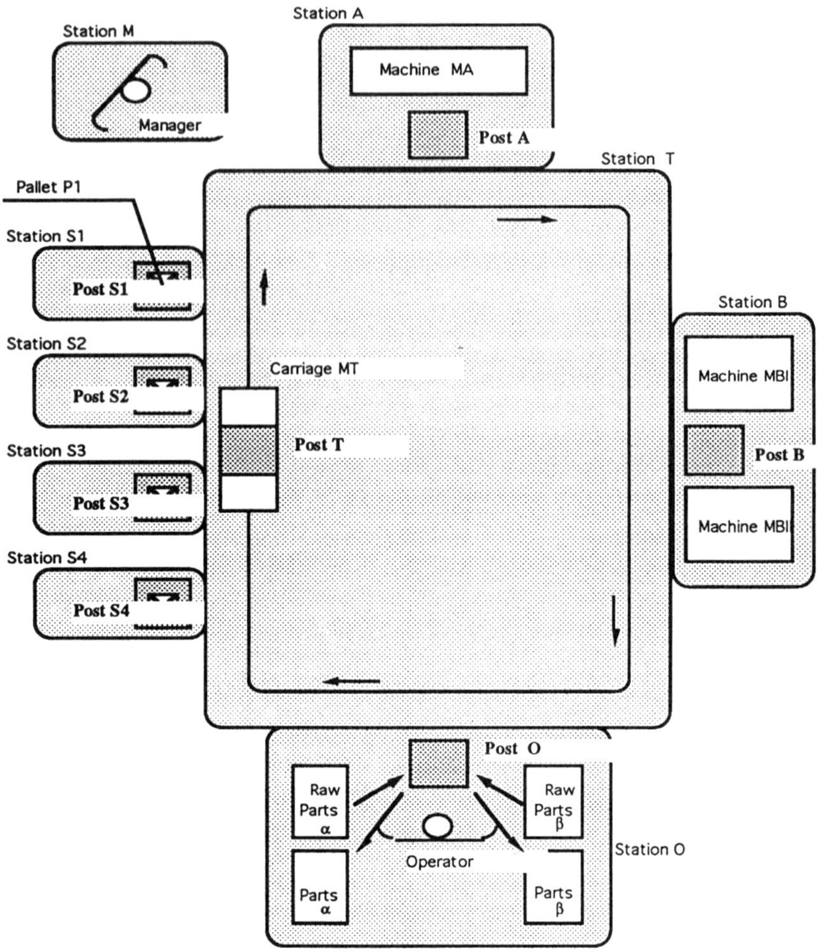

Fig. 7. Manufacturing cell

The cell has to produce two types of parts : α and β . The operating sequences are :

- for α : 001α --> OAα --> OB_I and OB_{II} --> OOuα
- for β : 001β --> OB_I and OB_{II} --> OAβ --> OOuβ

Remark. Only operations OB_I and OB_{II} are identical for both types of parts.

At start, we consider that the cell is in its "reference state" : all four pallets Pi are empty and situated at the waiting posts Si.

The manager's mission is to supervise the general operating modes. He triggers the system off from the "Stopped in Reference" mode to the "In Production IP" mode by giving the start order. He triggers the system off from the "In Production" mode to the "Return in Reference" mode by giving the stop order. When the system reaches the reference state, it enters automatically the "Stopped in Reference " mode.

In the IP mode, the pallets which are empty (at starting) or carry a finished part are systematically driven in the station O in order that the operator loads a new part. The operator may choose either part α or β.

The four pallets must share the stations A, B, O and T. There are conflicts for accessing these resources. We make some priorities to manage the conflicts : P1 > P2 > P3 > P4.

- Illustration of the sequence construct

It is a linear simple oriented graph, where nodes denote states and arrows denote transitions. The diagram easily specifies the order whereby the state of the system changes, or the state of a fragment of the system changes. In our case, we can describe by a sequence the evolution of the pallet P1 state when P1 leaves its initial state (empty, in the station S1), to go to the station O. See Fig. 8.

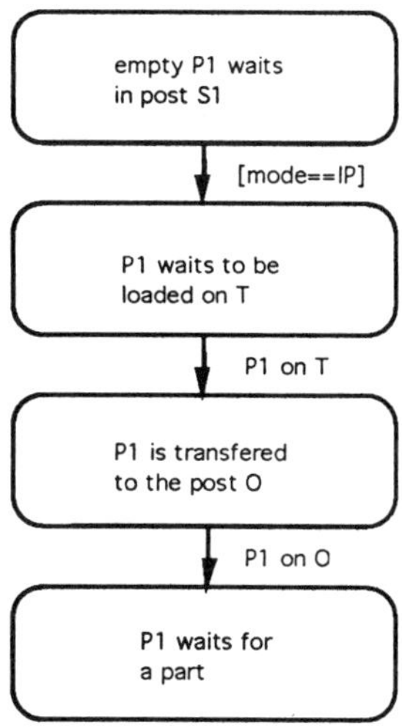

Fig. 8. Sequential operations

- Illustration of the alternative constructs

In statecharts, each transition will be labelled with an expression. When one wants to specify two evolutions from the same state, the construct of the transitions may differ depending on events and conditions combination. Let consider, in our example, the two following cases :

- when the pallet P1 is receiving a part in the post O, according to the choice of the operator, the loading operation OO1 terminates in event "End of OO1α " or "End of OO1β ". Depending on the event, the pallet will go to either post A or post B as shown in fig. 9.
- when the pallet P1 arrives on post A (event "P1 on A"), the machine MA has to execute either operation OAα or OAβ according to the type of the part fixed on P1. We have to introduce a variable pP1 which will indicate the part type. The evolution depends on the value of pP1 as shown in fig. 10.

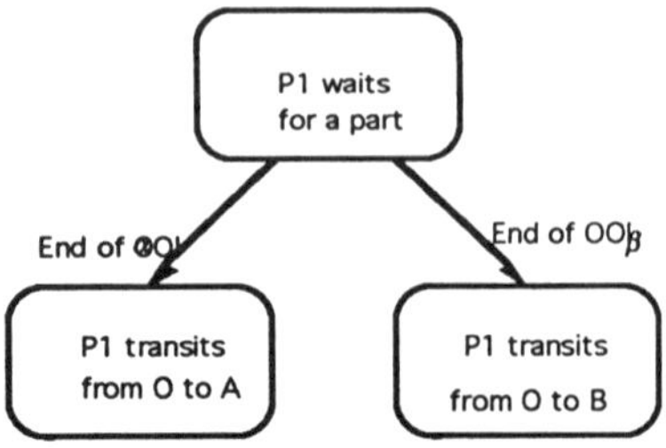

Fig. 9 Alternative construct

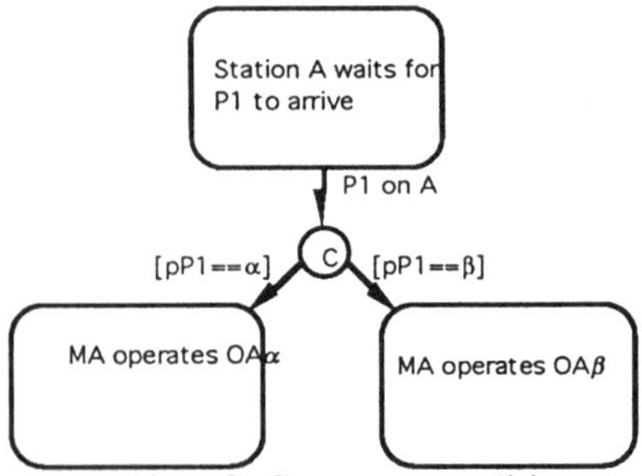

Fig. 10. Common condition

- Illustration of the loop construct

The loop construct is used whenever a sequence has to be repeated. The type of reasoning is the following : a condition is evaluate ; if it is true, a sequence is executed and the condition is re-evaluated ; this cycle starts again until the condition becomes false, at which point execution continues with the next part. Fig. 11 shows the corresponding construct using a conditional transition. In our example, the pallet P1 repeats the same production cycle while the mode selected by the manager is the IP mode as shown in fig. 12.

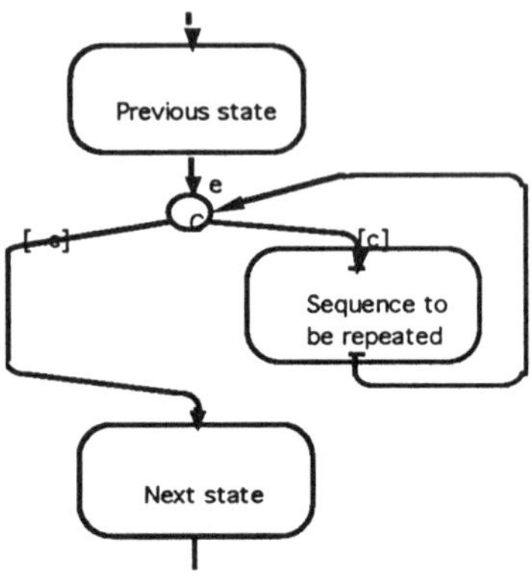

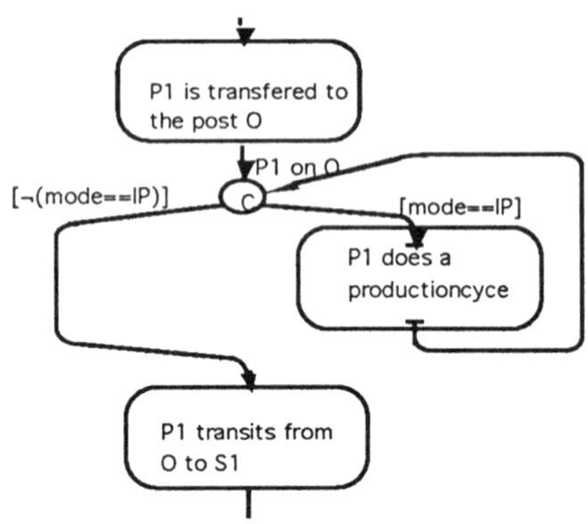

Fig. 11 While loop (common event) Fig. 12. While loop (common condition)

- Illustration of concurrency

Concurrency, noted orthogonality in statecharts, is a powerful construct for specifying simultaneous execution of more than one process. In statecharts, concurrency is introduced as communicating state-machines called AND-states. When decomposing a system, we have sub-systems which evolve independently and, in some cases, synchronise with each other. In specifying manufacturing systems, this construct is very important. In our example, we will consider the following sub-systems : pallets P1, P2, P3 and P4 and stations A, B, M, O, S1, S2, S3, S4 and T. Hence, we will have, at the top level, thirteen parallel diagrams, each corresponding to one specific equipment. as shown in fig. 13.

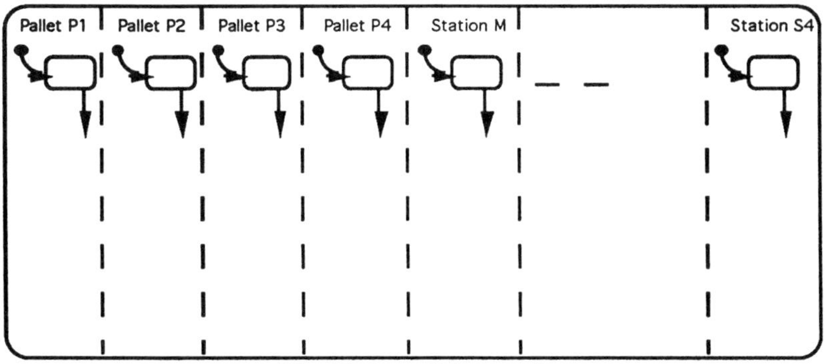

Fig. 13. Parrallel operations

By refinement of the station B, Fig.14 shows how we can detail the station state. States of both MB_I and MB_{II} machines become visible.

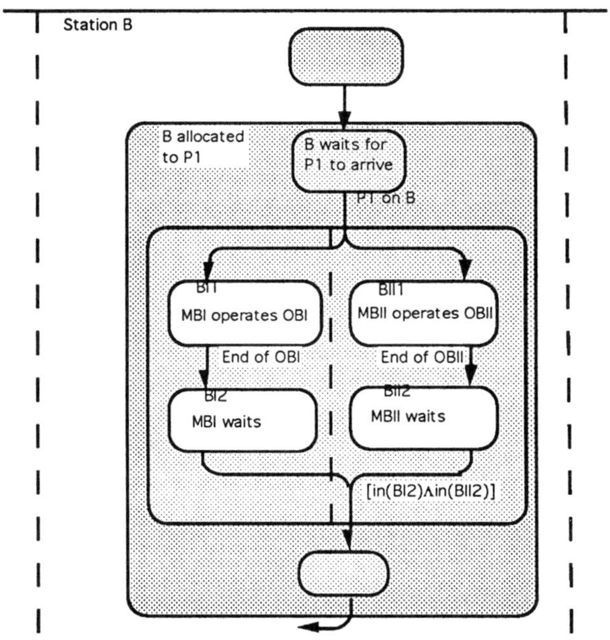

Fig. 14. State abstraction for station B dynamics

- Resource sharing specification

Among the sub-systems composing a manufacturing system, we have shared resources. So there must be a mechanism to allocate resources to the process users. We can specify this by Fig. 15.

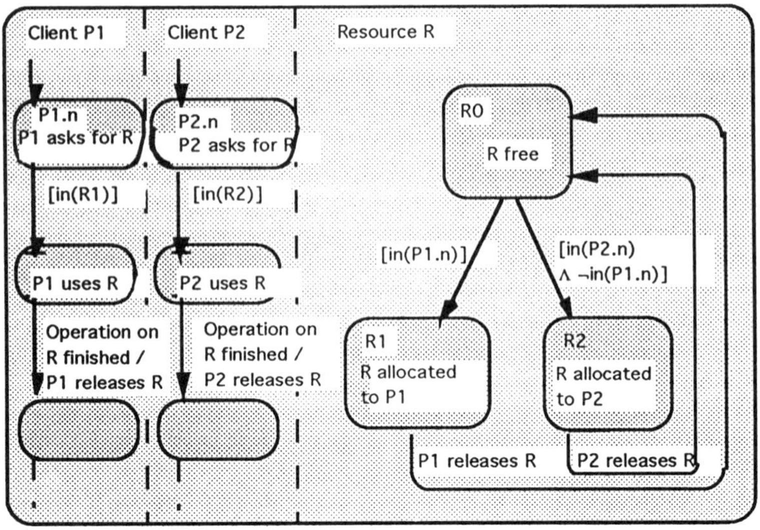

Fig. 15. Mutual exclusion for common resource

Interpretation :

- Each user Pi and the resource R are described as sub-systems using concurrency ;
- Each user diagram contains a state, denoted Pi.n, in which the user asks and waits for the resource R ;
- The resource diagram contains :

. a state, noted R0, which denotes that the resource is free and waits for an user ;

. as many states Ri as users ; Ri denotes the situation in which the resource is allocated to the client i. The transitions between R0 and all the Ri states expresse the resource sharing. The conditions supported by these transitions formalise the assignment logic. In Fig. 15, P1 has priority on P2.

Each user Pi has to release the resource when exploitation stops. The event "Pi release R" must be generated to produce the resource evolution to state "R free".

3.2.3 Properties assertion

The differences with Petri nets are for some the same as the ones discussed for Grafcet; the other differences are the abstraction and hierarchical representation, the event and state abstraction and historic constructs.

The expression of invariants in statecharts is carried out with temporal logic formulae [15].

The temporal specification language: The common temporal operators added on the ones used in propositional logic are ❑ and ◊ and defined as :

$P1 \rightarrow \Box P2$ If P1 is true then P2 will be true now and in the future

$P1 \rightarrow \Diamond P2$ If P1 is true then P2 will eventually be true in the future (there exists a future where P will be true).

* Invariance or safety properties : stating that some condition holds continuously throughtout the computation; this is summarised generally as "bad situations never happen ". The common example is the mutual exclusion problem, e.g. the resource will not be allocated to both users simultaneously which means that a collision never happens.

* Eventually or liveness properties: stating that under some conditions certain events must eventually happen in the future; the common example is fairness; this is summarised as " some good thing will happen ".

* Precedence properties, stating that an event always precedes another. Such property is easily handled using the Petri net based specification approach.

p U q *(p until q)*: either q holds or q will happen at some time in the future, and until that moment p will permanently hold.

O p *(next p)*: p holds at the step which immediately follows the current one.

Example: setting a formula which states that whenever there is a request for a resource, there will be followed by a grant event and consists in:

$$\Box(request \implies \Diamond grant)$$

The first two operators can be combined to define new operators as

* $\Box \Diamond$ p : *infinitely_often p*, at every future is followed by a later one in which p holds

* $\Diamond \Box$ p : *almost always p*, starting from some future, p will hold permanently.

The temporal logic language has been also used as a specification language and as assertion language for Petri nets based specification.

Summary on state based models

Major differences rely on semantics of expressivness and illustrated in table 1.

The state transition models are adequate for discret event modelling; they are embedded in advanced methods for complex systems; hence the following discussed methods are complementary tools to state transition models. Some extensions and applications have been proposed [3,11,16,17].

Construct	GRAFCET	PETRI NETS	STATECHARTS
Concurrency	Yes	Yes	Yes
State Abstraction	Yes	No	Yes
Event abstraction	No	No	Yes
History_Interrupt	No	No	Yes
Synchronous	Yes	Asynchronous	Yes
Synchronisation	Interpreted	Structural	Interpreted
Data type	No	High level Petri nets	No
Formal validation	No	powerful	Some properties
Software engineering applications	Discret event systems	Extension in SA-RT SA-RT/PN	activitycharts and module charts

Table 1 : summary for state transition models

IV. Specification methods

We discuss in this part the methods used mainly in sofware engineering framework; these methods consist of two or three specification languages where behavioural languages are state transition based models.

4.1 From state transition models to software specification methods

We show in this part that state transition models are not self sufficient to specify and modele complexe reactive systems or simple hybrid systems

Specifying manufacturing processes :

The often adopted control structure is made of five levels of abstraction [4,21]. This is illustrated in figure 16 along the computer systems configuration.

- Level 4 (planning): dealing with all aspects of management functions in the long-term and mid-term ,

- Level 3 (scheduling): function dealing with allocation of machines. This function is at first triggered off-line and then behaves as a dynamic task; the results are function of all parameters and available resources in the system; this function is often ommitted in certain cases.

- Level 2: coordination of all manufacturing cells,

- Level 1: real-time control of machines

- Level 0: Intellgent sensors and actuators, the level of interaction with the industrial process.

The hybrid nature of the system imposes a corresponding method. Petri nets is one of the mostly used specification language for this application; the main drawback was the lack of abstraction, expression of continuous aspect and context and in general the lack of a well established method.

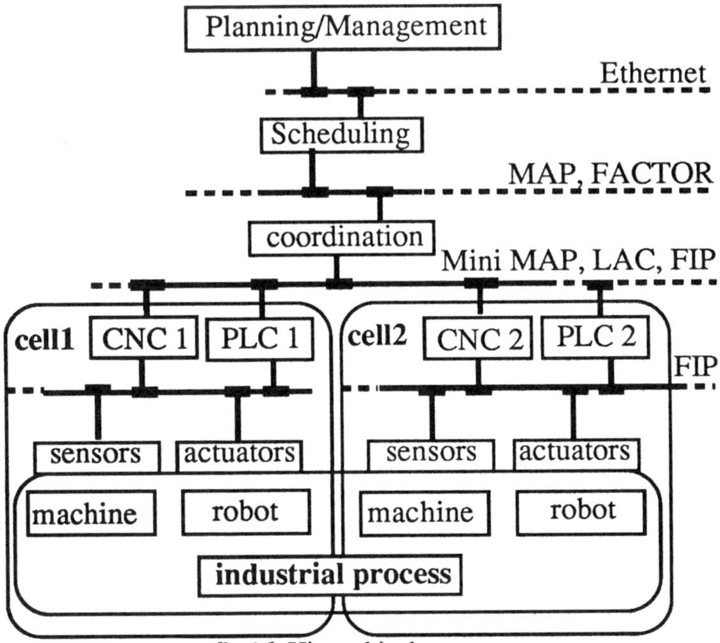

fig 16 Hierarchical structure

Hybrid systems

Most dynamic systems are hybrid; however, we find in some cases more discret event behaviour (*DES*) than continuous behaviour (*CS*) and vice versa for others and we denote these by *DES dominant* or *CS dominant* respectively. In the past and recently; in DES dominant we make abstraction of the CS part and in CS dominant , we make abstraction of DES part often denoted by *logic or control structure*.

Discret event systems are modelled using many available tools. The communities working on this range from specialists in *sequential logic* to ones in *discrete mathematics*. Corresponding formalisms were presented in part II.

Discret event systems are being studied massively since the last decade mainly applying transition system (Petri nets, Grafcet, Automate, ...) to control this type of systems.

There is an actual challenge in studying hybrid systems where CS and DES interact, for example : start-up and stop of a chemical plant in DES, the control loops are continuous systems; another example is a lift system :

- Calling lift, Starting and stopping at floors are automata based
- The control of drive is continuous;

CS concerns all aspects of continuous and discret time systems. However, we handle only discret time systems.

Among the main direction of studies, we propose and show how some methods used in software engineering discipline can be used to develop an approach to model a hybrid system.

We have to mention, however, the general confusion that classifies reactive systems as discret event systems because of the sporadic nature of events and transformational systems as continuous systems. Table 2 illustrates this.
For each feature, there are concepts and associated tool support :

- Abstraction for data and also data transformation
- Control modelling and associated expressive power

	Reactive systems	Transformationnal systems
Discret event systems	Logic controllers	simulation of systems
Continuous systems	control loops (PID, Adaptative)	optimisation, various computation

Table 2

Some proposals are available to improve the use of DFD methods and to set some rules and particularly SA-RT that enables these to be more expressive for specifying hybrid systems.

4.2 SA and SA-RT [6,10,23,24]:

Introduced in 1978 by DeMarco, traditional *Structured Analysis* (SA) [6] has been an industry standard method for software requirements analysis supported by numerous CASE tools. The popularity of structured specification techniques in industry is due to their use of intuitively appealing notations and concepts. These techniques are based on a functional view of the system, with the system being partitioned according to its functional aspects.

The basic model of this approach is the completed *Data Flow Digram* (DFD), which is developed by functional decomposition of the previous levels DFD or the *context diagram.* The context diagram presents the system as a single bubble representing the main purpose of the modeled system, connected to boxes called *terminators.* Terminators are used to represent active and passive entities situated in the environment of the modeled system.

The DFDs model the decomposition of the functions and interfaces of the system as physical data flow through the processes of the system. SA uses a *data dictionary*, which contains the definition of all of the data on the DFDs, and a pseudo-code, which specifies how basic processes transform input data to output data. The basic elements of DFD are the followings:

-*Data transformations* (*processes*): abstraction of an application's processing elements.

-*Stores:* abstraction of repositories of data.

-*Data flows*: abstraction of data communication among the above elements.

The DFD uses circles (bubbles) to represent processes, solid arrows to represent data flows, and pair of parallel line segments to represent stores. A labeled convention for bubbles and diagrams has been included to facilitate traceability between different levels of abstraction. At the highest level, the diagram context

specifies the context of the application with respect to the external environment. Figure 17 corresponds to a context diagram for a manufacturing cell

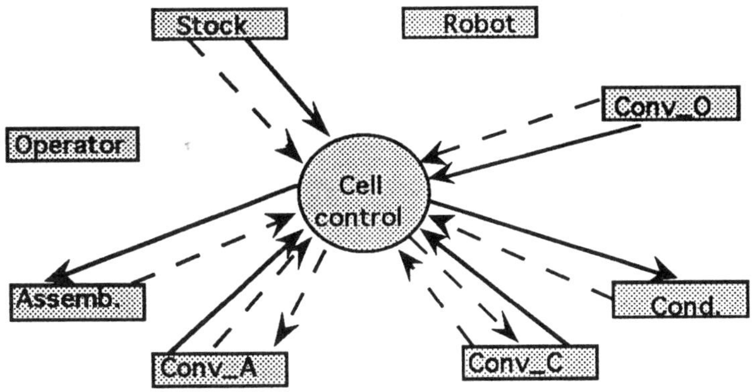

Figure 17 Context diagram

At the intermediate levels, The DFD diagram as the one below consists of data transformations (1,2,3) and also on control transformation (C). Figure 18 represents a DFD-CFD (Control Flow Diagram); the dotted circle is an abstraction of the control; when refined, an automata is obtained,which shows the activation of data transformation processes 1.0, 2.0 and 3.0.

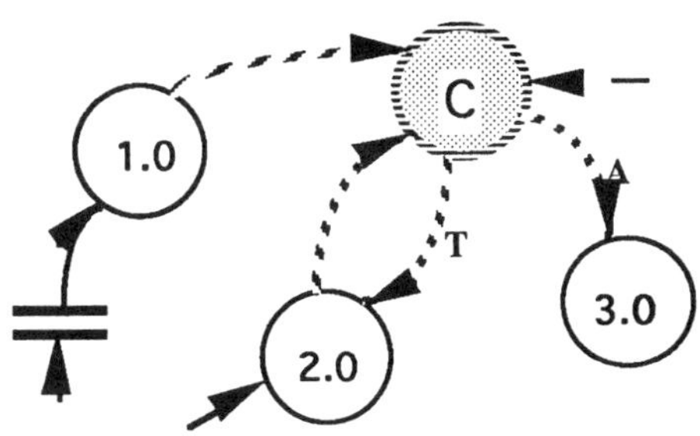

DFD / CFD

Figure 18 A DFD with control flow

At the lowest level, data transformation is represented by *Process Specifications* (P-Specs). These lowest level processes are called *primitives*.

The popularity of the DFD specification tool in industry is due to its use of intuitively notations and concepts. As a tool in the requirements specification phase of a system life cycle, the DFD has the following good points:

-it supports the creation of a modular and hierarchical structure to alleviate complexity.

-it is graphical and therefore enhances clarity.

-it has only a few primitives and concepts that are easily understood and used.

-requirements are expressed in terms of input and output data, so that they are testable.

-data flows are paths of information flow against which performance requirements can be assigned.

A number of researchers have proposed extensions to DFDs in order to update them to the real-time systems. The resulting method is called *Real-Time Structured Analysis* (SA-RT), it supports the specification of the timing, control, and behavioral aspects of the system.

Ward [24] has proposed as extension of DFDs a *Transformation Schemas* (TS), which are used as an integrated system by modelling both data flow and control flow structures. A TS is thus, at the same time, a DFD and a Control Flow Diagram (CFD). The TS notation is intended to be a modeling language for real-time systems. It models a system as a network of activities and connections. There are two types of activities according to their purpose. A *control transformation* (*control activity*) which models a portion of controlling work inside the system and determines when and for how long other activities will execute their actions, whereas a *data transformation (operational activity)* models a portion of operating work inside the system, which consists in processing data, either in uncontrolled way or under the control of a control activity.

There are two types of connections: *stores* and *flows*. Stores establish "non causal" relationships between the connected activities. A flow may be a *data flow* or a *control flow*. A control flow has no associated value and is either a *signal* or a *prompt*. A signal serves to indicate an occurrence of an event that is relevant to the behavior of the destination activity. A prompt models control imposed by one activity on another. There are three types of prompts:

-Trigger

-Activate:

-Deactivate

Each control activity in Transformation Schemas is described by a Meally-type Finite State Machine (FSM) with output - also known as a *State Transition Diagram* (STD) - in case of sequential logic, and activation or condition tables in case of combinational logic.

4.3. The statecharts-activitycharts method (SC-AC)

SC-AC offers three main languages statecharts, activity charts and modulecharts; only the the two first are considered here. This method apart from some other constructs has two types of languages : statecharts for the control part , called the behavioural view (equivalent to automata or state event matrix in SA-RT) and activitycharts for the functionnal view (as processes in a DFD for SA-RT). We can derive four different methods to develop specifications; some can be mapped directly from existing languages and associated methods.

An initial technique makes use of the control decomposition and some part of functional decomposition. In case where control is important (discret event features) , we use control decomposition to take benefit of abstraction (state and transition abstraction). Activity decomposition is equivalent to the existing SA-RT methodology.

This decision was based on heuristics and experience by trial and error. We point out that control decomposition involves only statecharts diagrams. Some experiences have been carried out using the present formalisms.

A statechart behaves as a task synchroniser (real-time executive) at the higher level of CIM hierarchy and is like a logic controller that can be implemented in PLC's at the lower level of the hierarchy.

Statecharts specification enable to supervise the execution of activities. These primitives behave as the ones used in real-time executives except message passing primitives which are as data flow between activities. These are summarised as follows in table 3.

REFERENCES	EVENTS	CONDITIONS	ACTIONS
state : S	entered(S), exited(S)	in(S)	
activity : A	started(A) stopped(A)	active(A) hanging(A)	start(A), stop(A) suspend(A) resume(A)
data : D, F	read(D), written(D) true(c), false(c)	D=F,D<F,D>F...	D:=exp m_true(c), m_false(c)
time	timeout(e,n)		schedule(a,n)

Table 3 event-action-condition primitives

We have been using these methods in a pragmactic way and common sense to specify hybrid systems [16,17].

V Case studies :

5.1 Case study on combining both models: We take study carried in the project in order to combine synchronous model [2] like statecharts and asynchronous model like Petri nets. This work has been carried jointly by LAAS-CNRS at Toulouse, LAIL at Lille and SPORTS at Nice (France) [1].

5.1.1 The environment : a manufacturing cell

Figure 19 shows a small manufacturing cell composed of two machines (M1, M2) and two robots (R1, R2). Parts arrive at 'FIFO - in' and when finished they leave the cell at 'FIFO-out'.

Each machine owns an input-buffer, an output-buffer, and a working place. As our approach is product-oriented the type of the machine is not distinguished. Only the operations which can be performed on a machine are given: M1 executes operations Op1 and Op2 while M2 only performs operation Op2.

The robots R1 and R2 perform transfer between FIFO IN, FIFO OUT and machines. Any of the robots can perform these operations. As for the machines, transfer operation are defined between the entities.

In order to obtain an interesting degree of flexibility, four types of products have been defined :

- G1 : operation1
- G2 : operation2

- - G12: operation1 followed by operation2
- - G21: operation2 followed by operation1

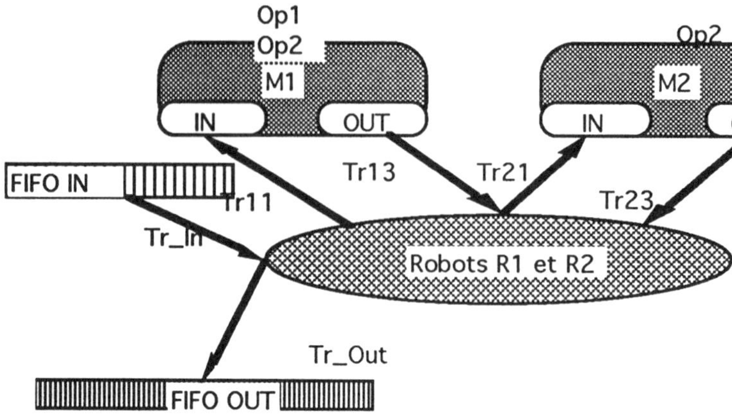

Figure 19 Cell configuration

Further we specify three user modes. An user mode specifies the types of products to be produced and their respective ratios. Thus the change from one user mode (UM) to another corresponds to the voluntary change of an operating mode of the system.
- UM 1 : Simultaneous production of G1 and G2 and their ratios are given.
- UM 2 : Simultaneous production of G1, G2 and G12. Respective ratios are given
- UM 3 : Simultaneous production of G1, G2 and G21.Respective ratios are given.

As it was mentioned above we limit the representation of the operating modes and subsequent of the user modes to the qualitative aspects: the ratios are left aside.

5.1.2 Petri Net specification

We have chosen a hierarchy of high level Petri nets for the modelling of the command. This work has been carried out at LAIL, Ecole Centrale, Lille, France.
* Operating Sequences modelling overall coordination
* Transport Command modelling resource synchronization
* Resource command modelling the behaviour of the production resources.

<u>Operating Sequences</u>

The generation of the Operating Sequences (OS) starts from a purely functional specification of the production process (Figures 20, 21) The operational aspect is introduced specifying the workstations where the operations are executed and the input and output buffers for each product. The accessibility relationships between these entities represent constraints to be respected.
The transitions of the Operating Sequences model either a transfer or a transformation,i.e., either a change of state or of location of the product, whereas the places define a state and a location for an object, i.e. the product modelled by

the token. The transport transitions translate the accessibility relationships between the components of the installation.

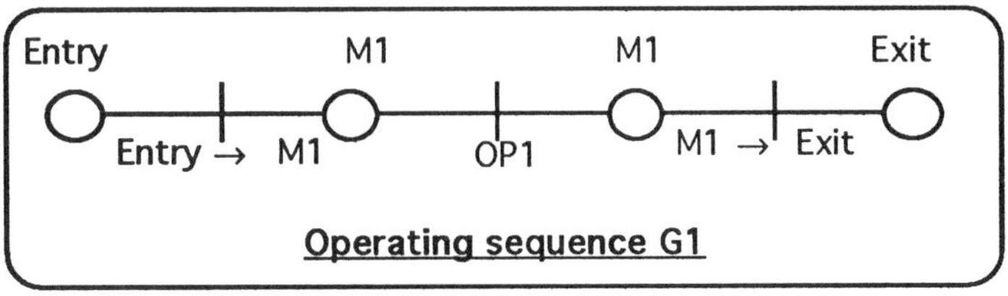

Fig 20 G1 operations

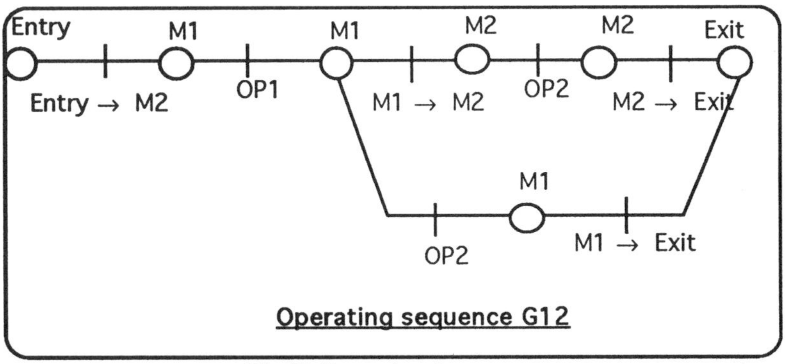

Fig 21 Operating Sequences

The command of the transport system synchronizes the production resources and its model should represent the entire flexibility of transports (figure 22). The flexibility results from the various accessibility relationships between production resources, which are the issue of direct accessibility relationships between characteristic locations. As an example we show PN model of the transfer from the input buffer to M1.

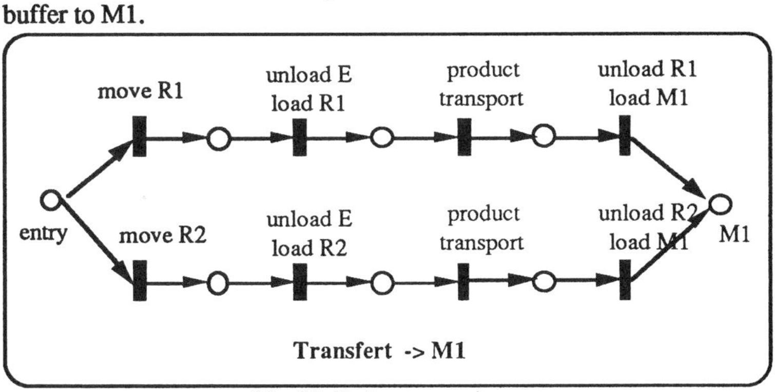

Fig 22 Robot operations

Only at this level of the command hierarchy the transportation tool performing the transfer is indicated.

The command models of the production resources

These models are the lowest level of the command hierarchy. They model the behaviour of the production resources and we could define a library for typical resources. Figure 23 shows the Petri nets of a robot and a machine.

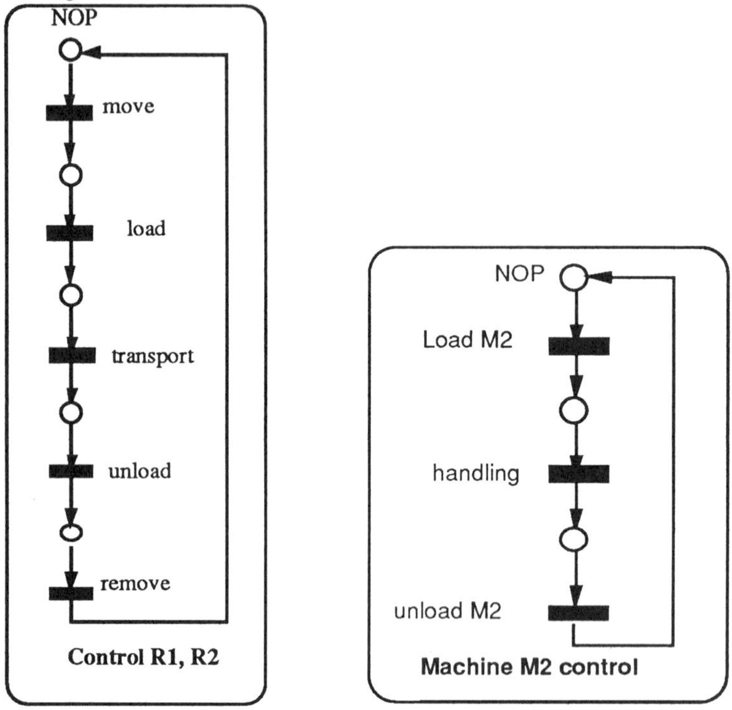

Fig 23 Robot and machine low level control

We now try to specify the operating modes and integrate to the control partially described in this part.

Operating modes specification

Flexible Manufacturing Systems are difficult to characterise by a steady state description, since the observation horizon is too short for the deduction of an asymptotic behaviour. The point in which the systems operates (operating point) is subject to current modifications. These changes can be due to a qualitative or quantitative modification of the required production or to a deterioration of the system performances.

An operating point is thus a certain type of production defined by the nature of the required products and their quantity.

The associated operating mode is the qualitative component of an operating point given by the different types of products to manufacture.

The deterioration of the performances changes the operating point if only the quantity of the products is touched, but if degradation inhibits the fabrication of a certain type of product the operating mode changes.

Thus the operating modes combine the top-down and bottom up characteristics of the information flow issuing from the conducting of the system and the monitoring (Kermad 1992). The operating mode management is modelled by StateCharts.

4.2. The Model of the Operating Modes

The model of the operating modes is hierarchical and its hierarchy is oriented on the one specified for the command model. We have to integrate into the model the fourth level which corresponds to the user modes.
An operating sequence is composed of transports and transformations in one word operations and one operation can be performed by several resources.

The StateCharts based specification

For the specification of the model we use StateCharts since they allow a fine representation of hierarchy, parallelism and the real-time reactive nature of the operating mode management.The functionnal aspects will be dealt with later.
The complete specification has resulted in about sixty charts. For conciseness, we will give a partial view of the hierarchy presenting the main charts.

- The General Supervision Monitor chart

In this chart (figure 24) the monitors of entities are composed in parallel, showing the fact that in reality they are considered independant.

Each entity has to be described to show its behavior in case of default, changing of user modes, freeze, thaw,... The descending or ascending propagation of the effects of theses changes are through the hierarchy. When the name of a state begins with an "@" this means that the chart has been developped by another chart.

- The ACTIVE_USER_MODE and OPERATING_SEQUENCE chart are presented in figures 25 - 27.

- The FEASIBILITY chart propagates ascendantly through the hierarchy (resources, operations, operating sequences, user modes) the defaults on resources that could happen.

- The FREEZE_THAW_USER_MODE deals with freeze and thaw states of user modes.

The top componnent represents the general observable behavior of the charts containing information for other charts that could communicate with it.

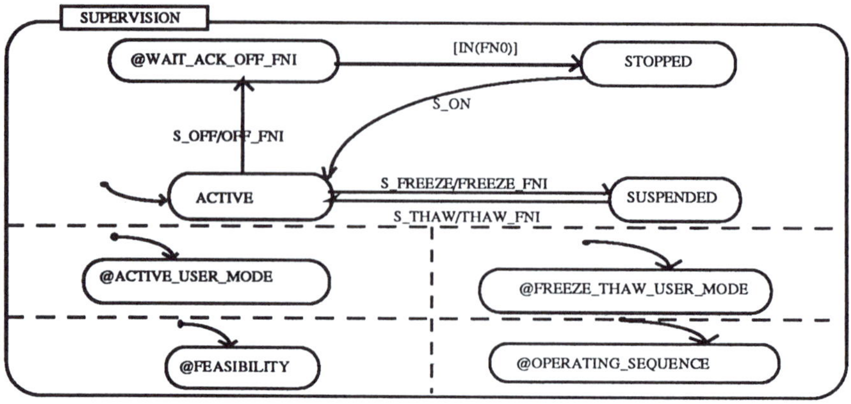

Fig 24 Supervision statechart

The Operation Sequence Monitor chart :

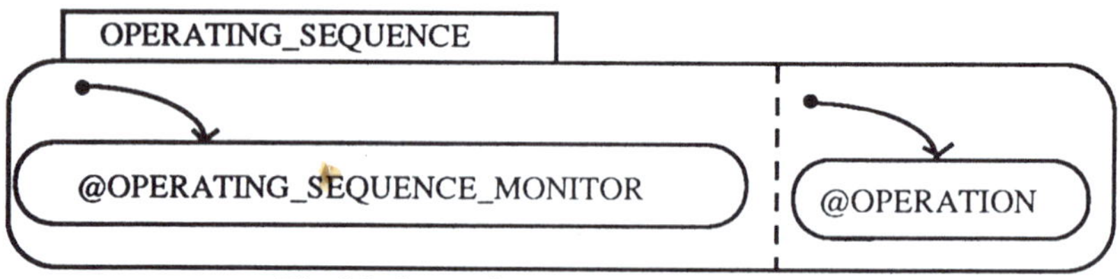

Fig 25 Operating_sequence statecharts

The figure 25 illustrates that the operations monitor is relevant only at the Operating sequences monitor level. The Operating Sequence Monitor subchart is described in fig 28, the other component is below.

- The Operation Monitor chart :

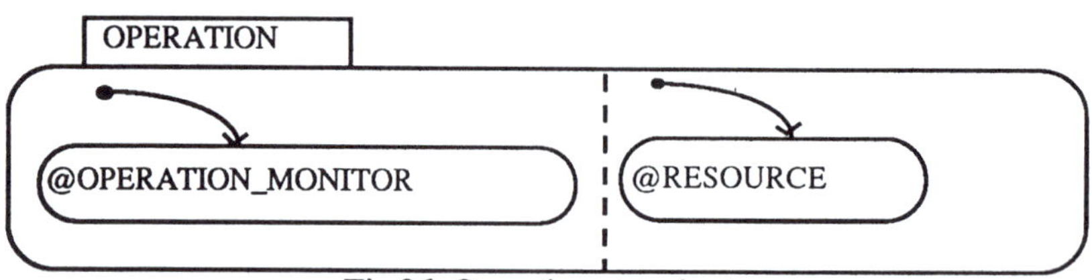

Fig 26 Operation statechart

The figure 26 illustrates that the resources monitor is relevant only at the Operation monitor level. The OPERATION MONITOR deals with the different states of an operation .

- The User Mode chart

The figure 27 describes the different user modes within which the cell can be. If the system changes its modes (descending event : F21) the change may not be done immediately for some cases, e.g. machining of a piece must be achieved. For this purpose, transition states (Tij) have been added to represent the waiting for acknowledge of terminaison. They also allow to launch the corresponding operating sequences.

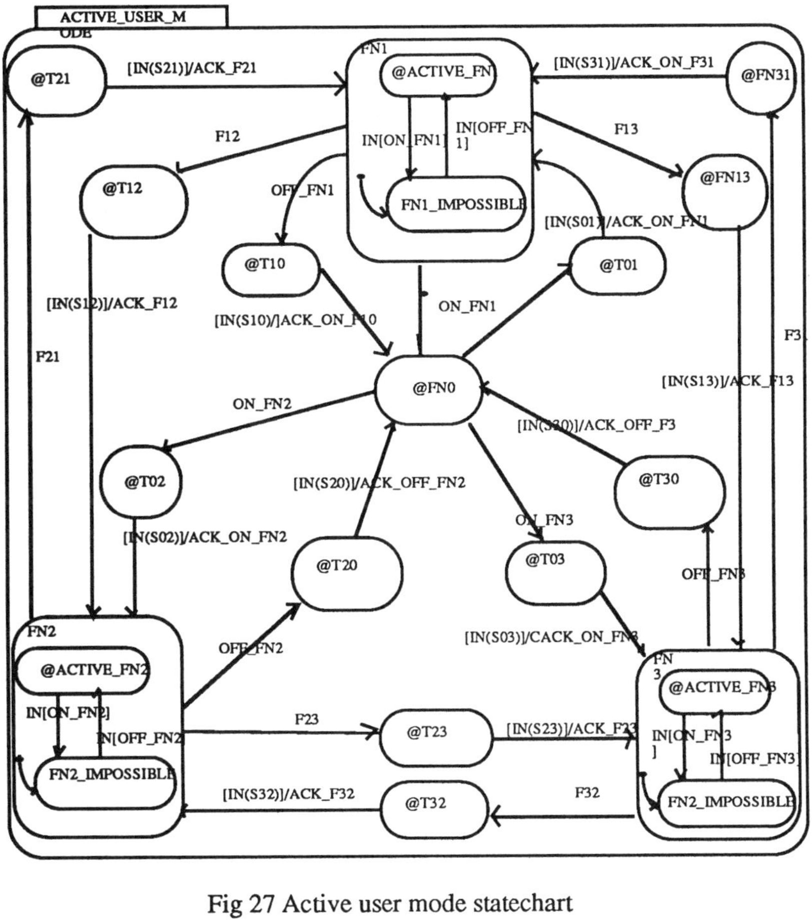

Fig 27 Active user mode statechart

*** The Operating Sequence Monitor chart**

Like other Monitors, figure 28 expresses the different states an operating sequence can be.

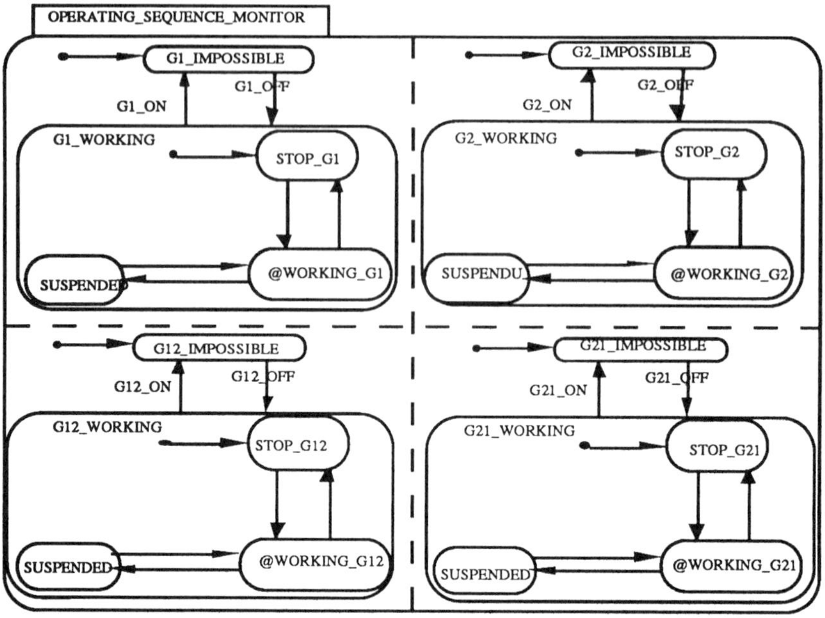

Fig 28 Three levels of abstraction for operating sequences

*** Validation**

The validation is being carried out through simulation. There are two simulation modes: batch and interactive. The batch simulation was carried out in our case in order to run all possible scenaris. This is done through the SCL simulation language, which enables to enter all possible execution specified by the user, timing states, variables values, statistical analysis etc ...

This step will give diagnostic on the correctness of the specification and to evaluate all operating modes, hence the specification can be improved until simulation results meet the requirement and, at this step, the design and implementation steps can be tackled.

Integrating the functionnal aspects

When specifying a reactive system, there are some entities to be expressed. The most common ones are :

- data (system definition),
- event (system change),
- data transformation (function),
- event transformation,
- time (related to all four aspects).

Each method focuses generally on some specific elements. For instance, Petri Net based methods, which are very adequate for discrete event systems, taking into account events and event transformation. SA and SADT methods are more

oriented to data transformation through DFD's and actigrams respectively,. SA-RT is one of the first approach to deal with all aspects in reactive systems specification and analysis.
Object oriented analysis approaches makes use of such methods by encapsulating data and data transformation into an object.
In the present problem in specifying operating modes of a reactive system, we must take into account explicity the functionnal aspect. The available model such as activity charts can be used in a coherent way with StateCharts (see part 2.2).
This part is limited to a preliminary approach. The top activitychart is given in Figure 29.

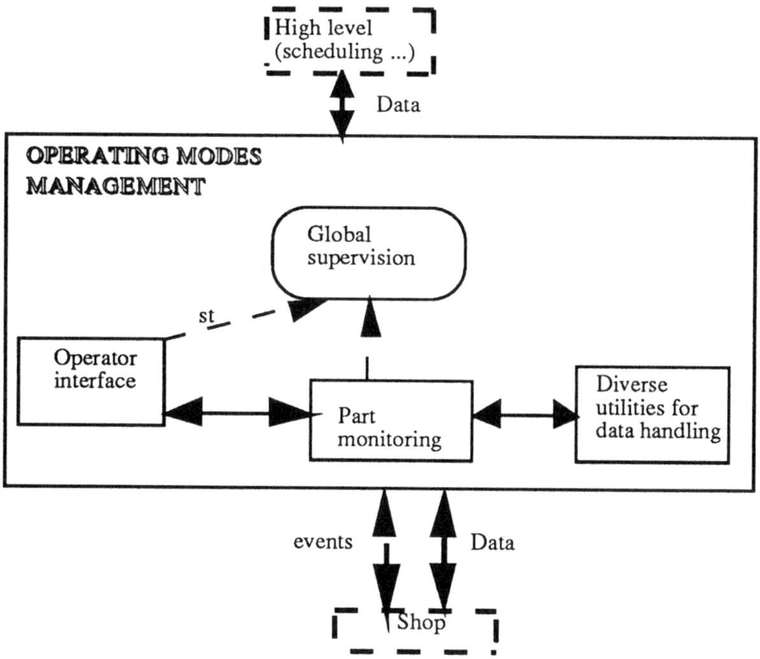

Fig 29 Operating modes activity chart

The data exchange bertween the activities involves all the information concerning the parts, tools and machines; the control exchange between activities and the control statechart deals with all the events generated by these activities. The control statechart can be decomposed as shown in Fig 30:

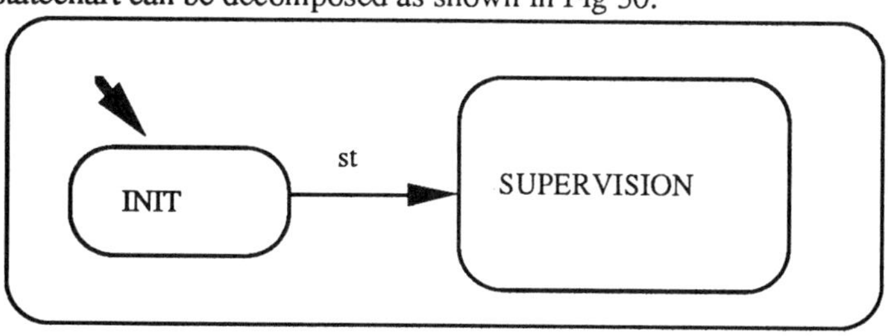

Fig 30 control statechart for operating modes

It was shown through the case study how to combine Petri nets and statecharts with respect to the synchrony hypothesis.

VI Research issues

Problems with actual methods :

DFD based methods are mostly used in complex projects as an alternative. A tradeoff must be found.

The main drawback of these methods is the lack of formal basis. This is an important issue when dealing with reactive systems. In the model presented in Section 2 , an attempt is made to introduce Petri nets as a language for specifying control structure; an interpreted extension of Petri net is proposed where real-time primitive (some based on POSIX standard) is proposed.

Undoubtedly, to implement a general system successfully the first thing to begin with is to establish a suitable model which can reflect the behavior of the system as much as possible. In this part of our work we propose a model for reactive system for manufactiring systems based on a modified version of Real-Time Structured Analysis (SA-RT) method and an extension of Petri Nets (PNs).

These tools, though useful, are not enough when control transformation deals with several tasks evolving concurrncy, synchronisation, etc. Additionally, in some cases the size of these representations may be very large and therefore their maintenance becomes impractical. For these reasons, we found that PNs are an appropriate tool in this respect owing to the following advantages:

-PNs supports parallelism, concurrency, and synchronism.

-the mathematical foundation, especially the concepts of tokens and markings, provides an excellent means of analysing the behavioral properties of the modeled system (deadlock, reachability, liveness, etc.).

In fact, for example, using deadlock detection, the analyst can determine what event sequences cause a system to loop or hang. Also, using reachability tests, the analyst can locate all possible event sequences that lead to a specific behavior in the system.

The differnce between these tools - PNs and STD - arises from the following facts:

-In PNs, the conept of tokens (marking) allows the execution of the model, and the simulation of its dynamic behavior.

-Abbreviations and extensions to PNs permit more condensed and powerful expressive specifications to be written.

Finally, one of the powerful argument for SA is the widespread availability of automated tools to support the method, PNs use very close notations as DFDs, so integrating PN modules in these automated tools will be carried out without any difficulty

Mapping from SA-RT/PN to Objects :

The general strategy is to use the SA-RT/PN specification: context diagrams, DFDs, PNs to identify objects (or classes) [14,22].

To identify the object (or class) of each:

-terminators in the context diagram.

-data stores and associated data transformations in a DFD.

-data flows in a DFD. which may require operations to manipulate the data.
Terminators can be easily mapped to objects and classes

Other work : some work is carried out in [18,20]; the main directions are
- Integrating timing specification in the proposed methods and
- Formalising SA-RT through VDM or Z. Formalising consists in translating a specification with SA-RT and translating it model with a formal semantics as VDM.

Acknowledgements : The author is indebted to the colleagues at LAAS-CNRS Toulouse, ENITarbes and ECLille for their diverse contributions to this work.

References

[1] C. Ausfelder et al : Combining Petri nets and statecharts to specify operating modes. 23 rd IEEE SMC (Science and cybernetics Conference, Lille, Sept 94.

[2] G. Berry, P. Couronne, G. Gonthier : Synchronous programming of reactive systems : an introduction to ESTEREL. *INRIA report No647, March 87.*

[3] Carvalho S., Noyes D., Sahraoui A.E.K, Soler F., 1991, Markov Statecharts, *3rd Int'l conference on informatics, cybernetics and systems, Baden-Baden, Germany.*

[4] Courvoisier M. et al, 1983, The SECOIA project, *LAAS-CNRS Research report, Toulouse.*

[5] R. David, H. Alla : Petri nets for the modelling of dynamic systems. *Automatica*, Vol 30, No.2, 1994, pp175-202.

[6]T. De Marco : Structured Analysis and System Specification. *Prentice-Hall, 1979.*

[7] D. Harel : Statecharts, a formalism for specifying complex systems, *Journal of science of computer programming, Vol. 8, No 1, 1987, pp 231-274.*

[8] D. Harel, A. Pnueli, J.P. Schmidt, R. Sherman : On the formal semantics of Statecharts,*IEEE LICS, Logic in computer science, 1987, pp 54-64.*

[9] D. Harel et al : Statemate, a working environment for the development of complex reactive systems. *IEEE Trans. Soft.Eng., Vol 16, No 4, April 90, pp 403-413.*

[10] Hatley D.J, Pirbhai I, 1987, Strategies for real-time specification, *Dorset House Publishing.*

[11] i-Logics, 1991, Statemate user manual, *Burlington, Mass.*

[12] H.R. Jorysz, F. Vernadat : CIM-OSA, total entreprise modelling and functional view. *Int. J. of CIM, 1990, Vol. 3, No 2-3.*

[13] L. Marce, P. Le Parc : Defining the semantics of Grafcet with synchronous languages. *IFAC-IFIP Workshop of real-time programming*, Bruges, June 92.

[14] [Mel-92] S. Mellor, S. Shlaer : Object life cycles : modelling the world into states. *Prentice Hall, 1992*

[15] A. Pnueli, Z. Manna : The temporal logic of reactive and concurrent systems. *Springer-Verlag, 1991.*

[16] A.E.K. Sahraoui, L. Gilhodes : Petri nets, temporal logic and Statecharts : a comparative study for specifying discrete event controllers, *European Control Conference ECC-91, Grenoble, 1991, Edi Hermes.*

[17] A.E.K. Sahraoui : An approach for monitoring discrete event systems, *AFCET APII (Automatique, Productique et Informatique Industrielle) Journal, Edi Dunod, Vol 26, 1992,No.2, pp 91-106*

[18] A.E.K. Sahraoui, N. Ould-Kaddour : A critical study on SA-RT and Statecharts methods to specify reactive systems. *Preprints of IFAC Congress, Sidney, Australia, July 93.*

[19] A.E.K Sahraoui, S. Brnusak : statecharts and its applications. Journal of advanced technics, Vol 1, No1, Dec 93.p1-53

[20] A.E.K Sahraoui : Specifying reactive complex systems : control and monitoring. Habilitation, LAAS technical report 94078, Feb 94, (*In French*).

[21] R. Valette and al : A Petri Net programmable logic controller. *Computer application in production engineering, CAPE'83.*

[22] N. Walters : Using Harel Statecharts to model object oriented behaviour. *ACM SIGSOFT, Vol 17, No4, Oct 92.*

[23] P. Ward, S.J. Mellor : Structured Development for Real-Times systems. *Yourdon Press, 1987.*

[24] P. Ward : The transformation schema : Control and timing. *IEEE Trans Soft.Eng,Vol-12, No2, Feb 86.*

12

Supervisory Control Specification and Synthesis

Mohsen A. Jafari
Department of Industrial Engineering
Rutgers University
P.O. Box 909
Piscataway, NJ 08855
email: jafari@gandalf.rutgers.edu

Abstract

The objective of this chapter is to give an insight to the problems of specification and synthesis of supervisory controller for manufacturing systems. The current practice in industry ranges from the low level ladder logic programming of PLCs to the high level graphical models which could automatically be translated into an executable code. For example, the use of GRAFCET and Sequential Function Chart to program PLCs have grown substantially in recent years. A main problem with all the existing models is that they do not provide the control engineer any means, other than simulation, of analyzing the functional and performance behavior of the supervised system. Also, with growing interest in lean manufacturing, there seems to be an increasing need for some formal methodology which takes the system requirements specification as an input and gives the supervisory control as its output. The existence of such a formal and analytical methodology would minimize the need for debugging and verification of the supervisory control implementation and at the same time would substantially reduce the time and cost of its development. This concept exists in theory but has very limited applications. In this chapter, we plan to explore some of this theory.

1.0 Introduction

The main theme of this chapter is to discuss the theory and the practice of specification and synthesis of supervisory controller for discrete event dynamic systems (DEDS), in general, and for manufacturing systems, in particular. Since the supervisory controller is eventually implemented as a computer program, one may find some similarities between this subject and the software requirements specification and synthesis. The general area of software requirements specification has been in existence for a number of years and many related articles have appeared in the literature mostly in the software engineering literature. In particular, there have been a number of related articles for process control systems (Cristian [1984], Jaffe et al. [1991], Nicollin et al. [1992], and Shaw [1992]). Generally speaking, these works employ finite automata to specify the system requirements. Some works suggest methodology for compiling the real time specification into a finite automata while others demonstrate the use of finite automata to represent concurrent real time processes. Some works also attempt to analyze the properties of the specified system, e.g., if the specified model follows some correct behavior.

The term *supervisory control* is somewhat loosely defined in the literature. In some occasions, it is mixed with such terms as *sequential control or coordination*, and *real time process control.* To understand this confusion, we note that in practice it is possible, but perhaps not efficient, to perform all these functions in one unit (e.g., a supervisory PLC). Here we will distinguish between the supervisory control and the real time process control functions. The real time process control function, by our definition, deals with local control of processes which are usually continuous in nature. The supervisory control function, however, deals with the state transitions of the individual processes as well as the coordination between these processes. One might say that the supervisory controller has a discrete view of the system, whereas, the real time process controller has a continuous view of the system. Supervisory control can be closed or open loop. In the case of the closed loop supervisory control the control actions are influenced by the input from the sensory devices monitoring the undergoing processes in the plant.

If we define a spectrum for the specification of supervisory controller such that a natural language based specification falls in the high extreme and machine level language in the low extreme of this spectrum, then the finite automata based specification is somewhere in the middle. Close to the lower end of the spectrum is ladder logic diagrams that have been commonly used in industry for programming the Programmable Logic Controllers (PLCs). These programs can be generated from a higher specification language, e.g, switching tables, state charts, or some type of natural language specification (see Bollinger and Duffie [1988]). More recently, GRAFCET has also been added to the list of the specification models from which one can generate PLC programs.

Finite automata are based on finite state machines and, therefore, are only a special case of Petri nets. Thus, the Petri nets provide a more general class of models for the specification of supervisory controller. As one might expect, Petri nets also fall somewhere near the middle of our spectrum. In addition to the original Petri nets (Petri [1962]), there are several extensions and variations of Petri nets some of which are intended for the specification of large and complex systems. An extension of Petri nets, namely, colored Petri net (Jensen [1983], Jafari [1992]) is specifically intended to model large scale systems by taking advantage of the various similarities that exists in the system.

There is a main body of research in the theory of supervisory control based on finite automata pioneered by Ramadge and Wonham [1987a,b]. The theory is mostly developed around the formal languages generated by finite automata. Though, the Petri net languages have been studied in great depth by Hack [1976] and later by others, the theory of supervisory controller based on Petri net languages is not fully developed. Generally speaking, the theory of formal languages was originally developed in computer science (see Hopcroft and Ullman [1979]). Later, it found its way to Petri nets and supervisory control formalism through the pioneering works of Hack [1976] and Ramadge and Wonham [1987a,b], respectively. Loosely speaking, a language generated by a Petri net or a finite automata is the set of all sequences of events that could be marked by that Petri net or finite state machine. An event can be representative of a single transition or a number of transitions. This Petri net or state machine could be the model of a plant or the supervisory controller for the plant. Our interest in the formal languages lie in the fact they provide a vehicle for specifying the supervisory controller. In other words, the designer of the control system may want to make sure that the plant or the system under control follows a sequence of actions according to a string in a specified language. On the other hand, one may want to make sure that the plant never reaches to some specific states (so called, *state avoidance problem,* Ramadge and Wonham [1987a]) or some events will never occur (so called, *string avoidance problem*). In this latter case, the language generated by the supervised plant is restricted to some specific language.

To better understand the issues involved in the specification and synthesis of supervisory controller, we shall first make an introductory discussion in section two on DEDS and their specification and analysis. The discussion on finite automata based supervisory controller will then follow in section three. In this section, we shall also give a brief discussion on finite automata and formal languages.

2.0 Discrete Event Dynamic Systems - Control Specification and Synthesis

Ostroff [1989] characterizes discrete event systems with their states and events, where events occur at discrete times and states have discrete values. The undergoing processes in the system are event-driven rather than clock-driven, have their own internal dynamic behavior, interact with the environment, operate concurrently, and communicate with each other. Typical examples of these systems are manufacturing systems, communication systems, transportation systems, etc.

There are usually functional or behavioral requirements which have to be satisfied for a correct (proper) and desirable operation of the system. These terms are defined loosely here. For example, filling a can which is moving on a conveyor, with some food ingredients must initiate and terminate at some specific time epochs. This is a required behavior in order for the system to produce an acceptable product. Another example could be that in a heat chamber or in an oven where some type of food product is being cooked, the temperature must remain within some pre-specified range, otherwise, the product will not meet the quality requirements. There are also performance requirements which have to be met. For example, there may be some minimum number of food packages that the above food packaging line must produce daily, or it may be required that the filling equipment are utilized at least for some specified proportion of time.

Some of the behavioral and performance requirements are met in the local control level, whereas, the others have to be met under the supervision of the supervisory controller. It is important that the closed loop behavior of the system (i.e., the plant) satisfies all the behavioral and performance requirements. In formal terms, this proposition can be written as

$$P \text{ and } C \Rightarrow S_P \tag{1}$$

where P is the uncontrolled behavior of the plant, C is the supervisor, and S_P is the specification of the plant operation. A main theoretical and practical issue here is to synthesis C such that relation (1) is satisfied. The common approach in the literature is to define S_P only in terms of behavioral requirements. But as we discussed above, performance requirements must also be included in the specification. For large scale and complex systems, it may be quite cumbersome to accommodate both behavioral and performance requirements in one step. As to how to solve this problem is yet a research issue.

In the above relation, the plant behavior P, the controller C, and the specification S_P must be expressed using some models. One paradigm could be to express P and

C using extended state machines and to use temporal logic to express S_P (Ostroff [1989]). Temporal logic is a branch of mathematical logic which also includes time. This paradigm lacks any algorithmic approach to synthesize the controller. However, temporal logic can be used to verify that S_P is satisfied by the plant-controller combination. *P* and *C* can also be specified using Petri nets. Another paradigm could be to express S_P in some high level model such as IDEF0 (Boucher and Jafari [1992] and Jafari and Boucher [1994]) and use Petri nets to express the plant and/or controller behavior. In this paradigm, the specification (e.g., IDEF0) is transformed into a Petri net which is then translated to a computer executable code (e.g., ladder diagrams for PLCs). Yet, in another paradigm one may just start with Petri nets and synthesize the controller using some bottom up or top down procedures (Jeng [1992], Zhou et al. [1991,1992], Darabi and Jafari [1994]). The advantage of using Petri nets is that some of the behavioral and performance properties of the system can be analyzed using the theory of Petri nets. In particular, the synthesis techniques ensure that some properties are maintained throughout the design process. The problem is that these synthesis techniques are very restricted and also that very few properties can be tested. In the case of using high level specification models, such as IDEF0, the specification is informal, and the analysis may lead to the study of the reachability set whose size increases exponentially with the size of the system. The same problem also exists when the discrete event system theory of Ramadge and Wonham [1987a,b] is used for the synthesis of the supervisory controller. In this paradigm, the specification is given using formal languages. The plant and the controller models are finite automata (state machines). Though, this theory seems quite attractive for small scale systems, it is yet to be explored for systems such as manufacturing systems where the number of states could be quite large. In section 3 we will concentrate on this theory. In section 4, we shall briefly describe some of the underlying theory of the supervisory control specification and synthesis based on Petri nets. Section 5 will address some future research problems.

3.0 Supervisory Control of Discrete Event Dynamic Systems - Finite Automata Approach

In this section, we shall present the theoretic aspects of supervisory control specification and synthesis. The discussion will center around the discrete event system theory pioneered by Ramadge and Wonham [1987a,b]. Since, the theory is heavily dependent on the theory of formal languages, we shall first give a brief introduction to this subject.

3.1 An Overview of Formal Languages and Finite Automata Theory

A *symbol* σ is an abstract entity that represents an event or a state transition taking place in a system. A *string s* is a finite sequence of symbols representing a sequence of events that take place in a system. The *length* $|s|$ of a string is the number of symbols in that string. The *empty* string λ is the string consisting of zero symbols. A *prefix* of a string is any number of leading symbols of that string. An *alphabet* Σ is a finite set of symbols. A *language L* is a set of strings of symbols from some alphabet Σ. Σ^* is the set of all strings over the alphabet Σ. $\bar{L}$ is the prefix closure of L, that is, it includes all the prefixes of the strings in L. For example, for $\Sigma=\{\alpha\}$, $\Sigma^*=\{\lambda,\alpha,\alpha\alpha,\alpha\alpha\alpha,\alpha\alpha\alpha\alpha,...\}$ and for $\Sigma=\{0,1\}$, $\Sigma^*=\{\lambda,0,1,00,01,10,11,000,001,...\}$.

A finite automaton is formally defined by a five-tuple $G = (Q,\Sigma,\delta,q_0,Q_m)$ where Q is the set of states q, Σ is the finite set of input alphabets or set of output symbols σ, $\delta : \Sigma \bullet Q \to Q$ is the transition function, $q_0 \in Q$ is the initial state and $Q_m \subset Q$ is the set of marked states. In general, δ is only a partial function, meaning that, for each fixed $q \in Q$, $\delta(\sigma,q)$ is defined only for some subset $\Sigma(q) \subset \Sigma$ that depends on q. G is equivalent to a directed graph with node set Q and an edge $q \to q'$ labeled σ for each triple (σ,q,q'), s.t. $q'=\delta(\sigma,q)$. Such an edge is called an event or state transition. Events are spontaneous, asynchronous (no reference to a clock) and instantaneous.

We denote by $L(G)$ the language generated by the automaton G. $L(G)$ is defined by

$$L(G) = \{s | s \in \Sigma^*,\ \delta(x_0,s) \text{ is defined}\}$$

Thus, $L(G)$ is the set of all possible sequence of events which take the initial state to some reachable state. We also denote by $L_m(G)$ the language accepted (or marked) by G. $L_m(G)$ is defined by

$$L_m(G) = \{s | s \in \Sigma^*, \delta(x_0,s) \in Q_m\}.$$

Only those sequence of events that take the initial state to some final state are considered as part of this set. These sequence of events could correspond to a sequence of elementary tasks leading to a job completion. One could interpret the strings in $L_m(G)$ as being recorded or marked for some specific reason. There are two types of finite automata: *Non-deterministic* (NFA) and *Deterministic* (*DFA*). In deterministic automaton, there is at most one transition from a state with the same symbol. In non-deterministic automaton, there can be more than one transition from a state with the same symbol. In fact, the automaton defined above is the deterministic one. In the non-deterministic case, the transition function δ maps into a set of states.

A finite automaton is said to be *trim* if all the states can be reached from the initial state (*accessible automaton*) and all the final states can be reached from all the other states (*co-accessible*). A trim generator is therefore both accessible and co-accessible. For the trim automaton, we have

$$\bar{L}_m(G) = L(G)$$

The reason is obvious: Any state can reach any final state. Therefore, a string of events leading to any state is a prefix to a marked string.

The languages accepted or generated by a finite automaton are easily described by *regular expressions*. Let L_1 and L_2 be the set of strings from Σ^*. The *concatenation* of L_1 and L_2, denoted by L_1L_2, is formed by choosing a string from L_1 followed by a string from L_2, in all possible combinations. The *closure* of L, shown by L^* is $\bigcup_{i=0}^{\infty} L^i$. That is, it has the strings constructed by concatenating any number of strings from L. *Regular expressions* over Σ are defined recursively as follows:

1) $\varnothing$ is a regular expressions and denotes the empty set.
2) λ is a regular expression, it represents an empty input.
3) For $a \in \Sigma$, a is a regular expression.
4) If r and s are regular expressions, then $(r+s)$, rs, r^* are also regular expressions.

An important result shown by Hopcroft and Ullman [1979] is that the language accepted or generated by a finite automaton is a regular language. They show that for every regular expression one can always find a non-deterministic finite automaton with λ-transitions which accepts or marks that expression. The proof is by the method of construction following an induction on the number of operators (union, concatenation, or closure) in the expression. To better understand this concept, we demonstrate an example from Hopcroft and Ullman [1979].

Example I

Construct an NFA for the regular expression $\alpha\beta^*+\beta$.

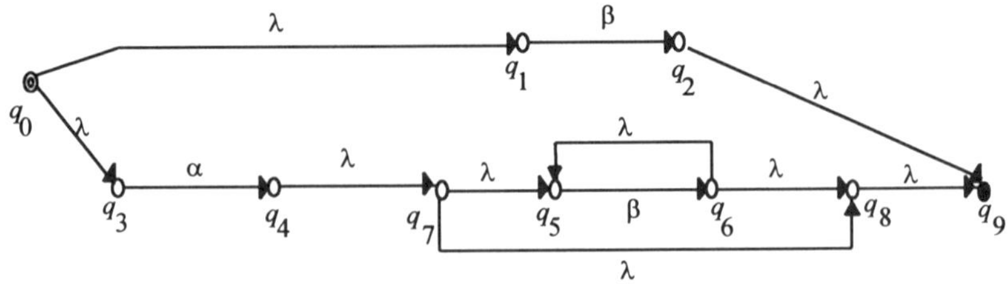

Figure 1: A Non-Deterministic Finite Automata

In Figure 1, the segment q_1q_2 corresponds to expressions β, q_3q_4 to expression α, everything between q_7q_8 to expression β^*. All the other connections are used to connect the above segments. We note that these connections are labeled λ, thus not affecting the language.

If L is a language accepted or marked by a DFA, then L is a regular language. The proof is on the fact that a path from the initial state to a final state can be described by a recursive relationship containing regular operators, union, concatenation, and closure. For every NFA we can construct a DFA in which each state corresponds to a group of states in the NFA. These two finite automata will have the same language. Also, for any NFA with λ-transition we can find an NFA with no λ-transition which accepts the same language.

Equivalence of Two Finite Automata

As it was shown by Hopcroft and Ullman [1979] there is an algorithm to determine if two finite automata are equivalent (i.e., they accept the same language). This is an important result needed to compute an efficient supervisory controller with a reduced size. With an arbitrary language L, we associate a natural equivalence relation R_L; denoted by $x\ R_L\ y$, iff for each string s, either both or neither of xs and ys is in L. In the worst case, each string is an equivalent class by itself. The number of equivalent classes (the index) is always finite for the regular language L.

There is also a natural equivalence relation on strings associated with finite automaton. Let $G = (Q, \Sigma, \delta, q_0, Q_m)$ be a DFA. For $x,y \in \Sigma^*$ let $x\ R_M\ y$ iff $\delta(q_0, x) = \delta(q_0, y)$. R_M divides Σ^* into equivalence classes, one class for each state reachable from q_0. Thus, for each state q reachable from q_0, the corresponding equivalence class $E_M(q)$ is the set of all strings which take G from q_0 to q. We note that if $x\ R_M\ y$ *then* $xw\ R_M\ yw\ \forall\ w \in \Sigma^*$. Such an equivalence relation is called right invariant w.r.t. concatenation. To understand the equivalence relations R_M and R_L we give the following example from Hopcroft and Ullman [1970].

Example II

Let $L = \alpha^*\beta\alpha^*$. L is accepted or marked by the DFA shown in Figure 2.

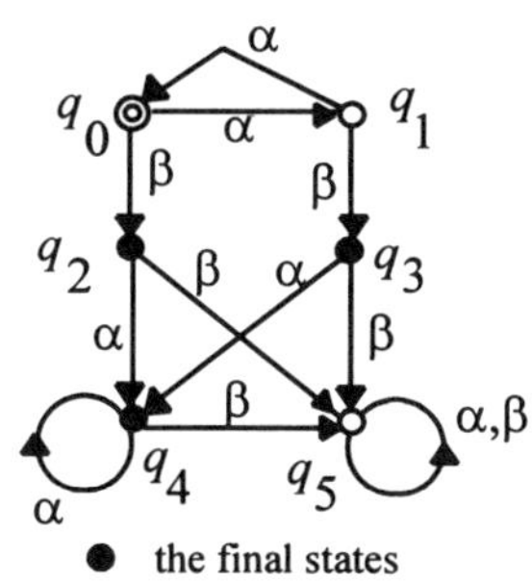

Figure 2: A DFA Marking $L = \alpha^*\beta\alpha^*$

For the relation R_M we have the following classes:

$$C_0 = (\alpha\alpha)^*,\ C_1 = (\alpha\alpha)^*\alpha,\ C_2 = (\alpha\alpha)^*\beta,\ C_3 = (\alpha\alpha)^*\alpha\beta,$$

$$C_4 = \alpha^*\beta\alpha\alpha^*,\ C_5 = \alpha^*\beta\alpha^*\beta(\alpha+\beta)^*.$$

Take for example C_0. This includes all the strings that bring the DFA from q_0 to itself. C_5, for example, includes all the strings that bring the DFA from q_0 to q_5. Note that $L = C_2 \cup C_3 \cup C_4$ as these classes include strings that take the automaton to one of the final states. To find the equivalence classes under relation R_L, we have to find the conditions necessary for either two strings xw and yw to belong to L or not to belong to L. From the above state diagram and also from the language L we can see that exactly one β is admissible in a marked string (leading to a final state) and that it can be proceeded or followed by one or more α's. More than one β in a string will not lead to a marked state. Less than one β in a string is admissible meaning that it is just a prefix. Three conditions exist:

i) x and y each have no β's,
ii) x and y each have one β, or
iii) x and y each have more than one β's.

Three equivalent classes emerge: $\zeta_1 = \alpha^*$, $\zeta_2 = \alpha^*\beta\alpha^*$, $\zeta_3 = \alpha^*\beta\alpha^*\beta(\alpha+\beta)^*$. The last class is basically all the strings that lead to state q_5. From R_L we may construct a DFA, $G' = (Q', \Sigma, \delta', q_0', Q_m')$, shown in Figure 3. There is a state for each equivalence class. The alphabet remains the same as that of G.

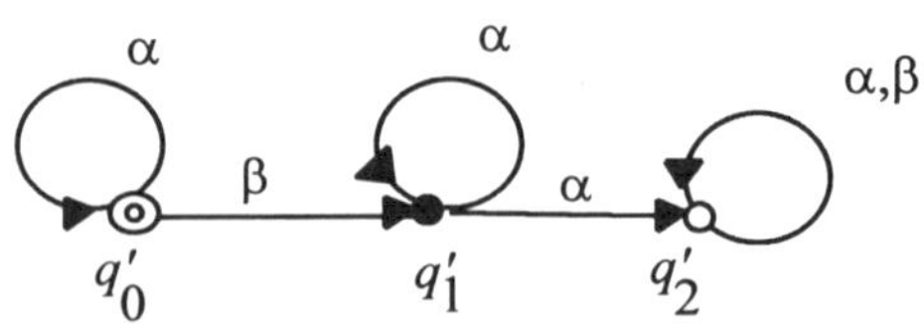

Figure 3: The equivalent DFA of Figure 2

q_0', q_1', q_2', respectively, correspond to $\zeta_1, \zeta_2, \zeta_3$. $\delta'(q_0', \beta) = q_1'$ because q_0' contains strings of form α^* and adding β to these strings gives us strings of the form $\alpha^*\beta$ which happens to be a string in q_1'. $Q_m' = \{q_1'\}$ is defined by the class ζ_2 which is the same as *L*. There is only a unique minimum state DFA for every regular language. Hopcroft and Ullman [1979] present an algorithm to compute this minimum DFA.

Each ζ_i relates to a set of states reachable from the initial state q_0. For example, ζ_1 relates to states q_0 and q_1. The states in each of these sets are equivalent such that given a sequence of events (a string) either the final states are reachable or not reachable regardless of which state is the starting state. Since, there is an equivalent class under relation R_M, ζ_i is union of one or more C_i 's. For the above example, we have $\zeta_1 = C_1 \cup C_0$, $\zeta_2 = C_2 \cup C_3 \cup C_4$, $\zeta_3 = C_5$. When the DFA of Figure 2 is in either of states q_0 or q_1 the DFA of Figure 3 is in state q_0'. Naturally, q_0' is the initial state for the DFA of Figure 3. The final state for this DFA corresponds to all the strings that are in *L*. Therefore, it relates to the final states in the DFA of Figure 2. It is now clear that any string that takes states q_0 or q_1 in the DFA of Figure 2 to its final set of states will also take q_0' in the DFA of Figure 3 to its final state.

Petri Net Languages

State machines are only a special class of Petri nets. Thus, the class of languages generated by Petri nets is more general than the class of languages generated by finite state machines. Every regular language is a Petri net language, but the reverse is not true. That is, the class of Petri net languages also include other types of languages. For a detail discussion on this see Peterson [1980].

3.2 Theory of Supervisory Control

The theory is developed around the following central issue: Given the uncontrolled behavior of the plant, a set of controllable events, and some specification of the desirable behavior, compute a supervisor which is maximally permissive. That is, no events are enforced and maximal flexibility is allowed. The supervisor only enables or disables the controllable events. It is of course possible that no supervisor can be obtained for some desirable behavior specification. Within this theory, the uncontrolled behavior of the plant is given by a finite automaton which is then modified to take into account the controllability of the events. The supervisor is also described by a finite automaton. The combination of the supervisor and the plant automata is a new automata. The theory includes an algorithmic solution to obtain the supervisory controller from this automata.

Controlled Discrete Event Processes: Definition and Terminology

Let G be a finite automata given by $G=(Q,\Sigma,\delta,q_0,Q_m)$. G may be non-deterministic. We shall denote by $L(G)$ and $L_m(G)$, respectively, the language and marked language generated by G. G can be considered the generator for a plant which does not admit external control. $L(G)$ is then interpreted as the open loop process language of the plant, and $L_m(G)$ as the marked language of the open loop process.

To close the loop for the control of the plant, one needs to define some means of enabling or disabling some or all of the events which could take place in the plant. Thus we distinguish between two subsets of Σ : Σ_c and $\Sigma_u = \Sigma - \Sigma_c$. We say an event (σ,q,q') is a controlled event if $\sigma \in \Sigma_c$. Let $\Gamma = \{0,1\}^{\Sigma_c}$ be the set of all binary assignments to the elements of Σ_c . Each assignment $\gamma \in \Gamma$, i.e., $\gamma : \Sigma_c \rightarrow \{0,1\}$ is a control pattern. Event $\sigma \in \Sigma_c$ is said to be enabled by γ if $\gamma(\sigma)=1$ or disabled by γ if $\gamma(\sigma)=0$. For $\sigma \in \Sigma-\Sigma_c$, $\gamma(\sigma)=1$. The *controlled discrete event process* (CDEP) is defined by generator $G_c=(Q,\Gamma\times\Sigma,\delta_c,q_0,Q_m)$ where

$$\delta_C : \Gamma \times \Sigma \times Q \rightarrow Q$$

$s.t.$

$$\delta_C(\gamma, \sigma, q) = \begin{cases} \delta(\sigma,q) & \textit{if } \delta(\sigma,q) \textit{ is defined and } \gamma(\sigma)=1 \\ \textit{undefined} & \textit{otherwise} \end{cases}$$

Ramadge and Wonham[1987a] describe G_c in the following way:

> *For each* $\gamma \in \Gamma$ *there is a generator* $G(\gamma)$ *formed by deleting from the graph of G those events* σ *with* $\gamma(\sigma) = 0$ *The external control action would consist simply in switching the control pattern through a sequence of elements* $\gamma, \gamma', \gamma'', \ldots$ *in* Γ.

G_c is the model of the plant which admits external control on some or all of its activities. $L(G_c)$ is the language generated by the plant if the disabling actions were absent and $L_m(G_c)$ is the uncontrolled marked language of the plant under the same conditions.

Supervisor: Definitions & Terminology

The function of the supervisor is to switch control pattern in such a way that a given CDEP G_c behavior satisfies some constraints. Formally, a supervisor S is a pair $S=(\chi,\phi)$ where $\chi = (X, \Sigma, \xi, x_0, X_m)$ is a deterministic automaton with (possibly infinite) state set X, input alphabet Σ, partial transition function $\xi : \Sigma \times X \rightarrow X$, initial state x_0 , and marker states X_m while ϕ: X $\rightarrow \Gamma$ is a total function that maps supervisor states x into control patterns γ; i.e., $\phi(x):\Sigma \rightarrow \{0,1\}$ with $\phi(x)(\sigma)=1$ $\forall$ $\sigma \in \Sigma - \Sigma_c$. ϕ is called *state feedback map*. Figure 4 gives a schematic of a closed loop system where the plant and the supervisory controller are coupled together.

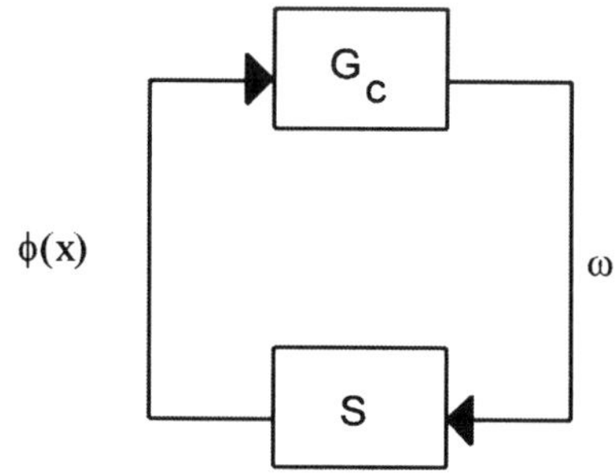

Figure 4: Closed Loop Control - Plant and the supervisory controller coupled

In the above closed system, the state transitions in S are forced by the events (ω) in G_c and G_c is constrained by successive control patterns determined by states of S.

Formally, we define the events in the coupled system by $\xi \times \delta_c$: $\Sigma \times X \times Q \rightarrow X \times Q$ where

$$(\sigma, x, q) \rightarrow (\xi(\sigma, x), (\delta_c(\phi(x), \sigma, q)). \tag{2}$$

The reader notes that this event is defined only if $\xi(\sigma,x)$, $\phi(x)(\sigma)=1$, and $\delta(\sigma,q)$ are defined. That is, by receiving input σ the plant and the supervisor make simultaneous transitions from their current states, x and q respectively, to new states defined by their respective transition functions. Of course, such transitions should have been possible and the event (labeled σ) should not have been disabled by the controller.

Supervised Discrete Event Process (SDEP)

Given G_c and S, we denote by $S|G_c$ the SDEP, such that G_c and S are coupled in the following way:

$$S|G_c = AC(X, Q, \Sigma, \xi, \delta_c, (x_0, q_0), X_m, Q_m)$$

where AC stands for accessible automaton (described in the previous section). That is, we are only interested in those states that are reachable from the initial state. In the above automaton, one basically couples the CDEP with the supervisor in a closed loop manner as shown in Figure 4. $S|G_c$ generates the language $L(S|G_c)$. This is the set of al possible finite sequences of events that can occur when S is coupled with G_c.

There are three languages that together describe the closed loop behavior:

$L(S|G_c)$= the language generated by the plant and supervisor coupled together,
$L_c(S|G_c)$=marked sequences of events of the uncontrolled process language that survive under supervision $= L(S|G_c) \cap L_m(G_c)$,
$L_m(S|G_c)$=marked sequences of events of the coupled system.

These languages are related:

$$L_m(S|G_c) \subseteq L_c(S|G_c) \subseteq L_m(G_c) \qquad (3)$$

We illustrate these results with an example from Ramadge and Wonham [1987a].

Example III

Let $\Sigma = \{\alpha, \beta, \gamma\}$. Suppose that the CDEP for a plant is represented by G_c shown in figure 5. Two supervisors are considered, S_1 and S_2 as shown in Figure 5. The coupling of the states is done according to the conditions associated with relation (2). We have

$$L_m(G_c) = (\alpha\gamma^*\beta)^*$$
$$L(S_1|G_c) = (\alpha\beta)^*(1+\alpha\gamma^*)$$
$$L_c(S_1|G_c) = L(S_1|G_c) \cap L_m(G_c) = (\alpha\beta)^* = L_m(S_1|G_c) \subset L_m(G_c)$$
$$\bar{L}_c(S_1|G_c) = (\alpha\beta)^*(1+\alpha)$$
$$L(S_2|G_c) = (\alpha\gamma)^*(1+(\alpha\gamma^*\beta)^*)$$
$$L_c(S_2|G_c) = (\alpha\gamma^*\beta)^* = L_m(G_c)$$
$$L_m(S_2|G_c) = (\alpha\beta)^*$$

We note that G_c under the supervision of S_1 fails to maintain all of its unsupervised marked strings, whereas, under the supervision of S_2 it keeps all those strings. This is due to the extra transitions of S_2 in state x_2 and also because c=1 in this state. Even if these extra transitions, i.e., α and β were possible in state x_2 of S_1, there would not be any change in $S_1|G_c$ because c=0 in state x_2 of S_1.

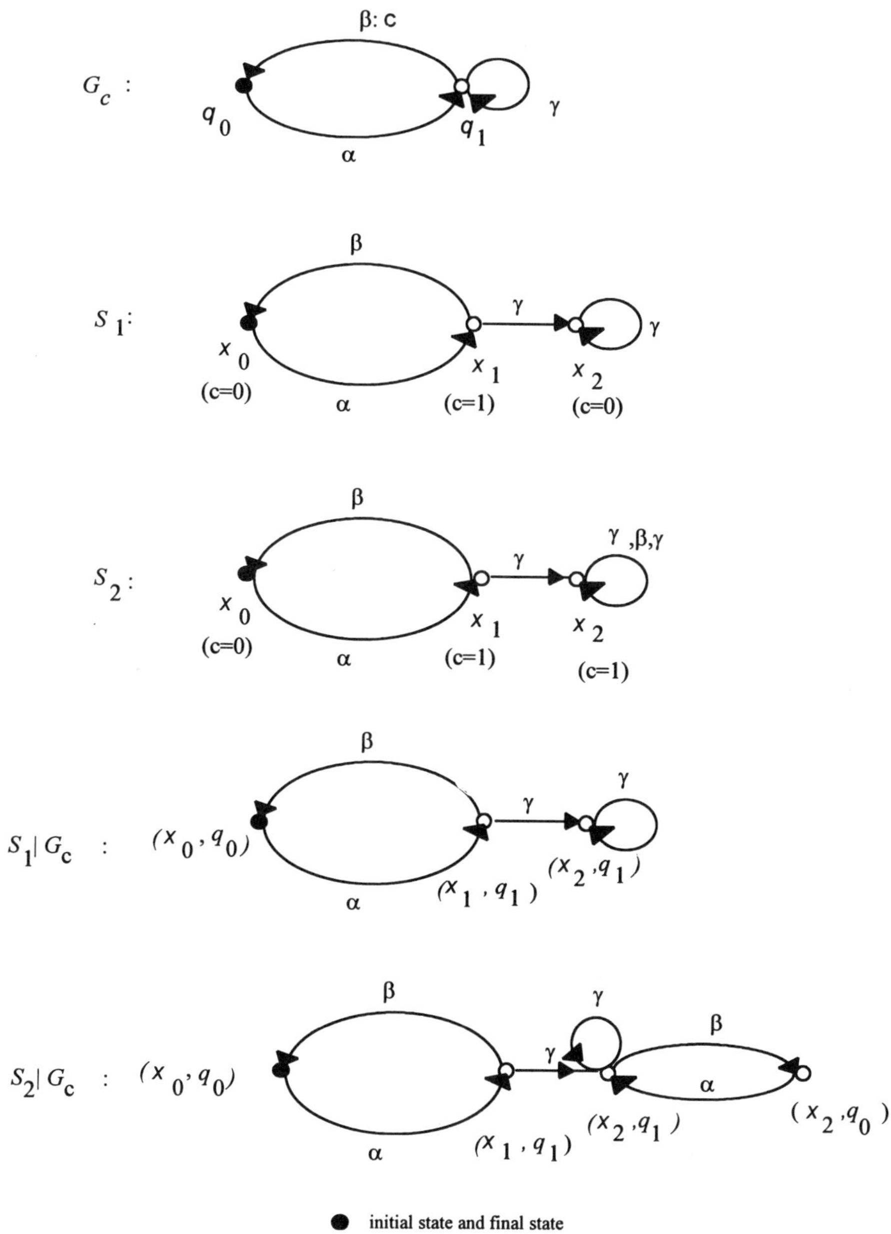

Figure 5: A Plant with two supervisors

To be of a real value, it is important that S is *complete* w.r.t. G_c. That is, if a string s is included in the closed loop behavior and σ is allowed to happen by the supervisor then $s\sigma$ must also be included in the closed loop behavior. Formally, we have:

If $s \in L(S|G_c), s\sigma \in L(G_c)$ and $s \in \Sigma$ $s.t. \phi(\xi(s,x_0))(\sigma) = 1$ then $s\sigma \in L(S|G_c)$.

There are two more desirable properties that Ramadge and Wonham [1987a] associate with a supervisor:

Non-Rejecting Property: $\bar{L}_c(S|G_c) = \bar{L}_m(S|G_c)$

Non-Blocking Property : $\bar{L}_c(S|G_c) = L(S|G_c)$

Non-rejecting property implies that the supervisor does not reject any behavior which leads to a completed (or a recorded) task. Suppose that there is a string $s \in \bar{L}_c(S|G_c)$ but $s \notin \bar{L}_m(S|G_c)$. This means that s can be completed to a task in $L_c(S|G_c)$ but not to a task in $L_m(S|G_c)$. We note that this latter set has all the strings that are marked (or recorded) in the coupled closed loop system, whereas the former set has all the strings that are marked in unsupervised process and are also acceptable supervised behavior. Therefore, the supervisor rejects some of the recorded markings that were originally intended for the plant. Non-blocking property implies that the supervisor does not block a task to be recorded or completed. Suppose that there is a string $s \in L(S|G_c)$ but $s \notin \bar{L}_c(S|G_c)$. This latter condition implies that s cannot be completed to a task in the supervised process. On the other hand, since G_c is trim, we have $L(G_c) = \bar{L}_m(G_c)$. Thus s can be completed to a task or recorded in the unsupervised plant. A supervisor which is complete, non-blocking, and non-rejecting is called *proper*. For the above example, supervisor S_1 is blocking and supervisor S_2 is rejecting. S_2 is rejecting because $S_2|G_c$ only marks $(\alpha\beta)^*$ while the complete set of tasks to be performed is $(\alpha\gamma^*\beta)^*$.

Next we will present, without proofs, some of the main results of Ramadge and Wonham [1987a]. We shall present these results under two main categories of *existence* and *synthesis*. It is assumed that G_c is a trim generator. Prior to representing these results, we shall give a few more definitions: Let $K \subset \Sigma^*$, $L \subset \Sigma^*$ be two arbitrary sublanguages. K is said to be

i) L-closed if $K = \bar{K} \cap L$

ii) controllable if $K \subset L(G_c)$ and $\bar{K}\Sigma_u \cap L(G_c) \subset \bar{K}$.

The second condition in (ii) implies that adding an uncontrollable event to an already physically acceptable behavior does not make the resulting behavior unacceptable.

The existence of Supervisory Controller

Let $K \subset L(G_c)$ *and* $K \neq \varnothing$. There exists a complete supervisor S such that $L(S|G_c) = K$ if and only if K is prefix closed ($K = \bar{K}$) and controllable. The supervisor S prevents the generation of strings in $L(G_c)$ that are not in K while making sure that all strings in K can be generated by the closed loop system.

Let $K \subset L_m(G_c)$ *and* $K \neq \varnothing$. (i) There exists a proper supervisor S such that $L_m(S|G_c) = K$ if and only if K is controllable. (ii) There exists a proper supervisor S such that $L_c(S|G_c) = K$ if and only if K is controllable and $L_m(G_c)$-closed.

The Supervisory Synthesis Problem

Let L_a and L_g be sublanguages of Σ^* such that $\varnothing \neq L_a \subset L_g \subset L_m(G_c)$. L_g can be considered as the legal behavior of the plant and L_a as the minimally acceptable behavior for the plant. Two problems are of interest:

Supervisory Marking Problem (SMP):

To construct a proper supervisor S for G_c such that $L_a \subset L_m(S|G_c) \subset L_g$.

Supervisory Control Problem (SCP):

To construct a proper supervisor S for G_c such that $L_a \subset L_c(S|G_c) \subset L_g$.

If one of the above problems is solvable then the other one is solvable too. It is desirable that the solution to either problem is minimally restrictive, i.e., $L_m(S|G_c)$ and $L_c(S|G_c)$ are as large as possible. To realize this, Ramadge and Wonham [1987a] define for $L \subset L(G_c)$

$$C_G(L) = \{K | K \subset L \; and \; K \; is \; controllable\}$$

$$F_G(L) = \{K | K \subset L \; and \; K = \bar{K} \cap L_m(G_c)\}$$

Both $C_G(L)$ and $F_G(L)$ are non empty and closed under union and each contains a unique supremal element with respect to inclusion. We shall denote by Sup $C_G(L)$ and Sup $F_G(L)$ these supremal elements, respectively. Wonham and Ramadge [1987a,b] characterize the supremal controllable sublanguage Θ of a given language L as the largest fixed point of a certain operator Ω. In the case where the languages involved are regular, they show that the fixed point Θ can be computed through an iterative scheme. Next we shall discuss computation of the supremal language.

Supremal Language

Let $L \subset L(G_c)$ be the *legal* language describing behavior that is both possible and acceptable. We recall that $L(G_c)$ is closed and represents the unsupervised *physically possible* behavior of the CDEP G_c, where G_c is a finite automaton. Let $\Omega : \mathcal{L} \rightarrow \mathcal{L}$ be an operator defined on the set $\mathcal{L}$ of all languages over Σ. For L and $L(G_c)$, $\Omega(K)$, where $K \subset L$, is given by

$$\Omega(K) = L \cap Sup\{T : T \subset \Sigma^*, T = \bar{T} \; and \; T\Sigma_u \cap L(G_c) \subset \bar{K}\} \tag{4}$$

We note that $\Omega(K)$ defines a prefixed closed sublanguage of L which is supremal w.r.t. inclusion, such that for every string $t \in \Omega(K)$ and $\sigma \in \Sigma_u$, such that $t\sigma$ is physically possible, $t\sigma$ is prefix of a string in K. The solution $K=\Omega(K)$ is a fixed point solution of Ω. It is worthwhile to know that Ω is monotone, i.e., if $A \subset B \subset \Sigma^*$ then $\Omega(A) \subset \Omega(B)$.

Wonham and Ramadge [1987b] show that for Θ=Sup $C_G(L)$, $\Theta=\Omega(\Theta)$ and that for every K such that $K=\Omega(K)$, $K \subset \Theta$. That is, Θ is the largest fixed point solution to Ω. To compute Θ, they propose an iterative solution which they prove it converges when the languages L and $L(G_c)$ are regular. In this case, Θ is also a regular language. The iteration works as follows:

Let $K_0 = L$ and $K_{j+1}=\Omega(K_j)$. Then $\Theta = \bigcap_{j=0}^{\infty} K_j$.

We note that K_js computed as such are monotonically decreasing, that is,

$$K_0 \subset K_1 \subset K_2 \subset \ldots$$

In the case of regular languages L and $L(G_c)$, the above sequence eventually converges to a fixed point. The computation of K_js using the above procedure is not straight-forward. Therefore, Wonham and Ramadge [1987b] proposed a more effective solution which we describe next.

Computation of the Supremal Language

Suppose that G_c is a trim automaton defined as before. Let $\Im_j = (X, \Sigma, \xi, x_0, X_m)$ be a trim automaton such that $L_m(\Im_j) = K_j$ where K_j is the sequence of regular languages defined by $K_0 = L$ and $K_{j+1} = \Omega(K_j)$, $j \geq 0$. It is necessary that $\Im_j$ refines G_c, i.e.,

$$\forall s,t \in \bar{L}_m(\Im_j),\ \xi(s,x_0) = \xi(t,x_0)\ \textit{implies that}\ \delta_c(s,q_0) = \delta_c(t,q_0)$$

This means that there is a correspondence between the states of $\Im_j$ and G_c. This correspondence can be actually described by a unique function

$$h: X \rightarrow Q\ \ \textit{satisfying}\ \ h \times \xi(s,x_0) = \delta_c(s,q_0),\ s \in \bar{L}_m(\Im_j)$$

The correspondence between the states in G_c and $\Im_j$ is shown in Figure 6.

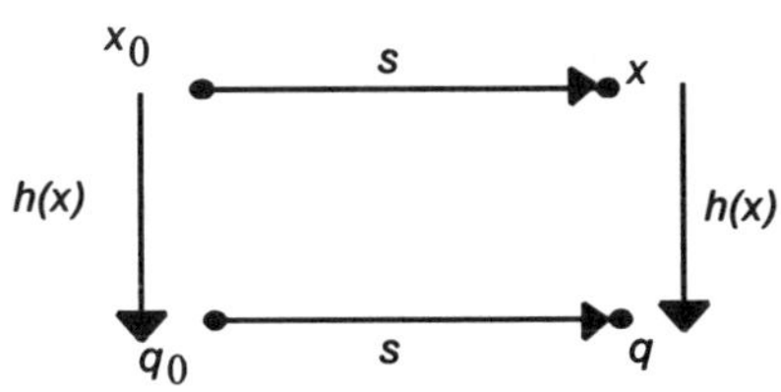

Figure 6: Graphical representation of refinement

In simple terms, refinement means that for any state in $\Im_j$ there is a state in G_c. Wonham and Ramadge [1987b] show that

$$w \in W(K_j)\ \textit{iff}\ w \in K_j\ \textit{and for each}\ u \in \overline{w},$$
$$x = x(u, x_0)\ \textit{implies that}\ \Sigma(\mathrm{h}(\mathrm{x})) \cap \Sigma_{\mathrm{u}} \subset \Sigma(x)$$

We recall that $\Sigma(h(x))$ is the set of all alphabets active in state q=h(x) of G_c and Σ_u is the set of all uncontrolled alphabets. What the above condition states is that the set of uncontrolled active events or alphabets in state $q = h \times \xi(u,x_0) = h(x) = \delta_c(u,q_0)$ has to be a subset of the active events in state $x = \xi(u,x_0)$ of $\Im_j$. If this condition is not true, then state x has to be dropped which means that string u has to be dropped which in turn means that string w should no longer be considered as a part of $\Omega(K_j)$. Wonham and Ramadge [1987b] put it this way:

To construct an automaton $\Im_j$ with $L(\Im_{j+1})=\Omega(K_j)$ simply remove those states of $\Im_j$ that fail to satisfy the active event constraint, namely,

$$(\Sigma(h(x)) \cap \Sigma_u) \subset \Sigma(x) \tag{5}$$

Next, we will demonstrate the above results with two examples in the order of their complexity. Both examples consider a manufacturing line with two machines and a single buffer.

Example IV

Consider a manufacturing system composed of two machines M_1 and M_2 in tandem separated by a buffer B. This is shown in Figure 7.

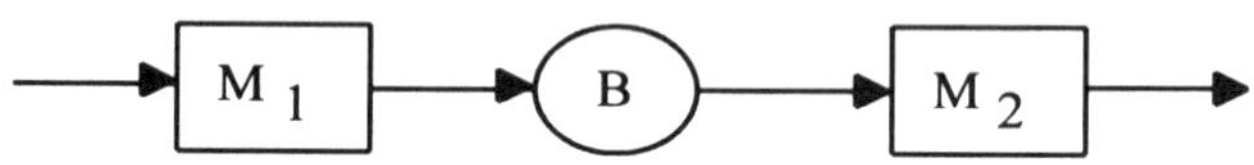

Figure 7: A manufacturing line

Both M_1 and M_2 are CDEPs with state diagrams shown in Figure 8.

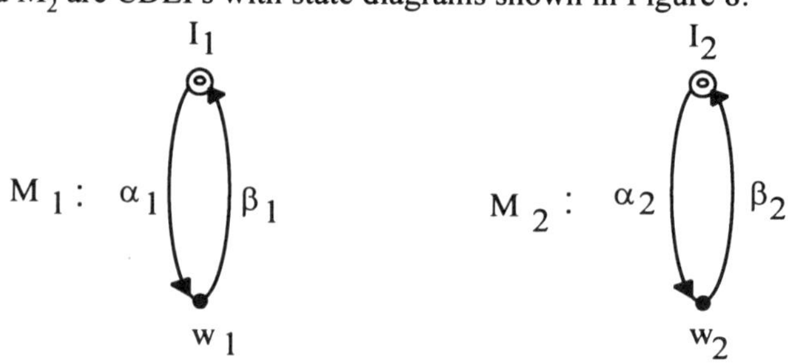

Figure 8: Automata of machines

We note that $\Sigma=\{\alpha_1,\alpha_2,\beta_1,\beta_2\}$ and $\Sigma_c=\{\alpha_1,\alpha_2\}$; I is the idle state whereas w is the working state. Buffer B has a single slot and can be modeled by an automaton with two states E (empty) and F (full). This is shown in Figure 9.

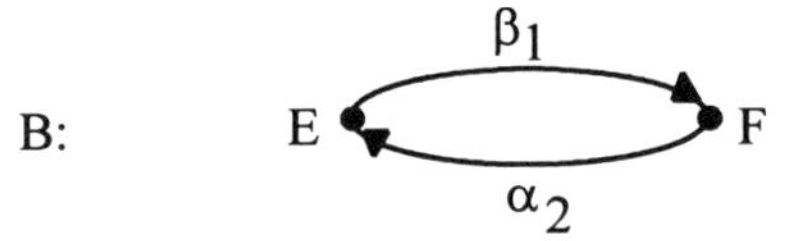

Figure 9: Automaton of buffer

Initially the system is in state (I_1, I_2, E). A work cycle $\alpha_1\beta_1$ of M_1 deposits a workpiece in B (driving B from E to F). M_2 then begins its work cycle by removing the workpiece from the buffer, thereby B returning to E. M is defined as a two dimensional finite automaton obtained from M_1 and M_2 in the following way:

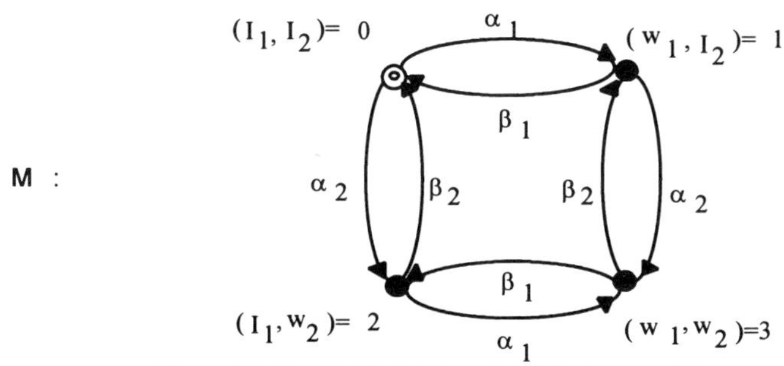

Figure 10: The shuffled Automata of the two machines

We note that M is obtained from M_1 and M_2 by combining or shuffling all the possible combinations of states of M_1 and M_2. In other words, by assuming that the two machines work asynchronously and independent. The simultaneous occurrence of an event in M_1 with an event in M_2, however, is ruled out.

Legal language L_g is defined by imposing the "safety" requirements determined by the capacity of the buffer: Let L_f be the "safety language" defined as below;

$$L_f = \{\rho : \rho \in \Sigma^* \; and \left|\alpha_2\right|(t) \le \left|\beta_1\right|(t) \le \left|\alpha_2\right|(t) + 1 \quad \forall t \in \bar{\rho}\};$$

where $|\sigma|(\rho)$ is the number of occurrences of σ in ρ. L_f contains those sequences of events where the number of workpiece in the buffer is either zero or one. An automaton for L_f is shown in Figure 11.

$\mathbf{L_f}$:

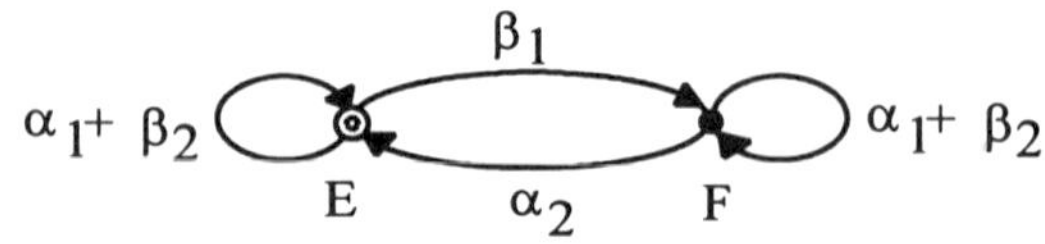

Figure 11: A generator for L_f

The legal language L_g is now defined by $L=L_f \cap M$ corresponding to the finite automaton with states (0E, 0F, 1E, 1F, 2E, 2F, 3E, 3F). Instead of state diagram, we will show L using the transition matrix with entities being the events:

$$
L: \begin{array}{c} 0E \\ 0F \\ 1E \\ 1F \\ 2E \\ 2F \\ 3E \\ 3F \end{array}
\begin{bmatrix}
 & & \alpha_1 & & & & & \\
 & & & \alpha_1 & \alpha_2 & & & \\
 & \beta_1 & & & & & & \\
 & & & & & & \alpha_2 & \\
\beta_2 & & & & & & \alpha_1 & \\
 & \beta_2 & & & & & & \alpha_1 \\
 & & \beta_2 & & & \beta_1 & & \\
 & & & \beta_2 & & & &
\end{bmatrix}
$$

Let $K_0=L$. We have

$$
\begin{array}{ll}
 & \begin{array}{cccccccccc} & & & & & & & & \Sigma(h(x))\cap\Sigma_u & \Sigma(x) \end{array} \\
K_0: \begin{array}{c} 0E \\ 0F \\ 1E \\ 1F \\ 2E \\ 2F \\ 3E \\ 3F \end{array} &
\left[\begin{array}{cccccccccc}
 & & \alpha_1 & & & & & & & \alpha_1 \\
 & & & \alpha_1 & \alpha_2 & & & & & \alpha_1\alpha_2 \\
 & \beta_1 & & & & & & & \beta_1 & \beta_1 \\
 & & & & & & \alpha_2 & & \beta_1 & \alpha_2 \\
\beta_2 & & & & & & \alpha_1 & & \beta_2 & \beta_2\alpha_1 \\
 & \beta_2 & & & & & & \alpha_1 & \beta_2 & \beta_2\alpha_1 \\
 & & \beta_2 & & & \beta_1 & & & \beta_1\beta_2 & \beta_1\beta_2 \\
 & & & \beta_2 & & & & & \beta_1\beta_2 & \beta_2
\end{array}\right]
\end{array}
$$

Initial state: 0E, Marked States: all states

We note that the states in L or K_0 have two dimensions, one from M and one from L_f. $h(x)$ is simply the first dimension which is from M. Both states 1F and 3F fail to satisfy the active event constraint and hence must be removed from the tableau. This yields;

$$
\begin{array}{ll}
 & \begin{array}{cccccccc} & & & & & & \Sigma(h(x))\cap\Sigma_u & \Sigma(x) \end{array} \\
K_1: \begin{array}{c} 0E \\ 0F \\ 1E \\ 2E \\ 2F \\ 3E \end{array} &
\left[\begin{array}{cccccccc}
 & & \alpha_1 & & & & & \alpha_1 \\
 & & & \alpha_2 & & & & \alpha_2 \\
 & \beta_1 & & & & & \beta_1 & \beta_1 \\
\beta_2 & & & & & \alpha_1 & \beta_2 & \beta_2\alpha_1 \\
 & \beta_2 & & & & & \beta_1\beta_2 & \beta_2 \\
 & & \beta_2 & & \beta_1 & & \beta_1\beta_2 & \beta_1\beta_2
\end{array}\right]
\end{array}
$$

Initial state: 0E, Marked States: all states

Since each state of K_1 satisfies the active event constraint, then $\Theta = \sup C_G (L_g)$. This is the least restrictive controllable language which satisfies the "safety requirement". The state diagram for Θ is shown Figure 12.

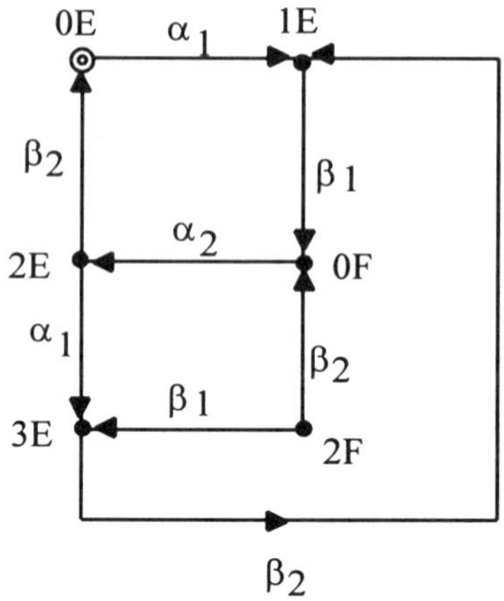

Figure 12: State diagram of Θ

Example V

Consider the manufacturing line of figure 7 again. But, this time we assume that machines break down and are repaired. The repair process is assumed to be controllable. A machine automaton is shown in Figure 13.

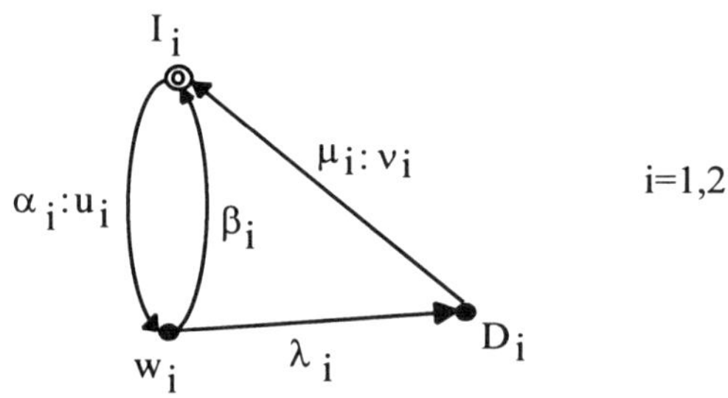

Figure 13: State diagram for machines M_1 and M_2.

M_1 and M_2 are CDEPS
Control u: enables / disables the transitions from I to w. (u=1 → accept a workpiece)
Control v: enables/ disables the transition from D to I (v=1 → M in state D is under repair)

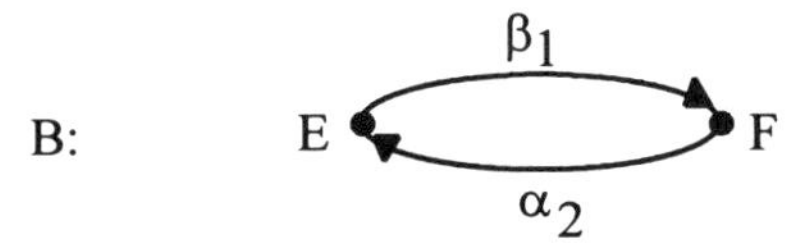

Figure 14: State diagram of buffer

Buffer is not a CDEP but an automaton driven by M_1 and M_2, shown in Figure 14. The following sequences of events are important to remember:
$\alpha_i\beta_i$: Successful completion of a workpiece
$\alpha_i\lambda_i\gamma_i$: machine failure and repair

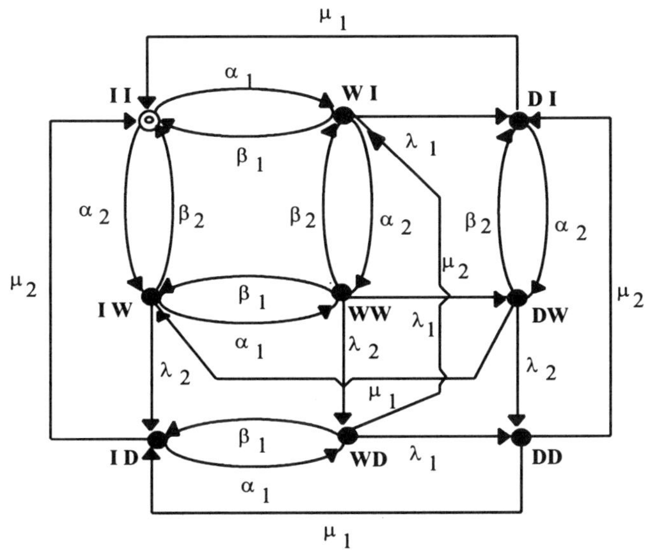

Figure 15: Shuffled automata (M_1M_2)=M

We note that M (shown in Figure 15) is obtained from M_1 and M_2. M is defined irrespective of buffer capacity or workpiece availability. Let L_b be the language defined by

$$L_b = \left\{ s : s \in \Sigma^* \ \textit{and}\ |\alpha_2|(t) \le |\beta_1|(t) \le |\alpha_2|(t) + 1 \quad \forall t \in \bar{s} \right\}$$

where $|\sigma|(s)$ is the number of occurrences of σ in s. L_b contains those sequences of events where the number of workpieces in the buffer is either zero or one. An automaton, B, for L_b is shown in Figure 16.

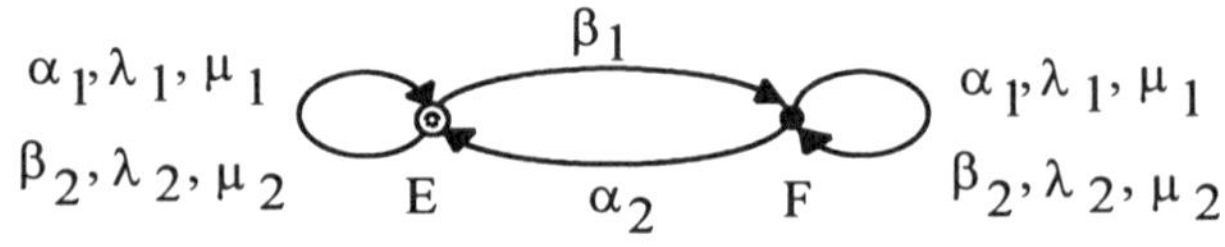

Figure 16: A generator for L_b

From L_b and M using the relation $L_1 = L_b \cap M$ we obtain a new language which is legal (satisfies the buffer requirement). The corresponding CDEP, G_c, has states {IIE, IIF, WIE, WIF, DIE, DIF, IWE, IWF, WWE, WWF, DWE, DWF, IDE, IDF, WDE, WDF, DDE, DDF} and the state diagram shown in Figure 17.

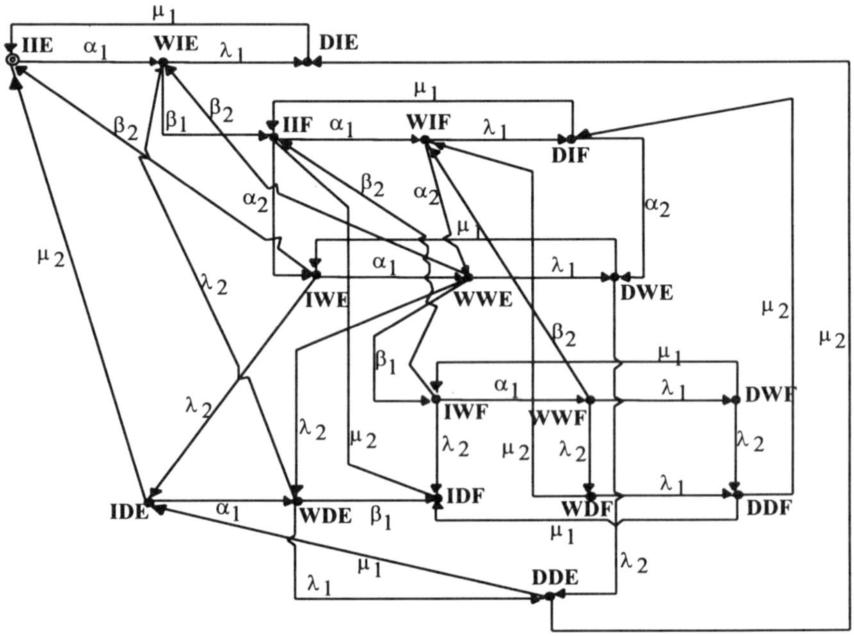

Figure 17: State diagram for G_c obtained from M and B such that $L(G_c) = L(M) \cap L_b$

Now we will add more requirements: (i) M_1 executes α_1 only if B is in E, (ii) M_1 cannot execute α_1 while M_2 is in D_2, (iii) If M_1 is in D_1 and M_2 is in D_2 then $\nu_1 = 0$. Let L_g be the legal language generated by the plant generator given that these requirements are satisfied. The state diagram of the automaton or the recognizer for L_g is shown in Figure 18.

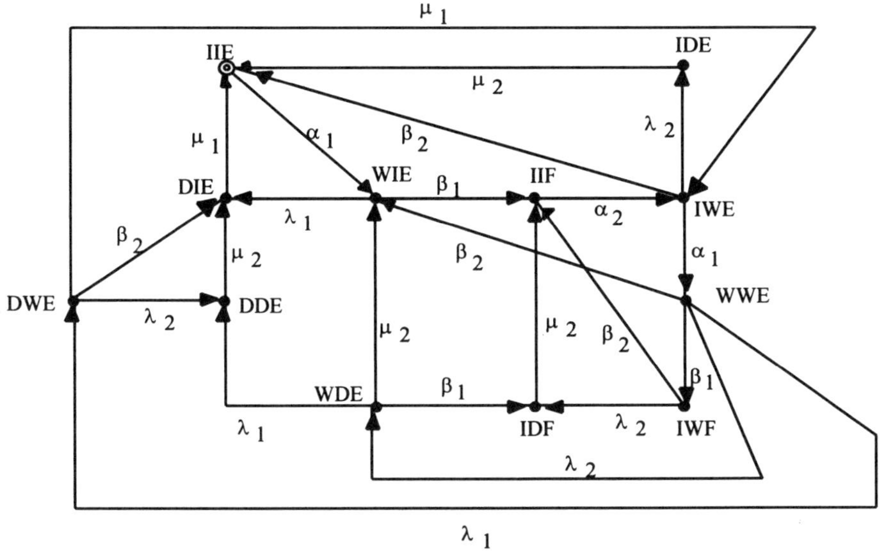

Figure 18: Recognizer for L_g

Comparing Figures 17 and 18 above, we note that the states {DWF, WDF, DDF, WIF, DIF, WWF } are eliminated from the automaton of L_g. The reason is obvious as we see that these states violate the above requirements.

To find the supremal language Θ=Sup $C_G(L_g)$ we again follow the iterative procedure described earlier. Let $K_0=L_g$. We note that $\Im_0$ s.t. $L_m(\Im_0)=K_0$ refines G_c. From the last two columns of the transition matrix for K_0 (not shown here) we note that $L_g=K_0$ is actually Θ. We know that L_g is the marked language generated by the state diagram of Figure 18. The strings in L_g are limited to the strings in $\bar{L}_g$ and $L_m(G_c)$. Therefore, L_g is $L_m(G)$ closed.

Now we have to construct a proper supervisor $S=(\chi,\phi)$ such that $L(S|G_c)=\bar{L}_g$ and $L_m(S|G_c)=L_g$. For this, we let $\chi=(X,\Sigma,\xi,x_0,X_m)$ be a trim recognizer for L_g. Figure 18 defines χ, where $X_m=\{x_0\}$. We also let $\phi:X \rightarrow \{0, 1\}^{\Sigma}$ to be defined as

$$\phi(x)(\mu_i \text{ or } \nu_i) = \begin{cases} 1 & \text{if } u_i \text{ or } v_i \text{ is issued from x} \\ 0 & \text{otherwise} \end{cases}$$

Ramadge and Wonham [1987a] give a formal proof for that S defined as such satisfies $L(S|G_c) = \bar{L}_g$ and $L_m(S|G_c) = L_g$. Here we give an informal discussion on why this happens. In order to show that $L(S|G_c) = \bar{L}_g$, we have to show that for any pair of physically possible states x and x' such that $x \in |\bar{L}_g|$ ($|L|$ defines the state diagram which generates L) and $x' \notin |\bar{L}_g|$, the event σ, resulting in a transition from x to x' is disabled by S. Also for any pair x and x' such that $x \in |\bar{L}_g|$ and $x' \in |\bar{L}_g|$ the event σ, resulting in a transition from x to x' is enabled by S. The way we defined $\phi(x)$ insures that the above conditions are satisfied. $L_m(S|G_c)=L_g=L_m(\chi)$ follows.

The supervisor S can be reduced in its number of states. This requires that we find the equivalent finite automaton which marks the same language L_g. Ramadge and Wonham [1987a] toss the term *control equivalence* for this purpose. Two strings $s, s' \in \Sigma^*$ are control equivalent if for all $\sigma \in \Sigma_c$, $s\sigma \in \bar{L}_g$ *iff* $s'\sigma \in L_g$. In other words, the control action following any two control equivalent strings is the same for every $\sigma \in \Sigma$. Having defined the equivalence relation, we can follow the procedure briefly described in section 3.1 to find the reduced supervisor, shown in Figure 19, for the above system:

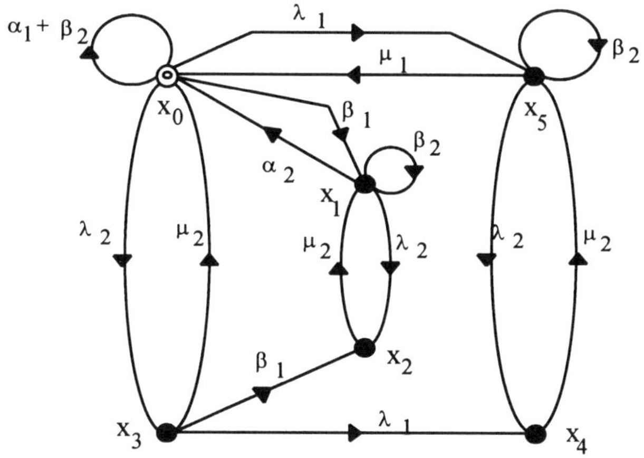

Figure 19: Reduced supervisor for the example

4.0 Supervisory control of Discrete Event Dynamic Systems - Petri Net Approach

Unfortunately, the theory presented in section 3 has not been fully extended to Petri nets. However, there are some related works, e.g., Sreenivas and Krogh [1992], which attempt to use this approach when the plant model or the supervisor is represented by Petri nets. Here, we shall not discuss these works. We will rather briefly discuss a different approach which has been taken mostly in Petri net literature to the problem of synthesizing supervisory controller.

One problem which has been around in the Petri net literature since its origination has been to analyze its various properties avoiding the reachability set. Various reduction and synthesis methods have been suggested (Murata [1989]). The synthesis techniques, in particular, have received considerable attention also from the point of view of the design and analysis of supervisory control. The works by Krogh and Beck [1986], Zhou et al. [1991,1992] or Jeng [1992], for example, attempt to synthesize the overall supervisory control structure by combining the simpler structures in a way that some desirable properties are maintained. The properties of major interest in these types of works are usually liveness, boundedness, reversibility, etc. In these works, there are no specific models for specification, even though, the design philosophy is still using relation (1) in some indirect way. Using the terminology of Ostroff [1989], these works use single language approach.

A slightly different approach has been adopted by Darabi and Jafari [1994]. Here, they use dual language approach. The requirements are represented in some high level language and Petri net is the control model. They present an algorithm which transforms this specification into a Petri net. The specification together with the transformation insures that the Petri net is correct and has some desirable properties. A similar approach has also been taken by Boucher and Jafari [1992] and Jafari and Boucher [1994]. They use IDEF0 as the specification model and present algorithms to transform the IDEF0 model into a Petri net.

5.0 Conclusion

Relation (1) posses an important problem both in theoretical and practical terms. It is important that the control engineers are provided with some software tool allowing them to specify, design, and analyze the controller in an efficient manner. The current state of technology in this area is quite distant from providing this environment. As we presented in this chapter, there are various theoretical models supporting this idea, but we believe they are far from being practical. The theory of Ramadge and Wonham or the other related works need to be explored and extended further to be practical. There has to be a simpler front end specification language. The same is true for the works on the Petri net synthesis or transformation models. In all cases, both behavioral and performance measures have to be considered in the synthesis and analysis.

References

Bollinger, J.G. and N.A. Duffie, Computer Control of Machines and Processes, Addison-Wesley, 1988.

Boucher, T.O. and M.A. Jafari, "Design of a factory Floor Sequence Controller from a High Level System Specification," *J. of Manufacturing Systems*, 11, 6, 1992.

Cristian, F., "Correct and Robust Programs," *IEEE Transactions on Software Engineering*, SE-10, 2, 1984.

Darabi, H. and M.A. Jafari, "Petri Net Transformation of a High Level Control Specification and the Related Theory," Working Paper, Dept. of Industrial Engineering, Rutgers University, P.O. Box 909, Piscataway, NJ 08855, 1994.

Hack, M.H.T., "Decidability Questions for Petri Nets," Ph.D. Dissertation, Massachusetts Institute of Technology, 1976.

Hopcroft, J.E. and J.D. Ullman, Introduction to Automata Theory, Languages, and Computation, Addison-Wesley, 1979.

Jafari, M.A., "An Architecture for Shop-Floor Controller Using Colored Petri Nets," *The Int'l J. of Flexible Manufacturing Systems*, 4, 159-181, 1992.

Jafari, M.A. and T.O. Boucher, "A Rule Based System for Generating a Ladder Logic Control Program from a High Level Systems Model," *J. of Intelligent Manufacturing*, 5, 103-120, 1994.

Jaffe, M.S., N.G. Leveson, P.E. Heimdahl, and B.E. Melhart, "Software Requirements Analysis for Real Time Process Control Systems," *IEEE Transactions on Software Engineering*, 17,3, 1991.

Jeng, M.D., "Theory and Application of Resource Control Petri Nets for Automated Manufacturing Systems," Ph.D. Dissertation, Rensselaer Polytechnic Institute, 1992.

Jensen, K., "High Level Petri Nets," *in Applications and Theory of Petri Nets,* A. Pagoni and G. Rozenberg (Eds.), Informatik-Fachberichte, 66, Springer-Verlag, New York, 1983.

Krogh, B.H. and C.L. Beck, "Synthesis of Place/Transition nets for Simulation and Control of Manufacturing Systems," *Proc. IFIP Symp. Large Scale Systems,* Zurich, Aug., 1986.

Murata, T., "Petri Nets: Properties, Analysis, and Applications," *Proc. IEEE*, 77, 4, 1989.

Nicollin, X., J. Sifakis, and S. Yovine, "Compiling Real Time Specifications into Extended Automata," *IEEE Transactions on Software Engineering*, 18,9, 1992.

Ostroff, J.S., Temporal Logic For Real Time Systems, Research Studies Press, 1989.

Peterson, J.L., Petri Net Theory and The Modeling of Systems, Prentice Hall, 1981.

Petri, C.A., "Kommunikation with Automaton," Ph.D. Dissertation, University of Bonn, West Germany, 1962.

Ramadge, P.J. and W.M. Wonham, "Supervisory Control of a Class of Discrete Event Processes," *SIAM J. Control and Optimization*, 25, 1, 1987.

Ramadge, P.J. and W.M. Wonham, " On the Supremal Controllable Sublanguage of a Given Language," *SIAM J. Control and Optimization,* 25, 3, 1987.

Shaw, A.C., "Communicating Real Time State Machines," *IEEE Transactions on Software Engineering*, 18, 9, 1992.

Sreenivas, R.S. and B.H. Krogh, "On Petri Net Models of Infinite State Supervisors," *IEEE Transactions on Automatic Control*, 37, 2, 1992.

Zhou, M.C. and F. DiCesare, "Parallel and Sequential Mutual Exclusion for Petri Net Modeling for Manufacturing Systems," *IEEE Transactions on Robotics and Automation,* 7,4, 515-527, 1991.

Zhou, M.C., F. DiCesare, and A.A. Desrochers,"A Hybrid Methodology for Synthesis of Petri Nets for Manufacturing Systems," *IEEE Transactions on Robotics and Automation,* 8,3, 350-361, 1992.

Index